Biochar for Sustainability
Production, Utilization, and Environmental Insights

Editors

Rosa Rodriguez
Instituto de Ingeniería Química, Facultad de Ingeniería,
Universidad Nacional de San Juan,
Grupo Vinculado al PROBIEN
(CONICET, Universidad Nacional del Comahue),
San Juan, Argentina

Germán Mazza
Instituto de Investigación y Desarrollo en Ingeniería de Procesos,
Biotecnología y Energías Alternativas,
PROBIEN (CONICET, Universidad Nacional del Comahue) Neuquén,
Argentina

CRC Press
Taylor & Francis Group
Boca Raton London New York

CRC Press is an imprint of the
Taylor & Francis Group, an **informa** business
A SCIENCE PUBLISHERS BOOK

Cover credit: Cover illustration reproduced by kind courtesy of the editors.

First edition published 2026
by CRC Press
2385 NW Executive Center Drive, Suite 320, Boca Raton FL 33431

and by CRC Press
4 Park Square, Milton Park, Abingdon, Oxon, OX14 4RN

CRC Press is an imprint of Taylor & Francis Group, LLC

Library of Congress Cataloging-in-Publication Data (applied for)

ISBN: 978-1-032-71384-7 (hbk)
ISBN: 978-1-032-72670-0 (pbk)
ISBN: 978-1-032-72669-4 (ebk)

DOI: 10.1201/9781032726694

Typeset in Times New Roman
by Prime Publishing Services

Preface

Humanity stands at a critical juncture, confronted by the intertwined challenges of climate change, environmental degradation, and the urgent need to transition from a linear, extractive economy to a circular, regenerative one. The pressure on natural resources, coupled with the escalating volume of waste generated by agricultural, industrial, and urban activities, demands innovative solutions that are not only technologically sound but also environmentally benign and economically viable. It is within this complex landscape that biochar --a porous, carbon-rich solid produced through the thermochemical conversion of biomass --has risen from a subject of niche scientific inquiry to a cornerstone of sustainable technology with multifaceted applications.

This volume, "Biochar for Sustainability: Production, Utilization, and Environmental Insights," is conceived as a comprehensive exploration of this dynamic field. It aims to provide a thorough synthesis of the state-of-the-art, bridging the gap between fundamental scientific principles and real-world implementation. The book is meticulously structured to guide the reader on a coherent journey from the foundational science of biochar production to its cutting-edge applications and systemic environmental evaluation.

The journey begins with an in-depth examination of the production pathways. The initial chapters delve into the heart of pyrolysis and gasification processes, dissecting the influence of critical operational variables such as temperature, heating rate, and feedstock composition on the yield and characteristics of the resulting biochar. Beyond empirical analysis, the book ventures into the realm of advanced modeling and simulation, showcasing how kinetic models, machine learning, and Computational Fluid Dynamics (CFD) are revolutionizing our ability to predict, optimize, and scale these processes efficiently. This solid foundation in production science is essential for tailoring biochar with specific properties suited for distinct applications.

The core of this work lies in its expansive and detailed review of biochar's diverse utilizations. The chapters systematically document its role as a powerful tool for environmental management. This includes its capacity to capture carbon dioxide from the atmosphere, treat wastewater by adsorbing a wide spectrum of organic and inorganic pollutants, and clean industrial gas emissions. Furthermore, the book dedicates significant attention to the agronomic benefits of biochar, exploring its function as a soil amendment to combat degradation, enhance fertility, and even as a key component in modern soilless cultivation systems, promoting water and nutrient efficiency.

The scope of applications extends into the realms of energy and materials science, highlighting biochar's potential as a solid biofuel, a precursor for energy storage devices like supercapacitors, and a sustainable additive that can enhance the properties of composite materials, from cement to biopolymers. This versatility underscores the principle of the circular economy, transforming low-value biomass waste into high-value products.

A defining and critical feature of this book is its commitment to assessing the true sustainability of biochar systems. A dedicated chapter on Life Cycle Assessment (LCA) provides a rigorous framework for evaluating the net environmental impacts of biochar production and use, ensuring that the proposed solutions are scrutinized holistically, from feedstock acquisition to end-of-life, avoiding problem-shifting and validating its green credentials.

Compiled by a multidisciplinary team of experts, this book is an indispensable resource for a broad audience --including students, researchers, engineers, environmental scientists, policymakers, and industry practitioners. Our aspiration is that it will serve not only as an authoritative reference but also as a source of inspiration. By presenting a consolidated vision of biochar's potential, we hope to catalyze further innovation, foster interdisciplinary collaboration, and accelerate the adoption of this remarkable material, contributing meaningfully to the building of a more sustainable and resilient future.

The Editors and Authors

Contents

Chapter 1

Pyrolysis and Gasification for Biochar Production. Influence of Operating Variables

Andrés Reyes-Urrutia[a*], Germán Mazza[a], Juan Pablo Capossio[a], Marcelo Echegaray[b] and Erick Torres[b]

[a]Instituto de Investigación y Desarrollo en Ingeniería de Procesos, Biotecnología y Energías Alternativas, PROBIEN (CONICET, Universidad Nacional del Comahue), Neuquén, Argentina

[b]Instituto de Ingeniería Química, Fac. de Ingeniería, Grupo Vinculado al PROBIEN (Universidad Nacional de San Juan, CONICET, Universidad Nacional del Comahue), San Juan, Argentina

1.1 INTRODUCTION

Lignocellulosic biomass is the most abundant source of organic material on Earth. It is considered a strategic renewable natural resource originating from different sectors such as forestry, agriculture, and industry (Mujtaba et al. 2023). It derives from the plant's photosynthesis process, converting solar energy into biodegradable, renewable organic matter. Lignocellulosic biomass is a sustainable source of energy and one of the most abundant available worldwide.

Due to its inability to be digested, lignocellulosic biomass is not part of the human food chain and is largely disposed of as waste. Therefore, unlike so-called energy crops, which are also of lignocellulosic origin, they do not

*Corresponding author: Andrés Reyes-Urrutia (andres.reyes@probien.gob.ar)

DOI: 10.1201/9781032726694-1

compete with crops intended for human consumption. Quantitatively, the main sources of resource generation are trees, wood waste used in construction, sawmill waste, and forest waste such as branches and fallen leaves in the mountains. Additionally, agricultural residues (rice straw, corn stalks, and sugarcane bagasse), cellulose waste (municipal solid waste such as pulp mill and sawmill residues), and so-called energy crops are significant contributors (Sahay 2022).

1.1.1 Chemical Composition and Physical-Chemical Properties of Lignocellulosic Bio-Waste

In general, lignocellulosic biomass primarily comprises cellulose, hemicellulose, and lignin. These are high molecular-weight organic polymers found in the cell walls of the constituent cells of lignocellulosic biomass, accounting for, on average, 75% of the weight of plant matter. Specifically, they provide structural strength to plants to maintain their upright position. In smaller proportions, there are also extractable organic species through solvents (oils, proteins, starch, sugar, among others) and some inorganic components that become ash during combustion (Wiselogel et al. 1996). Table 1.1 shows some examples of the composition of different types of residual lignocellulosic biomass reported in the literature:

Table 1.1 Biomass lignocellulosic waste compositions by mass fraction percentage on a dry basis

Biomass	Cellulose	Hemicellulose	Lignin	Source
Energy crops	36.60	16.1	21.9	Kaylen et al. (2000)
Rosehip husk	22.70	5.99	26.04	Torres-Sciancalepore et al. (2023)
Quince	58.57	12.37	23.86	Torres-Sciancalepore et al. (2022)

The physical properties of different lignocellulosic biomasses, such as calorific capacity, density, and thermal conductivity, among others, must be evaluated according to the specific need for that particular data.

1.2 BIOCHAR PRODUCTION TECHNOLOGIES

Biochar is produced by supplying heat to biomass in total or partial absence of oxygen. The technologies used to obtain biochar employ the process of pyrolysis, defined as the thermal treatment of biomass in the absence of oxygen, resulting in the production of solid products (charcoal), liquids (tar and an aqueous solution of organic compounds), and non-condensable gases. Thermochemical conversion methods include pyrolysis, hydrothermal carbonization, gasification, and torrefaction (Yaashikaa et al. 2020).

1.2.1 Pyrolysis

Pyrolysis is a process involving the thermal decomposition of organic materials in the absence of oxygen, carried out in the temperature range between 250°C and 900°C. It is considered an alternative option for transforming lignocellulosic biomass into higher-value products such as char (now referred to as biochar), synthesis gas, and bio-oil. Pyrolysis is intriguing as an independent process leading to the production of high-energy-density products and as an intermediate step in a gasification or combustion process (Maschio 1992).

While the yield of biochar during the pyrolysis process depends on the type and nature of the biomass used, if we are talking about the same residue, temperature is the main operational variable determining the efficiency of each product. It is important to highlight that different residues with similar ratios between hemicellulose/lignin and cellulose/lignin fractions will behave similarly during pyrolysis. Generally, the biochar yield decreases, and syngas production increases as the temperature increases during the pyrolysis process. Overall, a pyrolysis process aimed at biochar production requires low heating rates, with final temperatures not exceeding 400°C, and prolonged residence times.

1.2.1.1 Slow, Fast, and Intermediate Pyrolysis: Influence of Operating Variables on the Yield and Characteristics of Biochar

Pyrolysis may be classified as slow (or conventional), intermediate, and fast, according to the values or ranges of the operating variables such as pyrolysis temperature, Tp (°C or °K), heating rate, α (°C/sec), etc. (see Table 1.2). Flash pyrolysis is also essential and occurs at the highest heating rates and shortest residence times.

Slow pyrolysis is characterized by low heating rates, high residence time, larger particles, and lower temperatures on average. On the other hand, in fast pyrolysis, heating rates are higher, and residence times are shorter. The size of the particles is smaller, and the temperatures are higher on average. Intermediate pyrolysis has overall operating variables' ranging between the slow and fast types. Another important characteristic is that the feedstock's water content can be higher (Hornung 2013).

Regarding product yield, in slow pyrolysis, the biochar yield is the highest because there is sufficient residence time for volatile matter vapors to convert into charcoal in the solid matrix and non-condensable gases (Angin 2013). According to Rangabhashiyam and Balasubramanian (2019), at low heating rates, less than 10°C/min, weak chemical bonds are broken, and polymeric structures are slightly affected, allowing reorganization reactions to form a more stable solid matrix, thus inhibiting volatile compound formation. During the fast pyrolysis, volatile matter is released before reorganization reactions occur. This means that volatile components are released quickly and in large quantities, limiting the time available for reorganization reactions of volatile matter into more stable products that remain trapped in the solid matrix, thus decreasing biochar yield. Biochar produced at high heating rates has a low surface area and porosity due to rapid depolymerization on the biochar surface. Flash pyrolysis is primarily aimed at gas production. In this case, homogeneous gas-phase or heterogeneous gas-solid phase reactions predominate, primarily producing non-condensable gases, reducing bio-oil production and increasing synthesis gas yield (Sonil et al. 2014). Finally, intermediate-rate pyrolysis has a more balanced yield of char, tar, and gas.

Table 1.2 Approximate ranges of the different pyrolysis types (synthesis of several authors: Akinpelu et al. 2023; Ragula et al. 2019; Park et al. 2014)

Type	Pyrolysis Temperature (°C)	Heating Rate (°/seg)	Time (sec)	Particle Size (mm)	Impact on Yield
Slow	300–700	0.1–1	>500	5–50	Higher Biochar
Intermediate	600–700	1–20	10–60	1–2	Balanced
Fast	400–900	10–200	0.5–10	<1	Higher Tar
Flash	600–1,200	100–1,000	<1	0.25–1	Higher Gas

In the work of Prabir Basu (2010), a more formal criterion is set to differentiate the three types of pyrolysis, which is based on the relation between the heating time, t_h, and the reaction time, t_r. Consequently, in slow pyrolysis, the reaction time is much shorter than the heating time; conversely, in fast/flash pyrolysis, the heating time is much shorter than the reaction time due to the high heating rates. Finally, in intermediate pyrolysis, both times are similar to a certain extent.

- Slow pyrolysis $t_h >> t_r$ (1)
- Intermediate pyrolysis $t_h \approx t_r$ (2)
- Fast/flash pyrolysis $t_h << t_r$ (3)

Finally, it is important to point out that the criteria presented in equations 1–3 have to be used together with the information presented in Table 1.2 to determine the pyrolysis type because it may be misleading if it is taken solely.

1.2.2 Hydrothermal Carbonization

Hydrothermal carbonization (HTC) of biomass is a thermal conversion process carried out in a closed reactor in an oxygen-free environment, in the presence of water, self-pressurized, conducted at moderate temperatures (180–300°C), for several minutes to several hours (Lin et al. 2016). Due to its low-temperature operation and the elimination of the raw material pre-drying process, the HTC process is cost-effective for biochar production. The product obtained by the hydrothermal process is called hydrochar to differentiate the char obtained by dry techniques. As biomass is heated during the HTC or pyrolysis process, various chemical reactions occur, such as hydrolysis, dehydration, decarboxylation, aromatization, and recondensation, which modify its physical structure. Due to a lower activation energy compared to other possible reactions, hydrolysis predominates in HTC (Fang et al. 2018).

Temperatures above 400°C allow for the production of bio-oil in a process called hydrothermal liquefaction. Above 400°C, the process is called hydrothermal gasification and is focused on generating synthesis gas primarily composed of CO, H_2, CO_2, and CH_4 (Yaashikaa et al. 2020).

1.2.3 Gasification

Gasification is a thermochemical decomposition method for carrying out the partial oxidation of carbon-rich materials in the presence of gasifying agents, such as oxygen, air, steam, or combinations of air or oxygen/steam at high temperatures. The products of gasification are primarily gaseous species, forming the so-called synthesis gas, generally composed of CO, CO_2, CH_4, H_2, and traces of organic species with empirical formula C_xH_y. Although biochar is also generated, the yield is low, so it is considered a byproduct. The composition of synthesis gas depends on different factors such as biomass type, reactor type, and operating conditions, with the main ones being temperature, pressure, and gasifying agent. The generated char is generally composed of carbon, some hydrocarbons, and ashes. The ash fraction mainly depends on the type of biomass used and can be inferred from the proximate analysis of the lignocellulosic biomass.

Table 1.3 Main reactions in gasification (Basu 2010; Zhang et al. 2020a)

Type of Reaction	**Reaction**	**ΔH (1 atm, 298 K) kJ/mol**
R0 (devolatilization)	$C_mH_nO_p$ (biomass) + heat → gas + tar + solid (pyrolytic char*)	endothermic
	Heterogeneous Reactions of the Char	
R1 (Boudouard Reaction)	$C + CO_2 \leftrightarrow 2CO$	172
R2 (Carbon steam gasification)	$C + H_2O\ (g) \leftrightarrow CO + H_2$	131
R3	$C + 2H_2 \leftrightarrow CH_4$	–74.8
R4	$C + 0{,}5\ O_2 \leftrightarrow CO$	–110.5
R5	$C + O_2 \leftrightarrow CO_2$	–393.5
	Homogeneous Reactions	
Oxidation reactions		
R6	$CO + 0{,}5\ O_2 \rightarrow CO_2$	–283
R7	$CH_4 + 2O_2 \leftrightarrow CO_2 + 2H_2O\ (g)$	–802.3
R8	$H_2 + 0{,}5\ O_2 \rightarrow H_2O\ (g)$	–241.8
R9	$CO + H_2O\ (g) \leftrightarrow CO_2 + H_2$	–41.2
	Methanation Reactions	
R10	$2CO + 2H_2 \leftrightarrow CH_4 + CO_2$	–247.3
R11	$CO + 3H_2 \leftrightarrow CH_4 + H_2O\ (g)$	–206.1
R14	$CO_2 + 4H_2 \leftrightarrow CH_4 + 2H_2O\ (g)$	–164.9
	Steam Reforming Reactions	
R12	$CH_4 + H_2O\ (g) \leftrightarrow CO + 3H_2$	206.1
R13	$CH_4 + 0{,}5\ O_2 \rightarrow CO + 2H_2$	–35.7
	Tar Reaction (decomposition)	
R15	$C_aH_bO_c$ (tar) + heat → gas + C (solid)	endothermic

*pyrolytic char contains some volatiles and ash in addition to fixed carbon

It is possible to consider gasification as a process composed of different stages. Initially, the heat provided by the environment raises the temperature of the lignocellulosic biomass to reach the drying stage, around 100°C, and subsequently get the stage of thermal degradation of organic matter during pyrolysis in the temperature range between 150°C and 700°C (Ferreira et al. 2020).

During pyrolysis, condensable gaseous species such as water and tar are released, and non-condensable gases such as CO, CO_2, H_2, and light hydrocarbons like CH_4, among others, and the remaining solid called char are also generated. Subsequently, a series of homogeneous gas-phase reactions and heterogeneous reactions between the generated char and the present gaseous species occur. The set of mentioned reactions is presented below in Table 1.3 with their respective enthalpies under standard conditions.

1.2.3.1 Influence of Operating Variables on the Yield and Characteristics of Biochar

In gasification, it is necessary to operate with air in deficiency compared to what is required for complete combustion. The parameter ER, equivalence ratio, defines the relationship between the air supplied to the process and that required for the complete combustion of the biomass. ER is the variable that determines the yields of a gasifier and the quality of the gas generated, mediated by its heating value. For example, in pyrolysis, ER equals zero due to the absence of air in ideal inertization conditions, generating high yields of biochar and gas with low heating value. To ensure that combustion does not predominate in the gasification process, thereby avoiding gas with a high fraction of CO_2, ER values greater than 0.4 should not be used. In practice, ER values in the range of 0.2–0.3 are used. Within the specified range, increasing the equivalence ratio leads to a decrease in the H_2 content due to oxidation. The CO content also decreases within the specified range as ER increases; however, its change is not as pronounced as that of H_2 (Ghassemi and Shahsavan-Markadeh 2014).

Regarding the overall speed of the gasification process, a dominant parameter is the fixed carbon content of the biomass. The carbon conversion reactions into gases are relatively slow, so the fixed carbon content determines the size of the reactor and the performance of each phase (Basu 2010).

1.2.3.2 Primary Agents Used in Gasification

The gases used as reactants for the gasification of lignocellulosic biomass are water vapor, air, oxygen, or mixtures thereof. The composition of the gas generated during the process strongly depends on the agent used in the gasification as well as the heterogeneous gas-solid reactions (R1-R5 in Table 1.3). Typically, air and oxygen are used as the oxidizing agents of carbon to provide the necessary exothermic heat for the overall reaction process. On the other hand, water vapor acts as a hydrogen source to produce hydrogen-rich synthesis gas products (R2, R9, R12 in Table 1.2). In practice, species generated during pyrolysis or gasification, such as carbon dioxide and water vapor generated during drying, will also serve as gasification reactants in a gasifier (Zhang et al. 2020).

1.2.4 Torrefaction

Torrefaction is a mild pyrolysis process carried out in a temperature range of 230 to 300°C in the absence of oxygen, with a residence time of less than

30 minutes and heating rates lower than 50°C/min (Yaashikaa et al. 2020). Typically, the inert material used in laboratories to provide a non-oxidizing atmosphere is N_2. The original or raw lignocellulosic biomass is characterized by its high moisture content, low calorific value, ability to absorb and retain water from the environment (hygroscopic nature), and high volume with low apparent density (Chen et al. 2015). In this regard, as a result of the torrefaction process, biochar is generated, which, compared to the original material, exhibits improved energy density, higher calorific value, reduced moisture content, lower oxygen-to-carbon (O/C), and hydrogen-to-carbon (H/C) ratios, improved grinding capacity, and also reduced hygroscopic characteristics (Li and You 2022). Additionally, the depolymerization of cellulose generated in the process causes changes in the structural strength of the material, which, unlike the original, now becomes brittle, facilitating its size reduction.

1.3 DESCRIPTION OF THE DESIRABLE CHARACTERISTICS OF BIOCHAR

Biochar has two main characteristics: specific surface area (SSA, m^2/gr.) and porosity (ϕ, adim.). Other characteristics, such as water holding capacity (WHC), soil aggregate stability (SAS), and many others, are subsidiary and depend on the application. It is generally desired that both porosity and SSA are as high as possible.

Porosity is the ratio between the void volume (or total pore volume, TPV, cm^3/gr.) and the total volume of a porous medium or solid. The porous structure or network of a solid is formed by pores of different diameters called micropores (<2 nm), mesopores (2–50 nm), and macropores (>50 nm). The porous network has to be carefully selected according to the adsorbate molecules' diameter in environmental sorption applications such as wastewater treatment, soil remediation, and CO_2 capture (Leng et al. 2021). Alternatively, as an ameliorant or plant growth aid, the biochar's porosity plays an important role as the home for soil microorganisms, including bacteria and fungi as well as for augmenting the WHC. Further, SSA is defined as the total surface area of a solid per unit mass. It is important for nutrient fixation in soils used for plant growth, and when biochar is used as a catalyst, SSA is related to the active sites available for the substrate to bind.

On the other hand, the higher heating value (HHV, MJ/kg) quantifies the amount of heat released per unit mass or volume by the biochar once it is combusted and returned to ambient temperature. In combustion applications, biochars have to have good HHV, which, in turn, depends on the carbon content of the biochar as determined by the ultimate analysis.

1.4 INFLUENCE OF THE FEEDSTOCK AND OPERATING VARIABLES ON THE YIELD AND CHARACTERISTICS

Both porosity and SSA have a significant dependency on the choice of feedstock. The inorganic components (ash) of the feedstock cause pore blocking, while high organic carbon content helps to increase the micropore surface area. In particular, lignocellulosic biomass has lower ash content than manure or algal biomass. Also, its components, lignin, cellulose, and hemicellulose, have different thermal decomposition ranges. With its branched polymeric structure and short side chains, hemicellulose readily undergoes pyrolysis in the temperature range of 200–350°C. Cellulose is thermally stable due to the presence of ordered semi-crystalline chains linked together; it decomposes in the temperature range of 305–375°C. Finally, lignin, consisting of a complex structure of phenolic polymers, undergoes a slow, gradual decomposition, with the temperature range being 250–500°C according to Lu and Gu (2022), which, for other authors like Beusch (2021), is even wider. Thus, lignin contributes most significantly to char yield because it degrades slowly over a vast temperature range. Additionally, lignin-rich biomass has a higher specific surface area and porosity due to the stability of lignin, and biochars produced from lignin-rich feedstocks have a higher carbon content (Beusch 2021).

In summary, these factors indicate that lignocellulosic biomass is appropriate for biochar production with adequate characteristics: porosity, SSA, and HHV, according to the application. Considering this aspect, wood biomass is the best because it is lignin-rich and has the lowest ash content, followed by herbaceous and agricultural biomass.

The second most important variable influencing the biochar's porosity and SSA is pyrolysis temperature. Because pore development is also related to the release of volatile matter, at low pyrolysis temperatures (<400°C), porosity and SSA remain mostly constant or increase marginally. At temperatures over 500°C, the volatilization of biochar is at its highest, which is why the best SSA and porosity are in the range of 300–500°C (intermediate temperatures) (Collard and Blin 2014). But, at higher temperatures, the trend changes because, even though the pores may widen due to solid cracks, it is at the expense of lower SSA. This is also because temperature and ash content are directly correlated. So, ultimately, both porosity and SSA will decrease at higher pyrolysis temperatures.

As indicated above, the pyrolysis type has its biggest impact on the process's yield. However, although less influential, it still affects the biochar's characteristics, which are dictated by the heating rate that governs the heat and mass transfer in and out of the particles. Low heating rates produce biochar with higher carbon content, HHV, SSA, and lower ash content (Abdullah et al. 2023). On the other hand, high heating rates lead to fast depolymerization resulting in lower porosity and SSA. Lastly, it is important to point out that both high heating rates and long reaction times at high pyrolysis temperatures can damage the pore structure.

Alternatively, a sufficient reaction time and heating rate can accelerate the release of volatile matter and help develop the pore structure.

The size of the biomass particles affects the product yield, porosity, and SSA. Since particles with smaller diameters offer less opposition to the release of condensable gases, a higher liquid yield is expected. Also, higher pore volume and SSA are expected as well for the same cause. Finally, controlling the pressure of the inert atmosphere can lead to improvements in porosity and SSA. This is because higher pressure hinders the release of volatile matter and can implode the pore structure (Cetin et al. 2005). Consequently, atmospheric or even vacuum pressures are preferable. Figure 1.1 summarizes the impact each of the discussed variables has on the resulting yield and characteristics of the produced biochar.

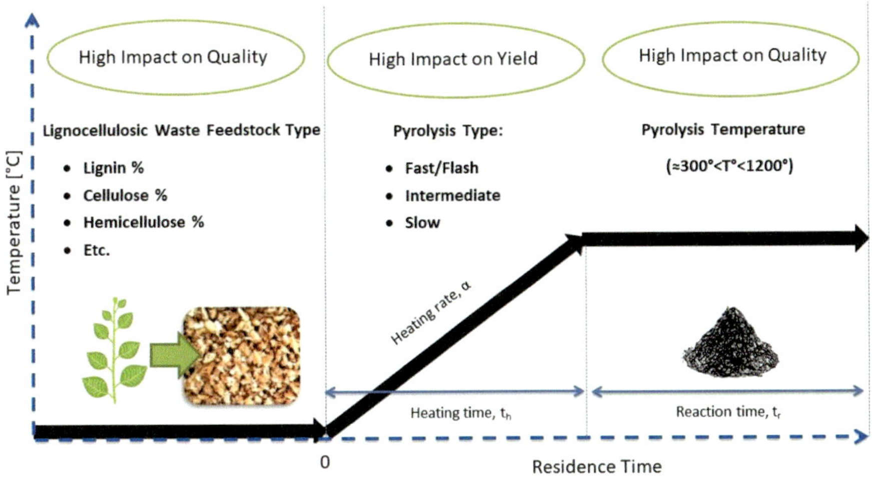

Figure 1.1 The relative impact of pyrolysis process variables on biochar yield and quality.

1.5 SOLAR ENERGY-ASSISTED PYRO-GASIFICATION

Solar energy-assisted pyro-gasification is emerging as a significant improvement in energy efficiency, economic viability, and environmental sustainability compared to conventional pyrolysis. This advanced method primarily utilizes renewable solar energy instead of electricity, which is often derived from non-renewable sources with detrimental environmental impacts. Using solar energy, the system provides the necessary heat for the endothermic reactions occurring during the conversion of lignocellulosic biomass into high-value by-products. Integrating solar energy in pyro-gasification demonstrated a promising pathway for cleaner and more sustainable bioenergy production.

A study compared the economic and environmental impacts of using concentrated solar thermal energy versus electric heating for the pyrolysis of date palm waste to produce biochar at approximately 400°C (Giwa et al. 2019). They found that solar pyrolysis reduced dependence on electricity, thereby lowering operational costs and exhibiting slightly higher economic indicators than conventional pyrolysis, with a return on investment and internal rate of return of 2.2% and 5% higher, respectively. The net present value was also higher for the solar process, indicating better long-term profitability. Furthermore, solar heating-based pyrolysis produced nearly 60% less CO_2 emissions, highlighting its significant sustainability and emissions reduction benefits.

Some studies have focused on solar pyrolysis utilizing only direct solar radiation (Morales et al. 2014). Generally, to achieve better control of process variables through a stable light source and to minimize the influence of external environmental factors, large-scale solar furnaces are simulated at the laboratory scale (Sobek and Werle 2019). Solar simulators employ concentrating mirrors or lenses to focus artificial light from high-power discharge lamps (typically xenon arc or tungsten halogen lamps) onto the sample placed inside a reactor. The heat flux density, corresponding to the temperature at the feedstock surface, is adjusted using shutters (Zeng et al. 2017).

Artificial solar pyrolysis typically achieves higher heating rates and temperatures than conventional pyrolysis. These variables, combined with the composition of the carbonaceous feedstock, are the primary factors influencing product yields, the composition of gases and bio-oil, and the morphological and chemical properties of the resulting biochar (Chintala 2018). The ability to reach higher temperatures and heating rates enhances the efficiency and effectiveness of the pyrolysis process, resulting in improved product quality and yield.

1.5.1 Influence of Operating Variables on the Yield and Characteristics of Biochar in Solar-Assisted Pyrolysis

Several studies have investigated the product distribution resulting from solar pyrolysis of various biomass feedstocks and the effects of temperature and heating rate. Solar furnaces typically produce two desirable products: bio-oils at lower temperatures, around 500°C, and gases at higher temperatures, approximately 1,200°C (Zeng et al. 2017).

For instance, Morales et al. (2014) investigated the design and construction of a parabolic-trough solar collector for the pyrolysis of orange peels in a borosilicate glass reactor. The solar pyrolysis system achieved an average irradiance of 12.55 kW/m^2 with a solar irradiation peak of 25.08 kW/m^2. They found that at an average reactor temperature of 290°C and a peak temperature of 465°C, approximately 79% of the orange peel biomass was volatilized, with the main product being bio-oil (77.6%), followed by char (20.9%) and non-condensable gases (1.4%). Additionally, they discovered that the chemical composition of the produced bio-oil included valuable compounds such as oleamide, squalene, and terpene, which are beneficial for the pharmaceutical and chemical industries (Morales et al. 2014).

A different study examined the slow solar pyrolysis of various waste biomass feedstocks, such as waste wood, waste straw, and sewage sludge. Indirect radiant heating was used with a 1.6 kW xenon lamp. The effective heat flux density at the reactor surface reached 172 kW/m^2, resulting in higher final process temperatures exceeding 800°C. Heating rates increased from 3.95 to 5.49°C/min. They found that for all three biomass types, bio-oil was the most yielding subproduct, with yields ranging from 68.4–70.2%, 63.8–61.4%, and 50.0–52.8% for waste wood, waste straw, and sewage sludge, respectively (Sobek and Werle 2021). On the other hand, in terms of char yields, an increase in heating rate caused a slight decrease in final char yields, with values generally between 21.2% and 39.5%. The poorest performance was observed in gas production, with values ranging from 8.6 to 11.5%.

At higher solar pyrolysis temperatures and heating rates, a decrease in biochar yield and an increase in liquid and gaseous products are expected. This is due to the thermal cracking of heavy hydrocarbon compounds and the volatilization of organic components in the biomass, which release more gases and vapors, consequently reducing the solid residue.

Ayala-Cortés et al. (2019) investigated the impact of solar pyrolysis temperatures ranging from 500 to 900°C and heating rates of 4 and 30°C/min on the yield of biochar derived from *Agave angustifolia* bagasse fibers biomass. The results indicated that slower heating rates led to an increase in char yield (24.6% at 4°C/min compared to 21.3% at 30°C/min). Conversely, an elevation in solar pyrolysis temperature from 500 to 900°C resulted in a reduction in char yield from 21.3 to 16.3%, respectively, attributed to the heightened crystallinity of the biochar. Furthermore, both variables significantly influenced the morphological structure, surface area, and electrochemical response of the biochar, leading to the formation of macropores and cracks at temperatures below 650°C.

Additionally, in a study conducted on solar intermediate pyrolysis of rice husk with a moderate heating rate of 160°C/min and temperatures ranging between 500°C and 800°C, it was observed that this process produced relatively higher liquid yields compared to conventional pyrolysis of the same biomass with lower heating rates (Weldekidan et al. 2018). The maximum values for bio-oil, biochar, and gas were 43.1% at 700°C, 43.0% at 500°C, and 25.5% at 800°C, respectively.

During fast solar pyrolysis, higher gas yields and energy content are achieved with faster heating rates (ranging from 10 to 500°C/s) and smaller particle sizes. This is due to the increased concentrations of combustible gases such as hydrogen, carbon monoxide, and methane. These gases can be effectively used as fuels in internal combustion engines or power plants, providing a renewable energy source with a high calorific value (Weldekidan et al. 2020). Temperature (ranging from 800 to 1,600°C) also has a significant impact on product distribution at a constant heating rate of 50°C/s. Higher temperatures favor gas production while reducing the yield of liquid and solid pyrolysis products (Chintala 2018; Weldekidan et al. 2020). This effect is attributed to the secondary decomposition of char residue at elevated temperatures.

Smaller biomass particles have a greater surface-to-volume ratio, enhancing heat transfer rates than larger particles. This can increase gas and bio-oil yields during pyrolysis processes with high heating rates (Weldekidan et al. 2020).

The existing literature suggests that solar energy represents a viable and sustainable option for biomass pyrolysis, particularly in regions characterized by high solar insolation. It underscores the potential of integrating solar energy with biomass pyrolysis to improve the efficiency and sustainability of renewable energy production and the generation of valuable chemical products. This approach harnesses renewable energy and mitigates greenhouse gas emissions compared to conventional methods, rendering it an environmentally friendly alternative.

1.6 MICROWAVE-ASSISTED PYRO-GASIFICATION

The widespread adoption of microwave pyrolysis technology has helped address some of the limitations of traditional pyrolysis plants. This technology enables fast and efficient volumetric heating within the core by directly interacting microwave energy with the fuel. As a result of microwave radiation, the heating rate of the fuel is substantially increased, leading to a significant reduction in conversion time from hours to minutes, decreased energy consumption, and an increased product yield. In addition, the power supply activates and deactivates instantly when turned on or off. Microwave pyrolysis operates at a lower temperature compared to conventional pyrolysis. This method also enables the production of high-quality products that can be used in various fields (Fricler et al. 2023).

Microwaves are typically defined as electromagnetic waves. These waves, or microwaves, comprise two main components: electric and magnetic fields, which are perpendicular to each other. In the electromagnetic spectrum, microwaves are characterized by frequencies ranging from 300 MHz to 300 GHz and wavelengths from 0.001 to 1 meter. Most microwave reactors used for chemical synthesis and domestic microwave ovens operate at a wavelength of approximately 12.25 cm, corresponding to a frequency of 2.45 GHz (Mohd Fuad et al. 2019).

Recently, there has been a heightened emphasis on controlling greenhouse gas emissions from industrial processes. Consequently, stakeholders have concentrated on the electrification of thermal processes as a crucial strategy for achieving industrial decarbonization. Microwave technology has the potential to significantly contribute to the electrification of thermal processes, allowing for the efficient utilization of intermittent renewable energy sources like solar and wind. This is especially relevant for small-scale, distributed chemical production systems. In addition to facilitating the electrification of thermal processes, microwaves offer several processing advantages over traditional thermal methods. These include rapid heating for quick startup and shutdown, enhanced reaction rates, and improved product selectivity. The advantages arise from the distinctive way microwaves transfer energy compared to conventional heating methods. Unlike traditional reactors that depend on heated fluidizing mediums or heating elements for heat transfer through convection and conduction, microwave reactors generate heat directly within materials. This is achieved through dielectric losses induced by mechanisms such as dipole rotation, ion conduction, and interfacial (Maxwell-Wagner) polarization when exposed to an electromagnetic field. The effectiveness of heating a material in a microwave field depends on its intrinsic dielectric properties (Ellison et al. 2023).

Microwave processing research explores converting carbonaceous feedstocks, including coal, biomass, and plastic waste, into chemical products through pyrolysis or gasification. This area has gained significant attention from researchers. While coal has been traditionally linked to higher atmospheric carbon emissions, recent studies have identified potential methods for utilizing coal in low-emission chemical production.

Pyrolysis temperatures are typically achieved through conventional heating, where heat is transferred to the material by conduction and convection. However, conventional heating has several drawbacks, including the need for small particle sizes to achieve rapid heat and mass transfer, poor temperature control, heat losses to the reactor walls, fluidizing beds (such as sand), and the surroundings. These limitations result in potentially high processing costs in large-scale systems due to the need for biomass pretreatment (such as milling) and reduced process efficiency, which limits the industrial application of conventional pyrolysis technology. Microwave heating can reduce costs by generating heat directly within the material. This is achieved through selective and volumetric heating, which minimizes system losses. Microwave pyrolysis may also eliminate the need for biomass grinding, as microwaves can penetrate large particle sizes, provided microwave energy is coupled with the material and an appropriate applicator configuration (Ellison et al. 2020).

1.6.1 Microwave-Assisted Pyrolysis Reaction Mechanism

In conventional heating processes, heat is transferred from an external source to the material's surface and then to its center through conduction, convection, and radiation. This makes the process slow, inefficient, and highly dependent on convection currents and the material's thermal conductivity. However, microwave heating involves directly converting electromagnetic energy into thermal energy, allowing the material to be heated more efficiently. Due to the volumetric heating of the feedstock, heat losses are minimized compared to conventional pyrolysis (Zhang et al. 2020b).

In microwave-assisted pyrolysis, microwave energy is absorbed by the biomass particle and converted into thermal energy, leading to a temperature gradient from the particle's interior to its exterior. The volatile materials released during the process diffuse from the inner core of the biomass feedstock to its outer surface at a lower temperature region. However, in conventional pyrolysis, the temperature gradient is from the outer surface (heating source) to the inner part of the biomass particle. In this case, heat is transferred from the heating source to the outer surface of the biomass particle and then to its inside. The released volatile materials, in this case, diffuse from the inside core of the biomass to its outside surface, which is at a higher temperature region (Chen et al. 2019). Therefore, the heating and reaction mechanisms of microwave-assisted pyrolysis offer several advantages over conventional pyrolysis due to improved heating mechanisms. The advantages of microwave-assisted processing include higher heating rates, non-contact heating, uniform energy transfer, lower thermal inertia, improved heating efficiency, the need for less pretreatment for feedstock, faster response rates, and greater control and safety. Additionally, microwave-assisted pyrolysis saves energy by accelerating thermochemical reactions and reducing process reaction time (Mohd Fuad et al. 2019; Zhang et al. 2016).

1.6.2 Parameters of Microwave-Assisted Pyro-Gasification

Effective microwave heating relies on the material's dielectric properties, including the dielectric constant and dielectric loss factor, which determine its ability to interact with an applied electric field. Microwave heating of low-loss materials requires the addition of a microwave absorber to promote dielectric heating in the bulk material. Various microwave absorbers have been documented in the literature for pyrolysis applications, including carbonaceous materials, metal oxides, other inorganic additives, and ionic liquids. Mixing char with biomass feedstocks has been found to improve the dielectric properties of the bulk mixture notably. When a microwave absorber is blended with the biomass feedstock, the absorber particles are selectively heated, transferring heat to neighboring particles and leading to bulk heating of the material. As the process commences within the biomass particles, they partially convert to char and become efficient absorbers themselves. Even the distribution of the absorber within the feedstock is essential to ensure consistent heating throughout the material (Antunes et al. 2018).

Ellison and co-workers investigated the effects of microwave power and percentage of residual carbon adsorbent on process temperature, pyrolysis product yield, and composition. The results showed that the addition of carbon significantly improved the dissipation of microwave energy compared to biomass alone, allowing pyrolysis temperatures to be achieved that could not be achieved with biomass alone. Interestingly, researchers found no significant temperature difference between 10% and 20% charcoal, indicating that using charcoal from biomass pyrolysis is sufficient to catalyze microwave heating (Ellison et al. 2020).

Ellison et al. (2023) investigated the effects of microwave and conventional pyrolysis and gasification on the char structure characteristics of different coal types. The characterization results revealed notable differences between the physical and chemical properties of chars produced by microwave and conventional heating. In most of the coal samples tested, conventionally generated chars had a higher surface area than those generated by microwaves after both pyrolysis and gasification. The hypothesis is that selective heating by microwaves creates localized hot spots within the char, causing rapid pore development followed by rapid collapse. The two heating methods showed a noticeable difference in the X-ray diffractograms. The microwave-generated chars exhibited evidence of crystalline carbon development, especially for the low-ash coal chars. This was not observed in conventionally generated chars. Microwave and conventional heating affect the thermal decomposition behavior and porous structure development of coal char in different ways. These differences can be attributed to the distinct heat generation and transfer mechanisms inherent to each heating mode. Conventional heating uniformly heats the coal bed, while microwaves selectively couple to the coal material based on variations in dielectric susceptibility within the coal bed. Although the bulk reaction temperature was maintained at 700°C during the experimental tests, parts of the coal bed may have exceeded this temperature during microwave selective heating.

Jiang et al. (2023) investigated the impacts of traditional pyrolysis and microwave pyrolysis on the evolution of products from the thermal treatment of wet and dried banana peel at 400°C and 600°C. The promotional effects of microwave pyrolysis on the cracking and gasification of volatiles resulted in more gases at the expense of bio-oil and biochar. The organic components in banana peel were thermally unstable, and combined with the high ash content, this led to a low degree of aromatization of the biochar in both microwave and traditional pyrolysis. Nevertheless, the biochar from microwave pyrolysis exhibited higher thermal stability due to enhanced cracking of oxygen-containing functionalities such as C=O and aliphatic structures on the surface of the biochar. These intense cracking reactions also produced biochar with a coarse and rugged surface during microwave pyrolysis.

In a study conducted by Mohit and Remya (2023), greywater-cultivated native polyculture microalgae were converted into biochar through microwave pyrolysis, achieving a maximum yield of 49.9% and a fixed carbon content of 68.7% under optimum conditions of 800 W microwave power and 8.6 minutes of reaction time. The optimized biochar exhibited low moisture content (5.9%) and volatile matter (20.2%), elevated fixed carbon content (68.7%), and a high heating value (28.3 MJ/kg). These characteristics align with coal standards, making the biochar suitable for biofuel applications.

According to the literature on the formation of solid byproducts from the thermochemical conversion of biomass, biochar is produced through the repolymerization of bio-oil constituents (Wang et al. 2017). Mechanistic explanations for biochar formation have focused on identifying the corresponding precursors and the thermodynamic behavior of potential precipitation reactions. These precursors for biochar formation were identified based on the changes in the relative composition of the bio-oil. Silos-Llamas et al. (2023) conducted a study to explore the reaction mechanisms involved in microwave-assisted fast pyrolysis to reveal the thermal decomposition of agricultural waste. They found that biochar formation during the microwave-assisted fast pyrolysis of pecan nutshells was due to the endothermic repolymerization of aromatic aldehydes (e.g., 2-hydroxybenzaldehyde and 3,5-dimethoxy-4-hydroxycinnamaldehyde) and aromatic ethers (e.g., 1,2,4-trimethoxybenzene and 1,2,4-trimethoxy-5-methylbenzene). Consequently, higher microwave power input and longer processing times improved the yield.

In contrast, biochar formation during the microwave-assisted fast pyrolysis of sugarcane bagasse and orange seeds resulted from the exothermic repolymerization of furans, leading to a lower biochar yield at higher microwave power inputs and longer processing times. Organic waste like sewage sludge, food waste, and livestock manure possesses several properties that make it challenging to use as a feedstock for microwave-assisted pyrolysis processes. Higher fixed carbon content in the organic waste significantly increases the gas yield and the H_2 and

CO contents, particularly enhancing char generation. Cellulose content negatively affects char and oil yields but positively impacts gas yields, while lignin and hemicellulose are strongly associated with increased char yields (Oh et al. 2024).

Several factors, including feedstock type, pyrolysis temperature, residence time, pressure, heating rate, susceptor, particle size, and microwave power, influence the conversion of agricultural waste biomass into biochar via microwave-assisted pyrolysis. Agricultural waste biomass with high fixed carbon and ash content tends to produce higher yields of biochar. Conversely, high pyrolysis temperatures and higher heating rates produce lower biochar yields. Additionally, biochar can be used as a susceptor in microwave-assisted pyrolysis to enhance heat and mass transfer (Potnuri et al. 2023). The ratio of volatile carbon in oil and gas and the proportion of carbon stabilized in biochar is influenced by the relative amounts of hemicellulose, cellulose, and lignin components. When pyrolyzed at temperatures around 500°C, feedstock with high lignin concentrations produces substantial yields of biochar and experiences minimal pyrolysis weight loss (Tripathi et al. 2016; Sharma et al. 2015).

Temperature significantly influences the reaction kinetics, thus impacting the yields and composition of final products in the microwave-assisted pyrolysis process. A lab-scale pyrolysis study was performed using pine, spruce chips, mixed larch, and softwood pellets to produce biochar at temperatures ranging from 350 to 550°C. The stability of the biochar increases with higher pyrolysis temperatures, but the yield of stable biochar decreases nearly linearly with temperature. Higher temperatures result in alkaline biochar, while biochar produced at lower temperatures is acidic (Ma et al. 2019).

The pressure inside the reactor plays a crucial role in determining the yield of pyrolysis products. Increasing pressure enhances the vapor residence time, resulting in the decomposition of organic matter and the production of secondary carbon. During the repolymerization process, the secondary carbon on the biochar surface enhances the production of the biochar (Manyà et al. 2013; Anand et al., 2022).

In microwave-assisted pyrolysis, the residence time is determined by the duration of the solid biomass processing, which typically ranges from minutes to hours. Prolonged residence times in microwave processing lead to the formation of non-condensable gases due to secondary cracking, which negatively impacts the yield of liquid bio-oil products and increases energy consumption. It is generally beneficial to employ short residence times to minimize the risk of secondary cracking of volatiles. Establishing a direct relationship between residence time and biomass can be complex, mainly due to influential factors such as temperature (Sharma et al. 2015; Siddique et al. 2022).

The choice of catalyst influences the distribution of pyrolysis products. Metals in raw biomass ash inhibit biochar synthesis, and both acidic and basic catalysts have similar effects. Certain catalysts increase biochar yield by reducing the liquid and gaseous product yield, while others boost liquid yield at the expense of gaseous product yield and biochar yield. Elements in the biomass, like ash, can act as catalysts, but they do not improve biochar output. Acid-base catalysts are believed to increase biochar production (Tripathi et al. 2016).

A study conducted by Kostas et al. (2020) was the first to demonstrate the successful pyrolysis of olive pomace using microwaves to produce biochar and bio-oil. The energy needed to pyrolyze up to 80% of the olive pomace was determined to be as low as 3.6 kJ/g. While microwave power did not affect the overall yields or chemical composition of the bio-oils obtained, it did alter the textural properties of the bio-chars and their ability to remove methylene blue dye. Optimal processing conditions were identified within the 3.6 kJ/g energy requirement, with a microwave power of 200 W and a processing time of 180 seconds. Under these conditions, the produced biochar had a surface area of 392.3 m^2/g, a micropore volume of 0.15 cm^3/g, and a methylene blue removal efficiency of 40 qMB mg/g. The results from this study demonstrate the advantages of microwave heating in pyrolysis systems and highlight a novel and promising approach for value-added recovery from natural waste resources.

1.7 CONCLUSIONS

This chapter addressed thermal processes applied to lignocellulosic residues for biochar production, highlighting the importance of this technology in the context of sustainability and renewable resource utilization. The chemical and physical fundamentals of lignocellulosic biomass were presented, along with predictive tools that allowed for estimating the calorific value and the behavior of residues during thermal processes.

Through the analysis of a range of technologies, from conventional pyrolysis to advanced methods such as microwave-assisted pyrolysis and concentrated solar energy, advancements in terms of energy efficiency and product quality were demonstrated. The effects of key operational variables, such as heating rate, temperature, and residue type, on biochar yield and properties were explored.

The chapter highlighted the versatility and practical applications of biochar in different contexts and underscored the need to continue innovating in cleaner and more sustainable technologies by integrating renewable energy sources. This approach established a solid foundation for the topics to be addressed in subsequent chapters, which will delve deeper into the implementation and optimization of these technologies on both industrial and small scales.

REFERENCES

Abdullah, N., R.M. Taib, N.S.M. Aziz, M.R. Omar and N.M. Disa. 2023. Banana pseudo-stem biochar derived from slow and fast pyrolysis process. Heliyon. 9(1): e12940.

Akinpelu, D.A., O.A. Adekoya, P.O. Oladoye, C.C. Ogbaga and J.A. Okolie. 2023. Machine learning applications in biomass pyrolysis: from biorefinery to end-of-life product management. Digital Chemical Engineering. 8: 100103.

Anand, A., V. Kumar and P. Kaushal. 2022. Biochar and its twin benefits: Crop residue management and climate change mitigation in India. Renewable and Sustainable Energy Reviews. 156: 111959.

Angin, D. 2013. Effect of pyrolysis temperature and heating rate on biochar obtained from pyrolysis of safflower seed press cake. Bioresource Technology. 128: 593–597.

Antunes, E., M.V. Jacob, G. Brodie and P.A. Schneider. 2018. Microwave pyrolysis of sewage biosolids: Dielectric properties, microwave susceptor role and its impact on biochar properties. Journal of Analytical and Applied Pyrolysis. 129: 93–100.

Ayala-Cortés, A., D.R. Lobato-Peralta, C.E. Arreola-Ramos, D.C. Martínez-Casillas, D.E. Pacheco-Catalán, A.K. Cuentas-Gallegos, et al. 2019. Exploring the influence of solar pyrolysis operation parameters on characteristics of carbon materials. Journal of Analytical and Applied Pyrolysis. 140: 290–298.

Basu, P. 2010. Pyrolysis and torrefaction. pp. 65–96. *In*: P. Basu (ed.). Biomass Gasification and Pyrolysis. Academic Press. ISBN 9780123749888.

Beusch, C. 2021. Biochar as a soil ameliorant: How biochar properties benefit soil fertility—A review. Journal of Geoscience and Environment Protection. 9: 28–46.

Cetin, E., R. Gupta and B. Moghtaderi. 2005. Effect of pyrolysis pressure and heating rate on radiata pine char structure and apparent gasification reactivity. Fuel. 84(10): 1328–1334.

Chen, W.H., J. Peng and X.T. Bi. 2015. A state-of-the-art review of biomass torrefaction, densification and applications. Renewable and Sustainable Energy Reviews. 44: 847–866.

Chen, L., Z. Yu, H. Xu, K. Wan, Y. Liao and X. Ma. 2019. Microwave-assisted co-pyrolysis of Chlorella vulgaris and wood sawdust using different additives. Bioresource Technology. 273: 34–39.

Chintala, V. 2018. Production, upgradation and utilization of solar assisted pyrolysis fuels from biomass–A technical review. Renewable and Sustainable Energy Reviews. 90: 120–130.

Collard, F.X. and J. Blin. 2014. A review on pyrolysis of biomass constituents: Mechanisms and composition of the products obtained from the conversion of cellulose, hemicelluloses and lignin. Renewable and Sustainable Energy Reviews. 38: 594–608.

Ellison, C.R., R. Hoff, C. Mărculescu and D. Boldor. 2020. Investigation of microwave-assisted pyrolysis of biomass with char in a rectangular waveguide applicator with built-in phase-shifting. Applied Energy. 259: 114217.

Ellison, C., V. Abdelsayed and M.W. Smith. 2023. Analysis of char structure and composition from microwave and conventional pyrolysis/gasification of low and middle rank coals. Fuel. 354: 129301.

Fang, J., L. Zhan, Y.S. Ok and B. Gao. 2018. Minireview of potential applications of hydrochar derived from hydrothermal carbonization of biomass. Journal of Industrial and Engineering Chemistry. 57: 15–21.

Ferreira, S., E. Monteiro, P. Brito and C. Vilarinho. 2020. A Comprehensive Review on Biomass Gasification Modified Equilibrium Models. *In*: P. Khumprom and M. Bošnjakovi (eds). Advances in Energy Research. Hyderabad, India: Vide Leaf.

Fricler, V.Y., G.S. Nyashina, K.Y. Vershinina, K.V. Vinogrodskiy, A.S. Shvets and P.A. Strizhak. 2023. Microwave pyrolysis of agricultural waste: Influence of catalysts, absorbers, particle size and blending components. Journal of Analytical and Applied Pyrolysis. 171: 105962.

Ghassemi, H. and R. Shahsavan-Markadeh. 2014. Effects of various operational parameters on biomass gasification process; a modified equilibrium model. Energy Conversion and Management. 79: 18–24.

Giwa, A., A. Yusuf, O. Ajumobi and P. Dzidzienyo. 2019. Pyrolysis of date palm waste to biochar using concentrated solar thermal energy: Economic and sustainability implications. Waste Management. 93: 14–22.

Hornung, A. 2013. Intermediate pyrolysis of biomass. *In*: L. Rosendahl (ed.). Biomass Combustion Science, Technology and Engineering. Woodhead Publishing. 172–186.

Jiang, Y., C. Li, L. Zhang, M. Fan, S. Zhang, W. Gao, et al. 2023. Pyrolysis of banana peel with microwave and furnace as the heating sources: The distinct impacts on evolution of the pyrolytic products. Process Safety and Environmental Protection. 173: 373–383.

Kaylen, M., D.L. Van Dyne, Y.S. Choi and M. Blase. 2000. Economic feasibility of producing ethanol from lignocellulosic feedstocks. Bioresource Technology. 72(1): 19–32.

Kostas, E.T., G. Durán-Jiménez, B.J. Shepherd, W. Meredith, L.A. Stevens, O.S. Williams, et al. 2020. Microwave pyrolysis of olive pomace for bio-oil and bio-char production. Chemical Engineering Journal. 387: 123404.

Leng, L., Q. Xiong, L. Yang, H. Li, Y. Zhou, W. Zhang, et al. 2021. An overview on engineering the surface area and porosity of biochar. Science of the Total Environment. 763: 144204.

Li, Y. and S. You. 2022. Biochar soil application: Soil improvement and pollution remediation. *In*: Daniel C.W. Tsang and Yong Sik Ok (eds). Biochar in Agriculture for Achieving Sustainable Development Goals. Academic Press. 97–102.

Lin, Y., X. Ma, X. Peng, Z. Yu, S. Fang, Y. Lin and Y. Fan. 2016. Combustion, pyrolysis and char CO_2-gasification characteristics of hydrothermal carbonization solid fuel from municipal solid wastes. Fuel. 181: 905–915.

Lu, X. and X. Gu. 2022. A review on lignin pyrolysis: Pyrolytic behavior, mechanism, and relevant upgrading for improving process efficiency. Biotechnology for Biofuels and Bioproducts. 15(1): 106.

Ma, Z., Y. Zhang, Y. Shen, J. Wang, Y. Yang, W. Zhang, et al. 2019. Oxygen migration characteristics during bamboo torrefaction process based on the properties of torrefied solid, gaseous, and liquid products. Biomass and Bioenergy. 128: 105300.

Manyà, J.J., F.X. Roca and J.F. Perales. 2013. TGA study examining the effect of pressure and peak temperature on biochar yield during pyrolysis of two-phase olive mill waste. Journal of Analytical and Applied Pyrolysis. 103: 86–95.

Maschio, G., C. Koufopanos and A. Lucchesi. 1992. Pyrolysis, a promising route for biomass utilization. Bioresource Technology. 42: 219–231.

Mohd Fuad, M.A.H., M.F. Hasan and F.N. Ani. 2019. Microwave torrefaction for viable fuel production: A review on theory, affecting factors, potential and challenges. Fuel. 253: 512–526.

Mohit, A. and N. Remya. 2023. Optimization of biochar production from greywater grown polyculture microalgae using microwave pyrolysis. Bioresource Technology. 388: 129666.

Morales, S., R. Miranda, D. Bustos, T. Cazares and H. Tran. 2014. Solar biomass pyrolysis for the production of bio-fuels and chemical commodities. Journal of Analytical and Applied Pyrolysis. 109: 65–78.

Mujtaba, M., L. Fraceto, M. Fazeli, S. Mukherjee, S.M. Savassa, G.A. de Medeiros, et al. 2023. Lignocellulosic biomass from agricultural waste to the circular economy: A review with focus on biofuels, biocomposites and bioplastics. Journal of Cleaner Production. 136815.

Oh, D.Y., D. Kim and K.Y. Park. 2024. A comprehensive comparative study on microwave-assisted pyrolysis products derived from raw and digested organic waste, with emphasis on sewage sludge, food waste, and livestock manure. Heliyon. 10(8): e29618.

Park, J., Y. Lee, C. Ryu and Y.K. Park. 2014. Slow pyrolysis of rice straw: analysis of products properties, carbon and energy yields. Bioresource technology. 155: 63–70.

Potnuri, R., D.V. Surya, C.S. Rao, A. Yadav, V. Sridevi and N. Remya. 2023. A review on analysis of biochar produced from microwave-assisted pyrolysis of agricultural waste biomass. Journal of Analytical and Applied Pyrolysis. 173: 106094.

Ragula, U.B.R., S. Devanathan and S. Subramanian. 2019. Modeling and optimization of product profiles in biomass pyrolysis. *In*: Recent Advances in Pyrolysis. IntechOpen. 43.

Rangabhashiyam, S. and P. Balasubramanian. 2019. The potential of lignocellulosic biomass precursors for biochar production: performance, mechanism and wastewater application-a review. Industrial Crops and Products. 128: 405–423.

Sahay, S. 2022. Lignin: Value addition is key to profitable biomass biorefinery. *In*: S. Sahay (ed.). Handbook of Biofuels. Academic Press: Cambridge, MA, USA. 207–232.

Sharma, A., V. Pareek and D. Zhang. 2015. Biomass pyrolysis—A review of modelling, process parameters and catalytic studies. Renewable and Sustainable Energy Reviews. 50: 1081–1096.

Siddique, I.J., A.A. Salema, E. Antunes and R. Vinu. 2022. Technical challenges in scaling up the microwave technology for biomass processing. Renewable and Sustainable Energy Reviews. 153: 111767.

Silos-Llamas, A.K., H. Sudibyo, V. Hernández-Montoya, W. Meredith and G. Durán-Jiménez. 2023. Fast pyrolysis of agricultural residues: Reaction mechanisms and effects of feedstock properties and microwave operating conditions on the yield and product composition. Journal of Analytical and Applied Pyrolysis. 175: 106217.

Sobek, S. and S. Werle. 2019. Solar pyrolysis of waste biomass: Part 1 reactor design, Renew. Energy. 143: 1939–1948.

Sobek, S. and S. Werle. 2021. Solar pyrolysis of waste biomass: A comparative study of products distribution, in situ heating behavior, and application of model-free kinetic predictions. Fuel. 292: 120365.

Sonil, N., M. Javeed, N.R. Sivamohan, A.K. Janusz and K.D. Ajay. 2014. Pathways of lignocellulosic biomass conversion to renewable fuels. Biomass Conversion and Biorefinery. 4: 157–191.

Torres-Sciancalepore, R., A. Fernandez, D. Asensio, M. Riveros, M.P. Fabani, G. Fouga, et al. 2022. Kinetic and thermodynamic comparative study of quince bio-waste slow pyrolysis before and after sustainable recovery of pectin compounds. Energy Conversion and Management. 252: 115076.

Torres-Sciancalepore, R., M. Riveros-Gomez, D. Zalazar-Garcia, D. Asensio, M.P. Fabani, R. Rodriguez, et al. 2023. Two-step valorization of invasive species *Rosa rubiginosa* L. husk waste through eco-friendly optimized pectin extraction and subsequent pyrolysis. Journal of Environmental Chemical Engineering. 11(5): 110802.

Tripathi, M., J.N. Sahu and P. Ganesan. 2016. Effect of process parameters on production of biochar from biomass waste through pyrolysis: A review. Renewable and Sustainable Energy Reviews. 55: 467–481.

Wang, S., G. Dai, H. Yang and Z. Luo. 2017. Lignocellulosic biomass pyrolysis mechanism: A state-of-the-art review. Progress in Energy and Combustion Science. 62: 33–86.

Weldekidan, H., V. Strezov, G. Town and T. Kan. 2018. Production and analysis of fuels and chemicals obtained from rice husk pyrolysis with concentrated solar radiation. Fuel. 233: 396–403.

Weldekidan, H., V. Strezov, R. Li, T. Kan, G. Town, R. Kumar, et al. 2020. Distribution of solar pyrolysis products and product gas composition produced from agricultural residues and animal wastes at different operating parameters. Renewable Energy. 151: 1102–1109.

Wiselogel, A., J. Tyson and D. Johonsson. 1996. Biomass feedstock resources and its composition. *In*: C.E. Wyman (ed.). Handbook on Bioethanol: Production and Utilization. Taylor and Francis. 105–118.

Yaashikaa, P.R., P.S. Kumar, S. Varjani and A. Saravanan. 2020. A critical review on the biochar production techniques, characterization, stability and applications for circular bioeconomy. Biotechnology Reports. 28: 570.

Zeng, K., D. Gauthier, J. Soria, G. Mazza and G. Flamant. 2017. Solar pyrolysis of carbonaceous feedstocks: A review. Solar Energy. 156: 73–92.

Zhang, B., Z. Zhong, Q. Xie, S. Liu and R. Ruan. 2016. Two-step fast microwave-assisted pyrolysis of biomass for bio-oil production using microwave absorbent and HZSM-5 catalyst. Journal of Environmental Sciences. 45: 240–247.

Zhang, Y., L. Wan, J. Guan, Q.A. Xiong, S. Zhang and X. Jin. 2020a. A review on biomass gasification: Effect of main parameters on char generation and reaction. Energy & Fuels. 34(11): 13438–13455.

Zhang, Y., Y. Cui, S. Liu, L. Fan, N. Zhou, P. Peng, et al. 2020b. Fast microwave-assisted pyrolysis of wastes for biofuels production–A review. Bioresource Technology. 297: 122480.

Chapter 2

Obtaining Biochar: Modeling and Simulation of Pyrolysis and Gasification

Daniela Zalazar-García*[a], Andrés Reyes-Urrutia[b], Marcelo Echegaray[a], Carla Lorenzo-Doncel[c] and Iside Mut[a]

[a]Instituto de Ingeniería Química, Fac. de Ingeniería, Grupo Vinculado al PROBIEN (Universidad Nacional de San Juan, CONICET, Universidad Nacional del Comahue), San Juan, Argentina

[b]Instituto de Investigación y Desarrollo en Ingeniería de Procesos, Biotecnología y Energías Alternativas, PROBIEN (CONICET, Universidad Nacional del Comahue), Neuquén, Argentina

[c]Departamento de Ingeniería Química, Fac. de Ingeniería, Universidad Nacional de San Juan, San Juan, Argentina

2.1 INTRODUCTION

Biochar is a solid product of pyro-gasification (thermal decomposition of biomass in the absence or deficiency of O_2). Its use has increased due to its properties as a soil improver, pollutant sequestration, and consequently increased agricultural productivity. In recent years, it has become a focus of study because it plays a fundamental role in climate change mitigation. These characteristics directly align with Sustainable Development Goal (SDG) 13: Climate Action, which calls for urgent action to combat climate change and its impacts.

In the technology field, computer modeling and simulation are essential for evaluating the performance of industrial-scale processes. Traditional modeling and

*Corresponding author: Daniela Zalazar-García (dzalazar@unsj.edu.ar)

DOI: 10.1201/9781032726694-2

simulation methods allow for the analysis of the behavior of operating variables. At the same time, the emergence of innovative computational approaches such as artificial intelligence (AI) and machine learning (ML) offers the possibility of anticipating the results of experiments performed in laboratory and pilot plant reactors. These models are widely used to support investment decisions before the design and construction phase of industrial facilities and equipment, highlighting the importance of their reliability for practical applications.

This chapter focuses on exergy analysis and computational methods such as artificial intelligence (AI), machine learning (ML), and fluid and transport modeling, which enable the simulation of multiphase flows in thermochemical reactors.

Exergy is a thermodynamic parameter that is holistically used to measure the efficiency and processes of environmental impacts. Thus, exergy measures the work potential (energy quality) in a determined system and its environment. Its main difference with energy is its conservation, whereas exergy degrades due to the irreversibilities of any nonideal process. Therefore, the decrease in energy quality can be measured using the concept of exergy, which is a crucial indicator in energy analysis and environmental performance.

The pioneers in the exergy approach application were Zoran Rant in the 1950s and Wall in the 1970s. They introduced exergy as an objective means of calculation to manage the use of energy and material resources, finding connections with ecological and economic systems. In biochar production, exergy analysis quantifies the associated efficiency and environmental impact. By evaluating the input and output streams of the system, it is possible to identify areas of energy and material deficiencies and thus improve the efficiency of the process. This is essential for developing more sustainable biochar production methods.

Furthermore, the exergy analysis allows for measuring the greenhouse gas emission (GGE) associated with biochar production. That is achieved by evaluating the exergy consumption throughout the life cycle of an agro-industrial system and identifying the emissions generated at each stage. This approach allows for comparing the environmental impacts of different processes and development strategies to minimize such emissions.

Despite its many advantages, the exergy approach is not always complete and needs to be complemented by other analysis methods, such as life cycle analysis. In this context, the integrated approach LCA-Exergy Analysis improves the sustainability assessment.

Regarding the computational approach to artificial intelligence, several algorithm types, including decision trees, neural networks, and regression, are commonly used. Statistics play a crucial role in evaluating the performance of these methods. Thus, the most used statistical tools are the F-test and the coefficient of determination (R^2). In biochar production, machine learning predicts process outcomes by analyzing large data sets from previous experiments and operational data. This approach helps to understand the implicit relationships between different process variables and the desired product quality and yield. By identifying crucial variables through sensitivity analysis obtained from model predictions, it is possible to optimize the process. Furthermore, by establishing the relationships between these variables, it is possible to evaluate design parameters, operating parameters, costs, and indices that assess the environmental sustainability of the process.

On the other hand, Computational Fluid Dynamics (CFD) simulation in biochar production addresses several aspects. First, it models the interaction between gases and solids, including gas flow and heat and mass transfer, which is fundamental for understanding system behavior. In addition, it facilitates the understanding of heat distribution and heat transfer within the pyrolysis reactor, a crucial factor for the efficiency of the process. It also represents the chemical reactions during pyrolysis, which are necessary for understanding how the products are formed and evolve. CFD makes it possible to control and optimize operating conditions by providing detailed information on the temperature distribution in the reactor. It also evaluates different reactor designs and operating configurations to improve the efficiency and quality of the biochar produced. Finally, it provides data on operating conditions and process parameters to adjust and optimize the quality of the final biochar. Overall, CFD simulation is a powerful tool to improve and optimize all aspects of the biochar production process, from heat transfer and fluid dynamics to final product quality.

In summary, this chapter will offer an in-depth insight into the involved processes and the potential for improved efficiency and sustainability. It will also examine the role of exergy analysis and advanced computational methods in optimizing biochar production (Figure 2.1).

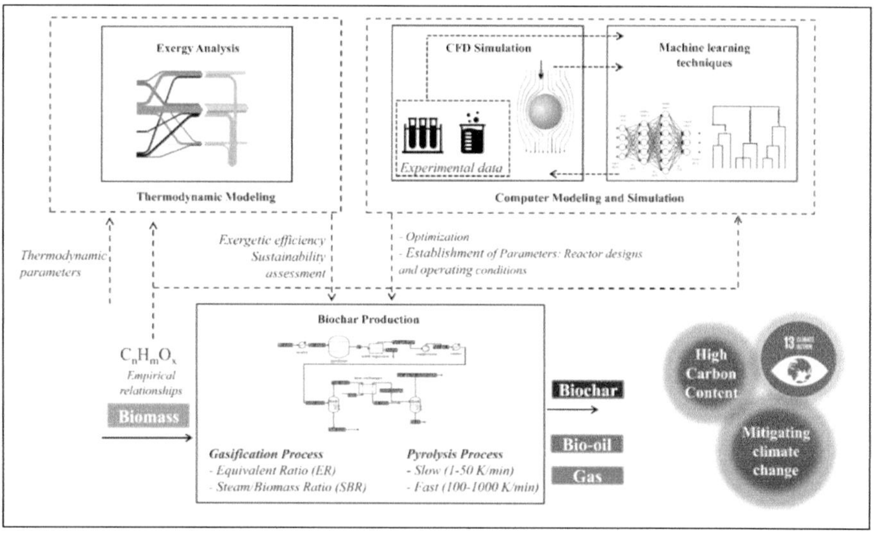

Figure 2.1 Types of process models addressed in Chapter 2.

2.2 EXERGY ASSESSMENT OF THE ENERGY SYSTEM IN BIOCHAR PRODUCTION

2.2.1 Exergy Analysis

A simplified empirical formula CH_aO_b (obtained through elemental analysis) is commonly used to represent biomass in the exergy modeling of biomass thermochemical processes. The chemical reaction for the decomposition of a generic biomass can be represented as follows:

$$CH_aO_b + wH_2O + m(O_2 + 3.76N_2) \rightarrow x_1CO + x_2H_2 + x_3CO_2 + x_4H_2O + x_5CH_4 + x_6CH_cO_d + m3.76N_2 + x_{bio\text{-}oil}\ CH_eO_f \quad (1)$$

Here, CH_cO_d is the empirical formula for biochar and CH_e is the empirical formula for the bio-oil. Bio-waste pyrolysis is the process in which this material is thermally decomposed by heating it in an inert atmosphere. This process results in the production of three distinct products: biochar, bio-oil, and fuel gas. The pyrolysis process can be conducted under a nitrogen (N_2) or carbon dioxide (CO_2) atmosphere.

Gasification occurs at high temperatures, where a gasifying agent such as oxygen, air, steam, carbon dioxide, or a mixture of these interacts with bio-waste feedstock. Biomass gasification involves several complex steps: drying, pyrolysis, oxidation, and reduction.

The pyro-gasification process occurs under steady-state conditions if the residence time in the reactor is long enough to achieve thermodynamic equilibrium (at atmospheric pressure, which is equal to 1 atm). When the system reaches chemical equilibrium, the Gibbs free energy of the reacting system is minimized, leading to the most stable composition.

The First Law of Thermodynamics (FLT), or energy balance, is widely used to analyze different power systems. In this context, all forms of energy are considered equal quality, and the loss of energy through transformation is not considered. Exergy is defined as a comprehensive measure of the quality of different energy forms in power systems that affect the environment. Exergy analysis assesses the quality of energy in power systems and the efficiency of processes. The initial step in power quality analysis is the identification and reduction of energy loss through exergy analysis. Consequently, the performance of process equipment and areas with the highest exergy are examined and highlighted. Exergy analysis is superior to energy analysis as it provides a better understanding of the irreversibility within the process. Unlike exergy analysis, energy analysis does not account for the degradation of transformed energy. Exergy analysis accurately identifies the locations where irreversibility occurs (Nanadegani and Sunden 2023).

Exergy can be defined as the maximum theoretical work that can be obtained when a system is brought into thermodynamic equilibrium with the environment through processes in which the system interacts solely with the environment (Piazzi et al. 2022).

Exergy analysis serves as a robust tool for evaluating energy conversion systems. Exergy encompasses four primary components: physical exergy, chemical exergy, kinetic exergy, and potential exergy. However, kinetic and potential exergies are typically neglected due to the minimal changes in height and velocity, rendering their values insignificant (Ding et al. 2021).

Physical exergy is the maximum work obtainable from a stream when its state changes from its initial temperature (T) and pressure (P) to the environmental state (T_0 = 298.15 K, P_0 = 1.013 bar). Chemical exergy measures the maximum work obtainable when a substance undergoes a change in chemical composition and transitions from the environmental state (at T_0 and P_0) to the dead state, also defined by T_0 and P_0. This is done while taking into account the species of common environmental components. The standard chemical exergy of a pure substance is available in various literature sources. When the chemical structure is unknown, such as for biomass and char, statistical correlations based on the elemental composition and heating value of the substance should be used (Piazzi et al. 2022).

Exergy analysis is a method that measures the degradation of available energy and identifies the causes, locations, and true magnitude of energy losses due to system inefficiencies. Exergy is a term that quantifies the available energy, availability, work capability, work potential, and related concepts. To achieve the greatest work or energy potential from input energies, operating the system reversibly undergoing various processes is necessary. However, in real processes, there are always losses that result in a loss of work potential, leading to irreversibility. These losses are generally due to irreversibility in the energy flow during different processes. Exergy consumption is referred to as irreversibility, loss of work potential, dissipated work, and dissipation (Arshad et al. 2019).

The exergy associated with the Second Law of Thermodynamics (SLT), which is due to irreversibility occurring during a process and cannot be recovered, can be examined in the following manner:

$$\begin{aligned}&\text{Exergy input} - \text{Exergy output} - \text{Exergy consumption}\\&= \text{Exergy accumulation}\end{aligned} \tag{2}$$

The consumption of exergy is a consequence of irreversibility, and this consumption is proportional to entropy generation.

During the flow process, irreversibilities are inevitable due to several factors, including frictional effects, sudden contraction or expansion, heat losses from finite temperature gradients, non-equilibrium compression and expansion processes, mixing, and chemical reactions. These factors collectively lead to increased entropy generation. Any phenomenon that generates entropy will result in exergy consumption or destruction, which is less than exergy accumulation. This is expressed below:

$$I = T_0 \cdot S_g \geq 0 \tag{3}$$

Thus, exergy destruction or accumulation is proportional to entropy generation. Furthermore, exergy destruction is a positive quantity greater than 0 for real processes and equal to 0 for reversible processes. Overall, the exergy destroyed in any thermodynamic system represents the loss of useful work potential, also known as irreversibility or lost work (Arshad et al. 2019).

The minimum amount of exergy in a system that cannot be processed due to existing technology and economic considerations, as well as efforts to prevent its destruction, is called unavoidable exergy losses. Avoidable exergy losses identify the greatest potential for process optimization. Accordingly, a system is analyzed for exergy, undesirable thermodynamic processes are identified, and the most feasible modifications are determined.

The input and output of exergy in a control volume involve three exchange processes: work, heat, and mass transfer. An exergy balance is formulated according to Equation (4) for a control volume in a steady state.

$$\dot{E}x_d = \sum_j \dot{E}x_{Q_j} - \dot{E}x_W + \sum_i \dot{E}x_i - \sum_k \dot{E}x_k \tag{4}$$

Here $\dot{E}x_d$ is exergy destruction caused by the irreversible rates on the control volume. $\dot{E}x_W$ is related to work exergy and $\sum_j \dot{E}x_{Q_j}$ is related to the concept of heat exergy, which is defined by Equation (5).

$$\dot{E}x_{Q_j} = \dot{Q}_j \left(1 - \frac{T_0}{T_j}\right) \tag{5}$$

In Equation (5), the quantity represented by the symbol $\dot{Q}_j$ is the heat transfer rate at a boundary with an instantaneous temperature of T_j (Rocco et al. 2014).

One of the principal findings of this analysis is the exergy efficiency of the process. To facilitate a more intuitive understanding of this concept, it is helpful to restate the exergy balance, assuming that the system is operating at a steady state. The exergy balance can be expressed as follows:

$$\dot{E}x_{\text{fuel}} = \dot{E}x_{\text{product}} + \dot{E}x_{\text{loss}} + \dot{E}x_d \tag{6}$$

$$\eta_{ex} = \frac{\dot{E}x_{\text{product}}}{\dot{E}x_{\text{fuel}}} = 1 - \frac{\dot{E}x_{\text{loss}} + \dot{E}x_d}{\dot{E}x_{\text{fuel}}} \tag{7}$$

In Equation (6), the waste or byproduct exergy ($\dot{E}x_{\text{loss}}$) represents the output streams that are not considered products. These fluxes are associated with an energy or material stream output, but their exergy is potentially recoverable. On the other hand, the exergy destruction ($\dot{E}x_d$) represents an irrecoverable loss.

The exergy analysis showed that the input exergy of nitrogen as an inert gas and the exergy of the produced water were minimal and were therefore excluded from the study. The energy and exergy balance for the biomass pyrolysis system can be expressed as follows in Equations (8) and (9), respectively:

$$\dot{E}_{\text{biomass}} + \dot{E}_{\text{inert gas}} + \dot{E}_{\text{process-heat}} = \dot{E}_{\text{biochar}} + \dot{E}_{\text{bio-oil}} + \dot{E}_{\text{gas}} \tag{8}$$

$$\dot{E}x_{\text{biomass}} + \dot{E}x_{\text{pyrolysis}} = \dot{E}x_{\text{biochar}} + \dot{E}x_{\text{bio-oil}} + \dot{E}x_{\text{gas}} \tag{9}$$

$\dot{E}_{\text{biomass}}$ and $\dot{E}_{\text{process-heat}}$ represent the input energy of the feedstock and the heat required for the pyrolysis process. $\dot{E}_{\text{biochar}}$, $\dot{E}_{\text{bio-oil}}$ and $\dot{E}_{\text{gas}}$ represent the output energy of biochar, bio-oil, and pyrolysis gas (Sakhiya et al. 2021). For this process, exergetic efficiency can be calculated as (Baghel et al. 2022):

$$\eta_{ex} = \frac{\dot{E}x_{\text{biochar}} + \dot{E}x_{\text{bio-oil}} + \dot{E}x_{\text{gas}}}{\dot{E}x_{\text{biomass}} + \dot{E}x_{\text{pyrolysis}}} \tag{10}$$

2.2.2 Sustainability Analysis

In every industrial process, sustainability analysis is of paramount importance, as it enables the minimization of the process's environmental impact on society while simultaneously integrating financial growth, social harmony, and environmental balance. The Sustainability Index (SI), which is calculated based on exergy efficiency, is employed to evaluate plant sustainability regarding fuel resources. A higher SI value indicates greater sustainability. In this context, the SI is employed to assess the long-term viability of the plant concerning its fuel sources. This can be expressed mathematically as:

$$\text{Sustainability Index (SI)} = \frac{1}{1-\eta_{ex}} \tag{11}$$

The potential for improvement has been evaluated by calculating possible enhancements to the system. This includes determining the destroyed exergy and exergy efficiency of the process, which can be expressed as follows in Equation (12).

$$\text{Improvement Potential (IP)} = \left[1-\left(\frac{Ex_d}{Ex_{in}}\right)\right] \cdot Ex_d \tag{12}$$

Here, Ex_{in} is the overall exergy potential of the process flow inputs.

The waste exergy factor is defined as the proportion of the process waste exergy value to the total input exergy value, $\eta_{\text{wasted exergy}}$. The environmental impact factor (f_{env}) was employed to assess the environmental impact of waste exergy or destroyed exergy (Trada et al. 2022).

$$f_{env} = \eta_{\text{wasted exergy}} \cdot \frac{1}{\eta_{ex}} \tag{13}$$

2.2.3 Exergy-Based Economic Analysis

Once the fuels and products associated with each process component have been identified, an exergoeconomic analysis can be performed to calculate the production costs for multiproduct processes. This analysis entails apportioning the overall process cost between the costs of the material streams and the equipment costs. The Specific Exergy Costing (SPECO) method comprises three principal stages: first, the identification of exergy streams; second, the definition of fuels and products; and third, the specification of cost streams. Fuels and products must be clearly defined for all equipment units. The cost of each specific equipment unit is distributed to the product flow based on modifying the exergy stream within that unit. The cost balance regarding exergy can be expressed as Equation (14). This equation represents the sum of all exergy cost streams entering unit k and the cost for the unit operation $\dot{Z}_k$ (which includes capital investment, operation, and maintenance of each component), and is equal to the sum of all cost streams leaving the unit. In this context, the symbols, c, $\dot{E}x_{\text{in, out}}$, $\dot{W}$, and $\dot{E}x_q$ represent the following: cost, input, and output exergy streams, work, and heat transfer, respectively. Using the SPECO method, a system of linear equations is derived from the cost balance of each component and then solved with auxiliary equations.

$$\sum_{\text{out}}\left(c_{\text{out}} \cdot \dot{E}x_{\text{out}}\right)_k + c_{w,k} \cdot \dot{W}_k = \sum_{\text{in}}\left(c_{\text{in}} \cdot \dot{E}x_{\text{in}}\right)_k + c_{Q,k} \cdot \dot{E}x_{Q,k} + \dot{Z}_k \tag{14}$$

When Equation (14) is applied to the pyrolysis process, the following balance is obtained (Torres et al. 2020):

$$\begin{aligned}&\sum_{\text{out}}\left(c_{\text{bio-oil}} \cdot \dot{E}x_{\text{bio-oil}}\right) + \left(c_{\text{biochar}} \cdot \dot{E}x_{\text{biochar}}\right) + \left(c_{\text{gas}} \cdot \dot{E}x_{\text{gas}}\right)\\ &= \sum_{\text{in}}\left(c_{\text{biomass}} \cdot \dot{E}x_{\text{biomass}}\right) + \left(c_{\text{inert gas}} \cdot \dot{E}x_{\text{inert gas}}\right) + \left(c_Q \cdot \dot{E}x_Q\right) + \dot{Z}_{\text{pyrolizer}}\end{aligned} \tag{15}$$

Ayala-Ruiz et al. (2022) presented a thermos-economic assessment of the fast pyrolysis process using banana waste. The study includes an exergoeconomic analysis to pinpoint components that require efficiency improvement. The drying and pyrolysis processes are identified as the main contributors to exergy destruction. They attributed the low efficiency to the irreversibility caused by the mixture of biomass with dry air and the evaporation of water contained within the biomass.

The slow pyrolysis of rice straw for biochar production was studied through a thermodynamic analysis. The results of the observations indicated that the pyrolysis temperature had a significant and positive effect on both the energy and exergy efficiency of the pyrolysis process. It was found that temperature played a crucial role in the thermodynamic analysis of the pyrolysis process. Both energy and exergy efficiencies increased as the temperature rose despite variations in individual product contributions. Two main reasons can be identified for this increase in efficiency at higher temperatures. Firstly, secondary thermal cracking of tar and char occurs at higher temperatures, converting solid and liquid products into gaseous products. Secondly, the rate of heat loss is relatively lower at higher temperatures due to the accumulation of heat within the reactor, resulting in higher energy and exergy efficiencies. It can, therefore, be concluded that high-temperature pyrolysis (600°C) results in a lower degree of thermodynamic irreversibility (Sakhiya et al. 2021).

In their study of the slow pyrolysis of lignocellulosic waste, Torres et al. (2020) identified two primary reasons for the loss of input exergy during the pyrolysis process. Firstly, some supplied heat is not utilized to produce high-value species with exergetic value (physical exergy). Secondly, exergy is lost due to mixing streams with different concentrations and reaction irreversibility (chemical exergy). Nevertheless, it was observed that the increase in gas exergy was sufficient to offset the elevated energy demand when the pyrolysis temperature was elevated.

2.3 COMPUTATIONAL FLUID DYNAMICS

Like other processes, the study of gasification or pyrolysis typically requires determining optimal operating conditions. Through multiple laboratory and pilot-scale experiments, critical process variables can be identified to adjust operating conditions according to established needs. Each of these experiments demands strict process control, precise measurement of each operational variable, and evaluation of products in terms of char yield, tar yield, non-condensable gases, and gas phase composition. In this regard, although experimentation is naturally a costly and time-consuming stage of development, the results of experiments provide highly reliable data on the process that can assist in designing larger-scale units. Additionally, data collected from processes such as gasification and pyrolysis can be used to validate computational models and assess the effect of changes in operating or geometric conditions of the units employed.

A pyrolysis or gasification reactor is a complex multiphase system in which a solid phase of particles continuously interacts with a moving gas that may or may not contain reactive species. In some cases, such as in fluidized beds, there are multiple solid phases where one is biomass and the other is the inert bed material that acts as a thermal buffer and improves the quality of mixing and heat transfer in the reactor.

The commonly employed models with computer-based resolution can be formulated at a local or global scale based on thermodynamic equilibrium, kinetics, neural networks, or computational fluid dynamics (CFD).

Each type of model presents different challenges. In particular, computational fluid dynamics (CFD) allows for the solution of the coupled Navier-Stokes differential equations using numerical methods on high-speed computers.

Three-dimensional, unsteady CFD simulations allow for the evaluation of the effect of geometric modifications in the reactive system of interest without the need for experimental evaluation. In this way, a deep understanding of the process can be achieved, and variables that are difficult to measure in experiments can be evaluated at a relatively low cost and within short time frames, even enabling the expansion of the range of operating conditions studied through experiments and the identification of operational limits related to process safety.

The software currently used can be commercial or open-source. There are differences between codes related to the available models, solution algorithms, interfaces, convergence methods, meshing methods, and other features. In general, obtaining similar results for the same process is possible. An important aspect from a comparative point of view is that when it comes to commercial software, the modifications allowed in the codes are limited, related to kinetics, heat and mass coefficients, boundary conditions, and drag, among others, but it is not possible to determine exactly how the code is implemented internally, which is possible in open-source software (Venier et al. 2019).

CFD offers two resolution strategies regarding global approaches, where complete reactors are simulated at a laboratory scale. On the one hand, Euler-Euler models or two-fluid models (TFM), in which all phases are continuous and interpenetrating, coexist and exchange mass, momentum, and heat, with the resolution for granular solids using granular kinetic theory. The properties of the particles are defined or calculated as averages over all particles in a control volume around the spatial location. On the other hand, discrete models (CFD-DEM) are more demanding in terms of computational resolution time due to the need to use particle search and tracking algorithms to resolve interactions between them. In this context, pseudo-continuous models are more efficient and most used to solve pyrolysis and gasification simulations.

In general, CFD codes include a set of submodels for the sequence of processes that occur in biomass particles, from the stage of receiving heat from a medium, the subsequent stage of moisture evaporation, then pyrolysis and gasification, depending on the case studied, and also gas-phase reactions.

The equipment commonly used for pyrolysis or gasification of lignocellulosic biomass are fixed bed units, entrained flow reactors, and fluidized beds (Luo et al. 2022). Additionally, for pyrolysis, screw reactors are also used (Uddin et al. 2018). Regarding the thermal effects in fluidized beds resolved through CFD schemes, a recent work by Reyes et al. (2022) demonstrates the capability of TFM-type models to solve such systems. The authors simulate different fluidized units and analyze the ability of OpenFoam and MFiX to resolve fluid dynamics and heat transfer to submerged objects. The cited work does not focus on resolving any chemical reaction; however, the resolution of reactors using CFD should undergo prior evaluation, especially when the medium that determines fluid dynamics, system temperature, and heat transfer to or from submerged objects is an inert material such as sand.

2.3.1 CFD Used in Biomass Pyrolysis

Regarding pyrolysis resolved with reactor-scale or global models, it is common in the literature to find studies where an adaptation of the Chemical Percolation Devolatilization (CPD) model (Fletcher et al. 1992) is employed. This model was originally developed to predict coal devolatilization. Adapting the model involves extrapolating the original model if biomass pyrolysis occurs as a weighted average from the weight fractions of its main components (cellulose, hemicellulose, and lignin). The employed scheme is shown in the following Figure 2.2, where the subscript "i" refers to each of the constituent species, and Y indicates the mass fraction of biochar generated in the reaction (Miller and Bellan 1997).

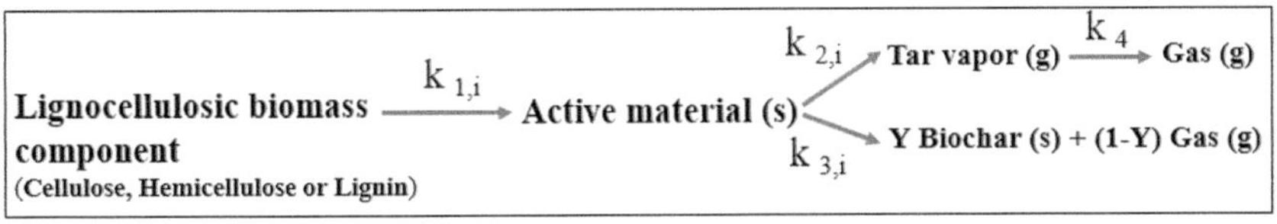

Figure 2.2 Kinetic scheme for biomass pyrolysis.

Xue et al. (2012) and Xiong et al. (2013) employed the CPD kinetic scheme to simulate biomass pyrolysis in laboratory-scale fluidized units using CFD, where mixing and heat transfer conditions ensure rapid pyrolysis. In the cited works, it was demonstrated that the kinetic scheme shown in Figure 2.2, with an Euler-Euler transient modeling approach, with two solid phases (inert material and continuously fed biomass) and the gas phase, allows obtaining yields similar to those obtained in experiments. The set of equations typical of the Euler-Euler modeling approach is also detailed in both works.

Multiple CFD-DEM studies on pyrolysis have incorporated the reaction mechanism developed by Debiagi et al. (2018), which describes the primary decomposition of biomass in terms of 32 first-order chemical reactions involving 29 solid species and 29 gaseous species. The model considers biomass composed of cellulose, three types of hemicellulose, three types of lignin, and two extractives. Tar is composed of 22 species and gas of six species. The kinetic parameters were determined through thermogravimetric analysis of 80 different biomass samples (Gao et al. 2021, Lu et al. 2021). Like the Euler-Euler modeling, the product yields resulted in similar experimental results.

The exact reaction mechanism was used by Gao et al. (2021) for the local simulation of pyrolysis in a single cylindrical particle with a diameter of 8 mm and a length of 19 mm, as well as for the simulation of a pilot-scale entrained flow reactor with polydisperse cylindrical particles, with a length-to-diameter ratio of 5 and diameters ranging from 140 to 1,791 μm. Isothermal and non-isothermal particle models were compared. For the local model, the non-isothermal model produced more accurate results, while in the case of the pilot-scale reactor, the isothermal characterization was more suitable.

Recently, Houston et al. (2022) added complexity to the primary decomposition model of Debiagi et al. (2018) by combining it with the complete secondary decomposition scheme developed by the Chemical Reaction Engineering and Chemical Kinetics Modeling group (CRECK) from Milan for gas-phase kinetics (Pelucchi et al. 2019). At its most detailed level, the secondary decomposition model contains 509 species and 20,240 reactions. The model was successfully used to predict the yields of biomass (switchgrass) pyrolysis in a laboratory-scale fluidized unit.

2.3.2 CFD Applications in Biomass Gasification

CFD resolution has also been applied in biomass gasification. Typically, the pyrolysis stage of gasification is simplified through a single reaction where biochar, non-condensable gases, and tar are generated. Subsequently, the emphasis is placed on resolving the set of heterogeneous reactions of biochar and gaseous species in the system, and, in some cases, a set of homogeneous reactions in the gas phase is also incorporated.

Liu et al. (2016) employed a discrete 3-D CFD model (MP-PIC—Multiphase Particle-In-Cell) to investigate steam gasification of almond pruning residues in a pilot-scale dual fluidized-bed gasifier. The pyrolysis reaction adopted by the authors is Equation (16), which excludes tar from the products. The authors hypothesize that it is completely converted into non-condensable gases to simplify the model.

$$\begin{aligned}\text{Biomass} \rightarrow\ & \alpha_1 CO + \alpha_2 CO_2 + \alpha_3 H_2 + \alpha_4 CH_4 + \\ & \alpha_5 C_2H_4 + \alpha_6 C_2H_6 + \alpha_7\ \text{Biochar}\end{aligned} \quad (16)$$

The model results showed that as the gasifier temperature increased, the concentrations of H_2 and CO increased, while the concentrations of CO_2 and CH_4 decreased in the synthesis gas. The study also demonstrated that a high gasifier temperature promoted synthesis gas production and increased the hydrogen content in the producer gas. Similar trends were observed in the study of the steam-to-biomass ratio (S/B). Given that it is a dual system, with a combustion chamber and a vapor gasification chamber, the effect of variations in the airflow on the synthesis gas composition was minimal.

Yin et al. (2020) utilized a hybrid model to study the gasification of 7.5 mm diameter biomass in an oxygen-enriched fluidized bed atmosphere. The pseudo-continuous TFM model resolved the fluid dynamics of the inert material bed (high-alumina bauxite), while a Discrete (DEM) scheme was responsible for determining the movement of biomass and chemical reactions. Pyrolysis was resolved through a single reaction, where, similar to the work of Liu et al. (2016), the generation of tar was neglected. The authors compared their results with experimental data (Praveen Kumar et al. 2006a) and concluded that the model could provide accurate predictions of synthesis gas composition.

Praveen Kumar et al. (2006) also employed a CFD-DEM model to study the gasification of different biomass types, including rubber wood, in a downdraft gasifier. In contrast to the previously mentioned authors, these authors considered a primary pyrolysis reaction as presented in Equation (17), where tar is present, along with a gaseous pseudo-species composed of C, H, O, S, and N. The decomposition of the volatile pseudo-species [Equation (18)] was also considered, where stoichiometric coefficients were evaluated with a specific algorithm programmed in C++. The CFD model was used to investigate the effect of the equivalence ratio on the composition of the outlet gas and gasifier temperature. Furthermore, different biomass feedstocks were tested for the gasification process, and it was concluded that rubber wood produced more synthesis gas than other biomasses.

$$\text{Biomass} \rightarrow \text{Volatile} + \text{Biochar} + \text{Ash} + \text{Tar} \quad (17)$$

$$\text{Volatile} \rightarrow X_1CO + X_2H_2 + X_3CO_2 + X_4CH_4 + X_5H_2O \ \sum X_i = 1 \quad (18)$$

2.4 MACHINE LEARNING APPLIED TO PYROLYSIS AND GASIFICATION PROCESS MODELING AND SIMULATION

Given the environmental importance of transforming biomass into such an attractive product as biochar, the need arises to create models capable of predicting the economic and relevant properties of this product, whether obtained by pyrolysis or by biomass gasification.

One way to simulate process flows and conditions to perform technical feasibility and risk analysis on processes and businesses is by using computational models (Sakheta et al. 2023b). Given the high complexity and several variables involved in thermochemical processes, it is essential to employ robust modeling for the design, optimization, and scale-up.

To obtain predictive models, one of the most widely used tools today is machine learning, a discipline that is part of the field of artificial intelligence. AI is the ability of machines to perform tasks that require reasoning characteristics, such as the ability to learn, the ability to self-correct, and the ability to make decisions without human intervention.

ML is a discipline that, through algorithms, equips computers with the ability to identify patterns in massive data and make predictions (predictive analytics). The supervised ML aim is to build a model that makes evidence-based predictions in the presence of uncertainty. A supervised learning algorithm takes a known set of input data and known responses to the data (output) and trains a model to generate reasonable predictions for the response to new data. Supervised learning uses classification and regression techniques to develop predictive models. Classification techniques predict categorical responses. Classification models classify input data. Regression techniques predict continuous responses, such as temperature changes or energy demand fluctuations. Typical applications include electricity load forecasting and algorithmic trading.

New machine learning techniques, such as neural networks (ANNs), are displacing traditional modeling methods, such as equilibrium thermodynamics, kinetic, and computational fluid dynamics (CFD) models. That is because machine learning models can predict outcomes in very complicated processes (Ascher et al. 2022b).

In property predictions, for their continuous values, unsupervised learning is used. This tool uses algorithms to identify patterns from input data, with which it can predict new events. The reader can note that, in contrast to mathematical models, predictive modeling is not simple to explain with an equation, so they are called black boxes.

The availability of these types of models makes it possible, based on the knowledge of a minimum amount of data on the characteristics of the raw material used and the pyrolysis conditions determined experimentally, to evaluate the properties of the biochar obtained from biomasses with different characteristics, i.e. to test the several raw materials (biomasses) or combinations of these (co-pyrolysis), without the need to experimentally carry out the test in a reactor (with the time and money costs that this entails). In other words, once the model is validated and trained, it can predict the output variables from the characteristics of the input biomass.

Simplified models are often based on limited data and cannot adequately accommodate variations in feedstock characteristics and pyrolysis conditions. In addition, most of these models treat biomass as a homogeneous entity or simply as the sum of its three main components without considering possible interactions between these components, inorganic or extractive species. Although some recent studies attempt to address these limitations, detailed information obtained from chemical analysis of biomass is regularly omitted or used indirectly to estimate the contents of the major components. That can introduce inaccuracies in pyrolysis predictions, as inorganic and extractive species in the biomass can influence both the yield and composition of the products obtained. In summary, failure to consider these additional biomass characteristics in kinetic models may affect the accuracy of predictions of yield and composition of biomass pyrolysis products.

2.4.1 ML Algorithm Selection for Biochar Production

Selecting the best machine-learning algorithm for a dataset is crucial in biochar production. No single algorithm works for all problems, but several factors can help narrow down the options. The choice depends on the need for interpretability and accuracy, as well as the nature of the dataset. Supervised learning can be used if the dataset includes both input and output variables. For example, with sustainability factors as inputs and biochar quality as the output, the production process can be optimized for environmental benefits. If there are only input variables, unsupervised learning should be used, such as identifying patterns in soil properties after biochar application. Different algorithms may be suitable depending on the need for accuracy and interpretability. For instance, to improve agricultural sustainability, accuracy would be prioritized over interpretability, using recurrent neural networks to provide outputs with sufficient accuracy for decision-making by researchers, farmers, and entrepreneurs.

Figure 2.3 shows a flowchart that outlines a structured approach to selecting the most suitable ML algorithm for a specific task.

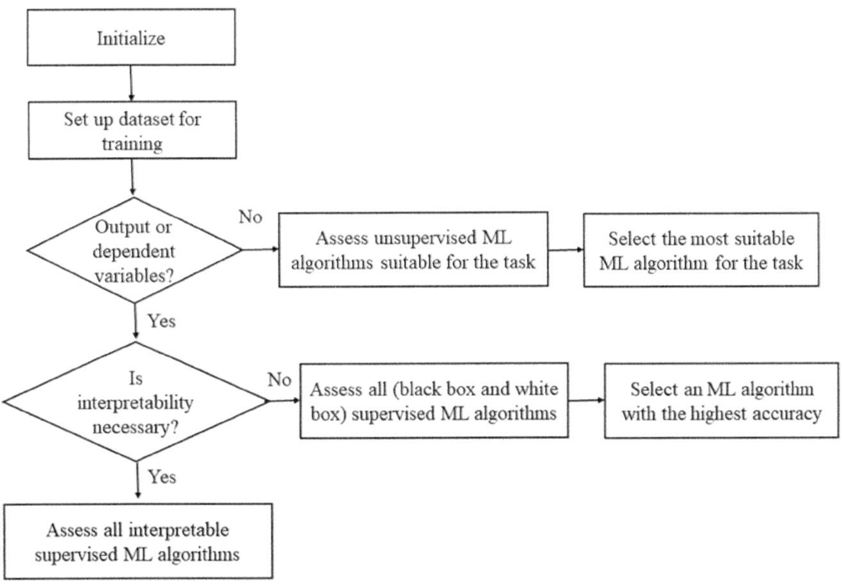

Figure 2.3 Steps for the best ML algorithm selection.

This approach can be applied to a variety of biochar production tasks, including optimizing production for enhanced carbon sequestration, improving biochar quality for specific agricultural applications, or identifying patterns in biochar properties following application to soil. In this sense, the available data determines the ML approach to be used.

Supervised learning is suitable when input and output or dependent data are available, while unsupervised learning is appropriate when only input data is available (Lee and Shin 2020). On the other hand, the desired level of interpretability versus accuracy needs to be considered. If understanding the decision-making process behind the ML model is crucial, interpretable models like decision trees or linear regression may be preferred. However, if high accuracy is paramount, more complex models like neural networks may be necessary (Lee and Shin 2020).

Regarding the efficiency and product quality improvement in pyrolysis and gasification processes, including biochar, significant challenges arise due to the complexity of these processes. AI-based approaches present a transformative option poised to revolutionize the bioconversion landscape.

Table 2.1 explores research based on AI approaches to predict outputs in pyrolysis systems. It highlights the diverse range of input variables, such as temperature, biomass composition, and particle size, that AI models can analyze to predict critical product yields (bio-oil, gas, and biochar) and properties (moisture content and H/C ratio).

Notably, the table showcases the high accuracy achieved by these models, with R^2 values exceeding 0.8 in many cases. The specific AI algorithms employed include Decision Trees, ANNs, and Random Forests, demonstrating the versatility of AI in tackling diverse prediction tasks within pyrolysis (Sakheta et al. 2023).

T is the highest temperature of the process, β is the heating rate, A is the preexponential factor, Ea is the activation energy, and n is the reaction order.

Like Table 2.1, Table 2.2 presents research about the AI approach's role in gasification systems. Here, the focus shifts to predicting gas products, H_2, methane (CH_4), and CO.

Additionally, the table delves into predicting system outputs like net output power and tar content, highlighting AI's potential for comprehensive system optimization. The table showcases an even wider range of AI algorithms being employed, including Least Squares Support Vector Machines (SVM), Nonlinear Autoregressive Exogenous models (NARXNNs), and online optimization routines. Notably, several studies report R^2 values exceeding 0.9, demonstrating the high accuracy of these AI-powered predictions.

2.4.2 Criteria Used to Evaluate the Performance of the Model

Various factors can be used to determine the quality and strength of engineering models. While some factors apply to all scales of models, there are also factors specific to certain modeling scales due to the differing scope and output of each

Table 2.1 Works were performed on lignocellulosic pyrolysis systems using AI tools

Reference	Input Variable	Predicted Variable	ML Algorithms	Metric for Assessment R^2
Hough et al. 2017	T, β, Carbon content in lignin fraction, Hydrogen content in lignin fraction in biomass	Solid yield, Moisture content of bio-oil, Gas composition, H/C ratio, Functional groups	Decision tree, ANN	0.982
Sunphorka et al. 2017	Lignin, hemicellulose, and cellulose contents in biomass	Kinetic parameters: A, Ea, and n	ANN	0.91
Zhong et al. 2020	T	bio-oil, gas, biomass distribution	CFD and ANN	>0.7274
Ullah et al. 2021	Ash content, Fixed carbon, Volatile matter, elemental analysis: C, H, O, N, Particle size, Nitrogen flow rate, β, and T	Bio-oil yield	Multi-linear regression, Decision tree, Random Forest, and Support vector machine,	>0.98
Khan et al. 2022	Moisture content, Fixed carbon, Particle size, Volatile matter, elemental analysis: C, H, O, N, cellulose, hemicellulose, lignin, T, β, Residence time, and Particle size	Biochar yield	ANN	>0.84
Dong et al. 2023	Moisture content, Fixed carbon, Particle size, Volatile matter, elemental analysis: C, H, O, N, T, β, heating time, Sweep flow rate, Residence time, and Particle size	Gas, bio-oil, and biochar	XGBoost, Adaboost, gradient boosting decision tree, and Random Forest	>0.78
Li et al. 2023	Ash content, fixed carbon, volatile matter, elemental analysis (C, H, O, N, S), retention time, β, T	Biochar yield, specific surface area, and total pore volume		>0.89
Fernandez et al. 2024	Elemental analysis: C, H, O, N form organic Matter, Paper content, high-density polyethylene content, low-density polyethylene content, polystyrene content, polypropylene content, polyvinyl chloride content, polylactic acid content, terylene content, β, and T	Mass left (wt.%) of Thermogravimetric Analysis curves	ANN	>0.82

Table 2.2 Sequential overview of AI-enabled advancements in lignocellulosic gasification systems.

Reference	Input Variable	Predicted Variable	ML Algorithms	Metric for Assessment
Mutlu and Yucel 2018	C, ER, H, O, fuel flow rate, temperature	H_2, CH_4, CO, CO_2, HHV	Least Squares SVM, Random Forests	Least squares SVM: RMSE < 1.5196
Yucel et al. 2019	Air flow rate, ash, C, ER, FC, H, MC, N, O, temperature, VM	H_2, CH_4, CO, CO_2, LHV	Nonlinear autoregressive exogenous (Mutlu and Yucel 2018), ANN	ANN: $R^2 > 0.9462$
Serrano and Castelló 2020	Ash, C, ER, H, moisture content, O, temperature	Tar	ANN	$R^2 > 0.9759$
Safarian et al. 2020	Air to fuel ratio, ash, C, FC, H, MC, N, O, S, temperature, VM	Net output power	ANN, Aspen Plus	$R^2 > 0.999$
Elmaz et al. 2020	Ash, C, ER, FC, fuel flow rate, H, MC, N, O, temperature distribution, VM	H_2, CH_4, CO, CO_2, HHV	Decision tree regressions, multilayer perceptron, Polynomial Regression, SVM	Decision tree regressions: $R^2 > 0.807$
Elmaz and Yücel 2020	Ash, C, ER, FC, H, MC, N, O, VM	H_2, CH_4, CO, CO2, HHV, temperature	Model predictive controller, NARXNN, online optimization routines, polynomial regression	$R^2 > 0.9839$
Safarian et al. 2021	Air to fuel ratio, ash, C, FC, H, MC, N, O, S, S/B, temperature, VM	Hydrogen-specific mass flow rate	Aspen Plus, ANN	$R^2 > 0.999$
Sezer and Özveren 2021	C, fuel flow rate, H, O, steam flow rate, temperature	Exergy	Aspen Plus, ANN	$R^2 > 0.996$
Ascher et al. 2022a	Ash, bed material, C, catalyst, ER, gasification agent, H, MC, operation mode, PS, reactor type, S, scale, temperature	C_2H_n, CH_4, CO, CO_2, Gas, H_2, LHV, N_2, tar and char yield	AdaBoost, ANN, SVM, Super Learner, Tree-based models	Most effective algorithm: Gradient boosting regression: $R^2 > 0.76$

modeling method. The main generic factor is model accuracy. Several approaches are used to measure the accuracy of models. The most common approach is to compare the model output with experimental data from the laboratory, pilot plant, or commercial scale plants using a common basis between the model and experiments. The error between the experimental reference data and data generated from the models can be quantified in different ways, such as root mean square error (RMSE), coefficient of determination (R^2), and error percentage (Sakheta et al. 2023).

Key validation parameters of the artificial neural network (ANN) model, including mean absolute error (MAE), root mean square error (RMSE), and determination coefficient (R^2), were employed to evaluate the accuracy of the model. A decrease in error between model predictions and experimental values, along with an increase in correlation, signifies the enhanced predictive capability of the model (Fernandez et al. 2024).

$$\mathrm{MAE} = \frac{1}{N}\left(\sum_{i=1}^{N}\left|Y_i - Y_{pi}\right|\right) \tag{19}$$

$$\mathrm{RMSE} = \sqrt{\frac{1}{N}\left(\sum_{i=1}^{N}\left(Y_i - Y_{pi}\right)^2\right)} \tag{20}$$

$$R^2 = 1 - \frac{\sum_{i=1}^{N}\left(Y_i - Y_{pi}\right)^2}{\sum_{i=1}^{N}\left(Y_i - \bar{Y}\right)^2} \tag{21}$$

$\underline{Y}$, Y_i, and Y_{pi} are the average of the experimental value, the experimental value, and the predicted value, and N is the predicted point number.

Additionally, calculation and model execution times should be considered when developing models. If the objective of the research is optimization or running sensitivity analysis, models with lengthy execution times will not be flexible enough to conduct such analysis since these analyses require multiple runs, iterations, and adjusting operating parameters for each run. (Kaczor et al. 2020).

Another criterion of high-quality models is their agility, i.e., their ability to model a range of feedstock, operating conditions used, and a number of features and target variables considered. Effective models can be used in place of experiments, which can be resource and time-consuming. When examining proposed feedstock or operating conditions, it would be more practical to use models that can extrapolate outcomes that include the proposed conversion conditions. The final factor is the availability of reliable input data required to develop the model. As the amount of data considered in the model development stage increases, the model's reliability and ability to identify normal values and exclude outliers increases (Sakheta et al. 2023).

2.5 CONCLUSION

Accurate modeling and exergetic evaluation enhance the efficiency and sustainability of gasification and pyrolysis processes and reduce costs associated with large-scale experimentation. Moreover, these processes yield biochar, a valuable co-product with applications in soil improvement and carbon sequestration. However, challenges persist in incorporating complex effects such as biomass variability and interaction among different components during conversion processes. Thus, the combination of exergy and computational modeling represents a powerful tool to advance research and development in biomass conversion technologies, enabling more efficient and sustainable gasification and pyrolysis systems design while producing biochar as a beneficial co-product.

REFERENCES

Arshad, A., H.M. Ali, A. Habib, M.A. Bashir, M. Jabbal and Y. Yan. 2019. Energy and exergy analysis of fuel cells: A review. Thermal Science and Engineering Progress. 9: 308–321.

Ascher, S., X. Wang, I. Watson, W. Sloan and S. You. 2022a. Interpretable machine learning to model biomass and waste gasification. Bioresource Technology. 364: 128062.

Ascher, S., I. Watson and S. You. 2022b. Machine learning methods for modelling the gasification and pyrolysis of biomass and waste. Renewable and Sustainable Energy Reviews. 155: 111902.

Ayala-Ruíz, N., D.H. Malagón-Romero and H.A. Milquez-Sanabria. 2022. Exergoeconomic evaluation of a banana waste pyrolysis plant for biofuel production. Journal of Cleaner Production. 359: 132108.

Baghel, P., A.K. Sakhiya and P. Kaushal. 2022. Influence of temperature on slow pyrolysis of Prosopis Juliflora: An experimental and thermodynamic approach. Renewable Energy. 185: 538–551.

Debiagi, P., G. Gentile, A. Cuoci, A. Frassoldati, E. Ranzi and T. Faravelli. 2018. A predictive model of biochar formation and characterization. Journal of Analytical and Applied Pyrolysis. 134: 326–335.

Ding, H., J. Li and D. Heydarian. 2021. Energy, exergy, exergoeconomic, and environmental analysis of a new biomass-driven cogeneration system. Sustainable Energy Technologies and Assessments. 45: 101044.

Dong, Z., X. Bai, D. Xu and W. Li. 2023. Machine learning prediction of pyrolytic products of lignocellulosic biomass based on physicochemical characteristics and pyrolysis conditions. Bioresource Technology. 367: 128182.

Elmaz, F. and Ö. Yücel. 2020. Data-driven identification and model predictive control of biomass gasification process for maximum energy production. Energy. 195: 117037.

Elmaz, F., Ö. Yücel and A.Y. Mutlu. 2020. Predictive modeling of biomass gasification with machine learning-based regression methods. Energy. 191: 116541.

Fernandez, A., D. Zalazar-García, C. Lorenzo-Doncel, D.M. Yepes Maya, E.E. Silva Lora, R. Rodriguez, et al. 2024. Kinetic Modeling of Co-Pyrogasification in Municipal Solid

Waste (MSW) Management: Towards Sustainable Resource Recovery and Energy Generation. Sustainability. 16: 4056.

Fletcher, T.H., A.R. Kerstein, R.J. Pugmire, M.S. Solum and D.M. Grant. 1992. Chemical percolation model for devolatilization. 3. Direct use of carbon-13 NMR data to predict effects of coal type. Energy & Fuels. 6: 414–431.

Gao, X., L. Lu, M. Shahnam, W.A. Rogers, K. Smith, K. Gaston, et al. 2021. Assessment of a detailed biomass pyrolysis kinetic scheme in multiscale simulations of a single-particle pyrolyzer and a pilot-scale entrained flow pyrolyzer. Chemical Engineering Journal. 418: 129347.

Hough, B.R., D.A.C. Beck, D.T. Schwartz and J. Pfaendtner. 2017. Application of machine learning to pyrolysis reaction networks: Reducing model solution time to enable process optimization. Computers and Chemical Engineering. 104: 56–63.

Houston, R., O. Oyedeji and N. Abdoulmoumine. 2022. Detailed biomass fast pyrolysis kinetics integrated to computational fluid dynamic (CFD) and discrete element modeling framework: Predicting product yields at the bench-scale. Chemical Engineering Journal. 444: 136419.

Kaczor, Z., Z. Buliński and S. Werle. 2020. Modelling approaches to waste biomass pyrolysis: a review. Renewable Energy. 159: 427–443.

Khan, M., Z. Ullah, O. Mašek, S. Raza Naqvi and M. Nouman Aslam Khan. 2022. Artificial neural networks for the prediction of biochar yield: A comparative study of metaheuristic algorithms. Bioresource Technology. 355: 127215.

Lee, I. and Y.J. Shin. 2020. Machine learning for enterprises: Applications, algorithm selection, and challenges. Business Horizons. 63: 157–170.

Li, H., Z. Ai, L. Yang, W. Zhang, Z. Yang, H. Peng, et al. 2023. Machine learning assisted predicting and engineering specific surface area and total pore volume of biochar. Bioresource Technology. 369: 128417.

Liu, H., R.J. Cattolica and R. Seiser. 2016. CFD studies on biomass gasification in a pilot-scale dual fluidized-bed system. International Journal of Hydrogen Energy. 41: 11974–11989.

Lu, L., X. Gao, A. Gel, G.M. Wiggins, M. Crowley, B. Pecha, et al. 2021. Investigating biomass composition and size effects on fast pyrolysis using global sensitivity analysis and CFD simulations. Chemical Engineering Journal. 421: 127789.

Luo, H., X. Wang, X. Liu, X. Wu, X. Shi and Q. Xiong. 2022. A review on CFD simulation of biomass pyrolysis in fluidized bed reactors with emphasis on particle-scale models. Journal of Analytical and Applied Pyrolysis. 162: 105433.

Miller, R.S. and J. Bellan. 1997. A Generalized Biomass Pyrolysis Model Based on Superimposed Cellulose, Hemicelluloseand Liqnin Kinetics. Combustion Science and Technology. 126: 97–137.

Mutlu, A.Y. and O. Yucel. 2018. An artificial intelligence based approach to predicting syngas composition for downdraft biomass gasification. Energy. 165: 895–901.

Nanadegani, F.S. and B. Sunden. 2023. Review of exergy and energy analysis of fuel cells. International Journal of Hydrogen Energy. 48: 32875–32942.

Pelucchi, M., C. Cavallotti, A. Cuoci, T. Faravelli, A. Frassoldati and E. Ranzi. 2019. Detailed kinetics of substituted phenolic species in pyrolysis bio-oils. Reaction Chemistry & Engineering. 4: 490–506.

Piazzi, S., F. Patuzzi and M. Baratieri. 2022. Energy and exergy analysis of different biomass gasification coupled to Fischer-Tropsch synthesis configurations. Energy. 249.

Praveen Kumar, D.G., H.U. Hebbar and M.N. Ramesh. 2006. Suitability of thin layer models for infrared–hot air-drying of onion slices. LWT - Food Science and Technology. 39: 700–705.

Reyes-Urrutia, A., C. Venier, N.J. Mariani, N. Nigro, R. Rodriguez and G. Mazza. 2022. A CFD comparative study of bubbling fluidized bed behavior with thermal effects using the open-source platforms mfix and openfoam. Fluids. 7(1): 1. https://doi.org/10.3390/fluids7010001.

Rocco, M.V., E. Colombo and E. Sciubba. 2014. Advances in exergy analysis: A novel assessment of the Extended Exergy Accounting method. Applied Energy. 113: 1405–1420.

Safarian, S., S.M. Ebrahimi Saryazdi, R. Unnthorsson and C. Richter. 2020. Artificial neural network integrated with thermodynamic equilibrium modeling of downdraft biomass gasification-power production plant. Energy. 213: 118800.

Safarian, S., S.M. Ebrahimi Saryazdi, R. Unnthorsson and C. Richter. 2021. Modeling of Hydrogen Production by Applying Biomass Gasification: Artificial Neural Network Modeling Approach. Fermentation. 7: 71.

Sakheta, A., R. Nayak, I. O'Hara and J. Ramirez. 2023. A review on modelling of thermochemical processing of biomass for biofuels and prospects of artificial intelligence-enhanced approaches. Bioresource Technology. 386: 129490.

Sakhiya, A.K., A. Anand, I. Aier, V.K. Vijay and P. Kaushal. 2021. Suitability of rice straw for biochar production through slow pyrolysis: Product characterization and thermodynamic analysis. Bioresource Technology Reports. 15.

Serrano, D. and D. Castelló. 2020. Tar prediction in bubbling fluidized bed gasification through artificial neural networks. Chemical Engineering Journal. 402: 126229.

Sezer, S. and U. Özveren. 2021. Investigation of syngas exergy value and hydrogen concentration in syngas from biomass gasification in a bubbling fluidized bed gasifier by using machine learning. International Journal of Hydrogen Energy. 46: 20377–20396.

Sunphorka, S., B. Chalermsinsuwan and P. Piumsomboon. 2017. Artificial neural network model for the prediction of kinetic parameters of biomass pyrolysis from its constituents. Fuel. 193: 142–158.

Torres, E., L.A. Rodriguez-Ortiz, D. Zalazar, M. Echegaray, R. Rodriguez, H. Zhang, et al. 2020. 4-E (environmental, economic, energetic and exergetic) analysis of slow pyrolysis of lignocellulosic waste. Renew Energy. 162: 296–307.

Trada, A., A. Chaudhary, D. Patel and D.S. Upadhyay. 2022. An alternative fuel production from sawdust through batch-type pyrolysis reactor: Fuel properties and thermodynamic analysis. Process Safety and Environmental Protection. 167: 332–342.

Uddin, M.N., K. Techato, J. Taweekun, M.M. Rahman, M.G. Rasul, T.M.I. Mahlia, et al. 2018. An overview of recent developments in biomass pyrolysis technologies. Energies (Basel). 11.

Ullah, Z., M. Khan, S. Raza Naqvi, W. Farooq, H. Yang, S. Wang, et al. 2021. A comparative study of machine learning methods for bio-oil yield prediction – A genetic algorithm-based features selection. Bioresource Technology. 335: 125292.

Venier, C.M., A. Reyes Urrutia, J.P. Capossio, J. Baeyens and G. Mazza. 2019. Comparing ANSYS Fluent® and OpenFOAM® simulations of Geldart A, B and D bubbling fluidized bed hydrodynamics. International Journal of Numerical Methods for Heat & Fluid Flow. 30: 93–118.

Xiong, Q., S.-C. Kong and A. Passalacqua. 2013. Development of a generalized numerical framework for simulating biomass fast pyrolysis in fluidized-bed reactors. Chemical Engineering Science. 99: 305–313.

Xue, Q., D. Dalluge, T.J. Heindel, R.O. Fox and R.C. Brown. 2012. Experimental validation and CFD modeling study of biomass fast pyrolysis in fluidized-bed reactors. Fuel. 97: 757–769.

Yin, W., S. Wang, K. Zhang and Y. He. 2020. Investigation of oxygen-enriched biomass gasification with TFM-DEM hybrid model. Chemical Engineering Science. 211: 115293.

Yucel, O., E.S. Aydin and H. Sadikoglu. 2019. Comparison of the different artificial neural networks in prediction of biomass gasification products. International Journal of Energy Research. 43: 5992–6003.

Zhong, H., Q. Xiong, L. Yin, J. Zhang, Y. Zhu, S. Liang, et al. 2020. CFD-based reduced-order modeling of fluidized-bed biomass fast pyrolysis using artificial neural network. Renew Energy. 152: 613–626.

Chapter 3

Slow Pyro-Gasification Kinetic Models

Anabel Fernandez*[a], Rosa Rodriguez[a], Germán Mazza[b], Yanina Baldán[a] and Rodrigo Torres-Sciancalepore[b]

[a]Instituto de Ingeniería Química, Fac. de Ingeniería, Grupo Vinculado al PROBIEN (Universidad Nacional de San Juan, CONICET, Universidad Nacional del Comahue), San Juan, Argentina

[b]Instituto de Investigación y Desarrollo en Ingeniería de Procesos, Biotecnología y Energías Alternativas, PROBIEN (CONICET-Universidad Nacional del Comahue), Neuquén, Argentina

3.1 INTRODUCTION

The structural composition of lignocellulosic biowastes enables the production of high-value energy products or their combustion for energy release. Pyrolysis, a promising conversion technology, decomposes these wastes in an inert atmosphere to produce bio-oil, gas, and biochar as by-products. This approach aims to recover energy and reduce the carbon footprint by lowering greenhouse gas emissions while addressing environmental contamination.

On the other hand, the design of any thermochemical plant must consider the sizing of the equipment and accessories, which requires knowledge of the reaction kinetics and the type of reactions that occur in the process. Therefore, it is essential to study the reactions and their kinetic parameters, such as activation energy (E) and the pre-exponential factor (A) (from the Arrhenius model), in addition to the functionality of the biomass conversion with the reaction rate (f(α), where α is the conversion degree). The kinetic parameters can be determined using model-free

*Corresponding author: Anabel Fernandez (anafernndez@unsj.edu.ar)

DOI: 10.1201/9781032726694-3

or model-fitting methods, as reported by authors such as Vyazovkin (2015), Torres-Sciancalepore et al. (2022), Rony et al. (2019) and Fernandez et al. (2024), among others.

To carry out the kinetics study, the use of TGA presents an advantage by furnishing insightful data on the thermal behaviors and pyrolytic traits of the feedstock. Additionally, TGA findings can be processed to infer, resulting in a deeper understanding of the underlying reaction mechanisms. Past research has successfully employed thermogravimetric data to ascertain different kinetic parameters and models for explaining these mechanisms. A comprehensive understanding of kinetic mechanisms is crucial in outlining the biomass conversion process during pyro-gasification. This comprehension is vital to designing, evaluating, and improving pyrolyzers or gasifiers since differing biomass feedstocks display distinctive characteristics under TGA analysis. A thorough understanding of TGA principles is primordial for successfully designing and operating industrial pyrolysis processes.

Recent advancements in machine learning, particularly in artificial neural networks (ANNs), have garnered significant interest in chemical engineering due to their efficiency and versatility in handling various operations and processes within the field. Despite their proven effectiveness in many areas, their application in specialized fields such as pyro-gasification, thermal analysis, and kinetic investigations is still in its early stages. ANNs, modeled after biological neural systems, offer several advantages, including the ability to approximate nonlinear functions and tolerate noisy data. However, in the kinetic analysis, most reviewed studies prefer data-driven methodologies (black box) over mechanistic or parametric models (white box). Consequently, artificial neural network models that rely on temperature and time variables face significant drawbacks, such as ambiguous interpretation and limited generalizability for predicting conversion rate data.

This chapter presents the kinetic characterization of various lignocellulosic waste pyro-gasification processes using thermogravimetric techniques. The isoconversional, master plots, and compensation effect methods were applied to determine the activation energy, reaction model, and pre-exponential factor. Furthermore, a comprehensive understanding of the kinetics study was achieved by integrating macro-TGA with advanced machine learning through ANN models, providing deeper insights into the process.

3.2 THERMOGRAVIMETRIC ANALYSIS: TGA AND DTG CURVES

Thermogravimetric Analysis (TGA) is a thermal analysis technique that involves monitoring the mass of a substance as a function of temperature or time. This is achieved by subjecting the sample to controlled temperature and atmospheric conditions. The experimental setup typically utilizes a thermogravimetric (TG) analyzer, within which the sample is carefully positioned on a precision balance inside a furnace. The furnace is responsible for heating the sample throughout the experiment. The heating process can be conducted under either isothermal or non-isothermal conditions. A purging gas flow (or carrier gas) is directed over the sample to control the environment. This gas may be inert or reactive. In conventional TGA experiments, inert gases are predominantly employed due to the pyrolytic nature of the process (thermal treatment in the absence of oxygen). The existing literature does not highlight any significant influence arising from the specific type of inert gas used.

While nitrogen is the most commonly used inert gas, argon and helium are viable options. Any changes in sample mass relative to temperature are meticulously recorded and analyzed throughout the experiment. TGA curves can be transformed into differential thermogravimetric (DTG) curves through differentiation. These DTG curves are derived from the first derivative of weight loss percentage (wt%) concerning temperature (dm/dT) or time (dm/dt). Valuable quantitative and qualitative information can be gleaned from these DTG curves. Qualitatively, they contribute to the material identification and differentiation of reactions. Quantitative analysis involves measuring the peak height and its corresponding temperature for the segments of maximum weight loss within a curve.

3.2.1 Slow Pyro-Gasification of Lignocellulosic Biowastes

Pyrolysis and gasification are recognized as highly efficient methods for the production of biofuels from biomass. These advanced thermochemical techniques enable the conversion of waste into valuable products: biochar, bio-oil, and biogas, thereby improving the feasibility of storage and transport. The specific process parameters and the characteristics of the biomass have a strong influence on the yield and quality of the desired products. For instance, bio-oils are predominantly produced through fast pyrolysis, which involves rapid heating rates and short residence times, optimizing the liquid yield. Conversely, syngas, a mixture of hydrogen and carbon monoxide, is primarily obtained through gasification, a process that operates at higher temperatures and often in the presence of a controlled amount of oxygen or steam. Biochar production is maximized by slow pyrolysis, which employs lower temperatures and longer residence times, resulting in a solid residue rich in carbon. This solid fraction is notable for its diverse applications, including its use as a renewable fuel, an effective soil conditioner, or an efficient adsorbent for pollutants.

The efficiency and output of slow pyro-gasification are influenced by factors such as temperature, residence time, and the nature of the feedstock. The inherent composition of lignocellulosic biowastes, which includes cellulose, hemicellulose, and lignin, also plays a crucial role in determining the quality and quantity of the final products.

Overall, slow pyro-gasification of lignocellulosic biowastes offers a sustainable approach to managing these wastes, reducing greenhouse gas emissions, and generating renewable energy and materials. This approach contributes to the development of a circular bioeconomy.

3.3 CHARACTERISTIC INDICES

Bioenergy indices (BI) represent a set of crucial parameters for biomass characterization. These indices permit the evaluation of the calorific value in terms of energy density, thereby enabling the comparison of fuel capacity and CO_2 emissions of biomass with those of traditional fossil fuels. The indices calculated for both types of waste include bioenergy density (BD), fossil fuel equivalent volume (FFEV), and potential CO_2 retention (PCOR).

Bioenergy density (BD) indicates the energy released during the complete combustion per unit volume of biofuel:

$$BD = LHV \times \rho \tag{1}$$

Here, ρ is the bulk density (kg m^{-3}) (ASTM 2013). This methodology is followed to determine how tightly packed or compacted the particles are within a specific volume of material.

Fossil fuel equivalent volume (FFEV) provides information on the amount (in liters) of fossil fuel (average value) required to produce the same amount of energy as one cubic meter of biofuel (Alves et al. 2020). FFEV depends on the ratio between the bioenergy density of biofuel ($BD_{biofuel}$) and fossil fuel ($BD_{fossil\ fuel}$). The reference values for different fossil fuels can be found in the literature (Protásio et al. 2013).

$$FFEV = \frac{BD_{biofuel}}{BD_{fossil\ fuel}} \times 1{,}000 \tag{2}$$

The potential CO_2 retention (PCOR) enables the calculation of the mass of CO_2 released from combustion that could potentially be suppressed by replacing certain fossil fuels with their equivalent in biofuel:

$$PCOR = EF \times FFEV \tag{3}$$

Here, EF is the emission factor of different fuels reported in the literature (Protásio et al. 2013, Yaro et al. 2023).

Table 3.1 shows the results concerning the bioenergy indices for biowaste reported by Fernandez et al. (2022) and Torres-Sciancalepore et al. (2022). The biowaste studied was apple pomace (AP), grape marc (GM), grape stalk (GS), and quince waste (QW), which was previously subjected to a procedure to extract and quantify pectin, thus generating another depleted residue: pectin-free quince waste (PFQW). It is important to note that the cited authors used the same methodology. This allows for a proper discussion of the results and their corresponding conclusions.

Table 3.1 demonstrates that the BD and FFEV values obtained for various biowaste are not significantly different. The biowaste under study yielded PCOR values exceeding 2 tons of CO_2 released per cubic meter of biomass, avoidable compared to petroleum, diesel fuel, and gasoline.

The PCOR value obtained is lower than that of fuel oil and would enable the avoidance of approximately 1,650 kg of CO_2 per cubic meter of biomass. The

Table 3.1 Bioenergy indices for different biowaste

Bioenergy Indices	AP	GM	GS	QW	PFQW
BD (GJ/m^3)	24.05 ± 0.18	22.62 ± 0.29	21.99 ± 0.16	22.14 ± 0.13	22.79 ± 0.12
$FFEV_{petroleum}$ ($L_{fossil}/m^3_{biomass}$)	649 ± 0.18	610 ± 0.29	593 ± 0.16	597.87 ± 0.23	615.35 ± 0.12
$FFEV_{diesel\ fuel}$ ($L_{fossil}/m^3_{biomass}$)	663 ± 0.18	623 ± 0.29	606 ± 0.16	610.40 ± 0.14	628.24 ± 0.31
$FFEV_{fuel\ oil}$ ($L_{fossil}/m^3_{biomass}$)	602 ± 0.18	566 ± 0.29	550 ± 0.16	554.45 ± 0.11	570.66 ± 0.25
$FFEV_{gasoline}$ ($L_{fossil}/m^3_{biomass}$)	737 ± 0.18	693 ± 0.29	674 ± 0.16	678.70 ± 0.18	698.54 ± 0.20
$PCOR_{petroleum}$ (kg CO_2)	2,228 ± 0.18	2,095 ± 0.29	2,037 ± 0.16	2,050.71 ± 0.33	2,110.65 ± 0.14
$PCOR_{diesel\ fuel}$ (kg CO_2)	2,340 ± 0.18	2,201 ± 0.29	2,140 ± 0.16	2,154.72 ± 0.22	2,217.70 ± 0.18
$PCOR_{fuel\ oil}$ (kg CO_2)	1,771 ± 0.18	1,665 ± 0.29	1,619 ± 0.16	1,630.09 ± 0.29	1,677.74 ± 0.30
$PCOR_{gasoline}$ (kg CO_2)	2,905 ± 0.18	2,732 ± 0.29	2,656 ± 0.16	2,674.09 ± 0.16	2,752.25 ± 0.11

biowaste under study exhibits favorable conditions for heat generation per unit of biomass volume, reduced transportation costs, and enhanced storage capacity.

3.4 KINETICS ANALYSIS: ISOCONVERSIONAL METHODS

The differential equation can express the reaction rate for a single-step mechanism:

$$\frac{d\alpha}{dt} = k(T) f(\alpha) \tag{4}$$

Here, α is the conversion of the solid reactant that can be defined as a function of the mass loss as follows:

$$\alpha = \frac{m_0 - m_t}{m_0 - m_\infty} \tag{5}$$

Assigning m_0 to the initial mass, m_t to the mass at time t, and m_∞ to the residual mass after the heating process finishes. The term $f(\alpha)$ in Equation (5) corresponds to the kinetic model, which is an exclusive function of the conversion. On the other hand, k is the specific rate constant, which can be defined according to the empirical Arrhenius equation:

$$k(T) = A \exp\left(\frac{-E}{RT}\right) \tag{6}$$

The main objective of the kinetic analysis is to determine the kinetic triplet: E (activation energy), A (pre-exponential factor), and $f(\alpha)$ (functionality of the biomass conversion with the reaction rate) to be able to predict the decomposition rate of biowaste by pyrolysis. For non-isothermal determinations, Equation (5) can be rewritten as follows:

$$\frac{d\alpha}{dT}\frac{dT}{dt} = k(T) f(\alpha) \tag{7}$$

If the heating rate is constant, then $dT/dt = \beta$. Combining Equations (6) and (7) yields:

$$\frac{d\alpha}{dT} = \frac{A}{\beta} \exp\left(\frac{-E}{RT}\right) f(\alpha) \tag{8}$$

By taking the derivative of Equation (5):

$$\frac{d\alpha}{dt} = \frac{-1}{m_0 - m_\infty}\frac{dm}{dt} \tag{9}$$

Dm/*dt* is the derivative thermogravimetric (DTG) data, which may be obtained through numerical differentiation, such as Lagrange's second-order numerical operator.

Integral isoconversional methods are used to determine E, which are obtained from the integrated form of Equation (8):

$$g(\alpha)=\int_0^{\alpha}\frac{d\alpha}{f(\alpha)}=\int_{T_0}^{T}\frac{A}{\beta}\exp\left(\frac{-E}{RT}\right)dT=\frac{AE}{\beta R}\left[P(u)\right] \quad (10)$$

where $u = E/RT$ and $P(u)$ is the approximation of the temperature integral. Depending on the function $P(u)$ used to approximate the integral, various integral isoconversional methods arise, among which the FWO (Flynn-Wall-Ozawa), KAS (Kissinger-Akahira-Sunose), and Starink methods stand out (da Silva et al. 2020).

FWO:
$$\ln\beta_j = \text{Constant} - 1.0516\left(\frac{E}{RT}\right) \quad (11)$$

KAS:
$$\ln\left(\frac{\beta_j}{T^2}\right) = \text{Constant} - \frac{E}{RT} \quad (12)$$

Starink:
$$\ln\left(\frac{\beta_j}{T^{1.92}}\right) = \text{Constant} - 1.0008\,\frac{E}{RT} \quad (13)$$

E can be obtained from the slope of the regression analysis line in any of the three integral methods by plotting $\ln(\beta_j/T^m)$ vs $1/T$ where $j = 1,\ldots, J$, with J being the number of heating rates used in the experiment.

The variability of E with the conversion is important to determine whether it is possible to model the pyrolysis kinetics considering a single step or if a multiple-step approach is needed instead. A significant variation of E with the conversion could mean the existence of complex kinetic processes, such as the occurrence of multiple parallel reactions (Vyazovkin et al. 2011).

To determine E, conversion intervals were selected with a step size of $\Delta\alpha = 0.05$, spanning from 0.15 to 0.85. This range was chosen to avoid noise in the results at low decomposition rates of biowaste, which can lead to unrepresentative E values (Vyazovkin and Wight 1999). The selection of the appropriate isoconversional model was based on an analysis of the coefficient of determination (R^2) of the corresponding linear fits. Fernandez et al. (2022) and Torres-Sciancalepore et al. (2022) studied the kinetic behavior of quince waste (QW), pectin-free quince waste (PFQW), grape marc (GM), grape stalk (GS), and apple pomace (AP). These authors observed that, among the three models previously presented, the model that demonstrated the better statistical parameter values in the linear fit was the FWO model, and the results are presented in Table 3.2.

It may be observed that the activation energy of PFQW, a residue that was subjected to an acid extraction of pectins, was considerably lower than that of the other wastes. These findings can be attributed to two factors: (a) bioactive compounds are embedded within the cell walls of lignocellulosic biomass, located between the hemicellulose and cellulose chains. Their extraction modifies the biomass structure, resulting in changes to the system's activation energy; and

Table 3.2 E and R^2 from isoconversional methods for biowaste pyrolysis

α	QW		PFQW		GM		GS		AP	
	E [kJ/mol]	R^2	E [kJ/mol]	R^2	E [kJ/mol]	R^2	E [kJ/mol]	R^2	E [kJ/mol]	R^2
0.15	73.01	0.971	54.58	0.992	71.31	1.000	62.31	0.995	70.80	0.993
0.20	69.36	0.964	48.33	0.996	70.72	0.993	66.75	0.967	62.08	0.990
0.25	65.63	0.965	44.75	0.999	76.63	1.000	78.08	0.997	67.39	0.960
0.30	62.25	0.959	41.36	0.995	63.79	0.991	74.43	0.987	64.44	0.980
0.35	59.10	0.962	39.07	0.996	67.08	0.939	64.27	0.991	62.45	0.984
0.40	57.24	0.962	36.76	0.997	91.53	1.000	69.73	0.991	55.75	0.981
0.45	54.71	0.964	35.11	0.991	64.84	0.950	82.36	0.999	66.77	0.980
0.50	53.29	0.967	33.78	0.992	69.49	0.960	96.15	0.995	77.84	0.958
0.55	51.13	0.966	32.42	0.994	96.40	0.986	111.41	0.936	67.57	0.958
0.60	48.97	0.970	31.19	0.995	89.81	0.992	104.91	0.925	75.5	0.939
0.65	47.42	0.978	29.96	0.995	117.08	0.998	101.83	0.940	70.27	1.000
0.70	48.01	0.999	29.56	0.993	106.43	0.997	105.6	0.998	78.77	0.925
0.75	94.89	0.992	29.14	0.996	65.82	0.999	75.61	0.916	81.13	0.981
0.80	44.86	0.994	24.63	0.984	67.84	0.998	78.88	0.953	93.97	0.940
0.85	36.56	0.990	21.76	0.986	89.07	1.000	76.52	0.980	70.19	0.904
Average	57.76		35.49		82.47		85.51		76.21	

(b) the acid-based extraction process for bioactive compounds leads to partial hydrolysis of the polysaccharides (hemicelluloses and celluloses) in the biomass. This hydrolysis increases the chemical accessibility and reactivity of these polysaccharides, thereby lowering the activation energy. In contrast, lignin is not significantly affected, so its activation energy remains largely unchanged (Fernandez et al. 2024).

3.5 ESTIMATING FREQUENCY FACTOR

For a series of closely similar reactions, different authors have reported the existence of the kinetic compensation effect (KCE), which can be evidenced in a linear relationship between the logarithm of the pre-exponential factor and the activation energy (Liu and Guo 2001).

$$\ln A = a + bE \tag{14}$$

Here, a and b are the fitting parameters corresponding to the ordinate to the origin and slope, respectively. They emerge from the process of fitting the experimental data to the model.

The physical origins of the KCE are not entirely certain, although it is believed that it is related to the thermodynamic or enthalpy-entropy compensation (EEC) through the transition state theory since the activation energy and the pre-exponential factor are related to these thermodynamic parameters (Garvín et al. 2017).

$$A = k^{\ddagger} \exp\left(\frac{\Delta S^{\ddagger}}{R}\right) \tag{15}$$

$$E = \Delta H^{\ddagger} + RT \approx \Delta H^{\ddagger} \tag{16}$$

Distinguishing that in both expressions, the symbol '‡' represents the transition state, $\Delta S^{\ddagger}$ and $\Delta H^{\ddagger}$ represent the entropy and enthalpy change between reactants and the transition state and $k^{\ddagger} = \frac{k_B T}{h}$ according to Eyring's theory (k_B is Boltzmann's constant and h Planck's constant). It is possible to determine the pre-exponential factor using the compensation effect by considering different kinetic models, $f(\alpha)$, representing the different closely similar reactions (Vyazovkin et al. 2011). From the Coats and Redfern linear approximation (Coats and Redfern 1964), it is possible to calculate the activation energy and pre-exponential factor corresponding to different kinetic models ("i" denotes each model).

$$\ln\left(\frac{g_z(\alpha_i)}{T^2}\right) = \ln\left(\frac{A_{i,z} R}{\beta E_{i,z}}\right) - \frac{E_{i,z}}{RT} + \ln\left(1 - \frac{2RT}{E_{i,z}}\right) \tag{17}$$

Since $2RT/E_{i,k}$ is usually much lower than unity, $\ln\left(1 - \frac{2RT}{E_{i,z}}\right) \to 0$, then:

$$\ln\left(\frac{g_z(\alpha_i)}{T^2}\right) = \ln\left(\frac{A_{i,z}R}{\beta E_{i,z}}\right) - \frac{E_{i,z}}{RT} \tag{18}$$

In Equation (18), the subscript z represents the reaction model used. Using multiple kinetic models, g_z, (see Table 3.3), it is possible to perform a linear regression analysis to get, from the slope and intersect, the pair ($E_{i,z}$; ln $A_{i,z}$) from Equation (18). The pairs obtained from each model are plotted in a scatter diagram and linear regression analysis is performed to obtain the compensation parameters a and b of Equation (14). Thus,

$$A_i = \exp\,(a\bar{E}_i + b) \tag{19}$$

Here, $\bar{E}_i$ the mean value of the activation energy obtained by the isoconversional methods previously described.

Table 3.3 Conversion function, $f(\alpha)$, and integral rate, $g(\alpha)$, for different reaction models

Reaction Model	***f*(*α*)**	$g(\alpha) = \int_0^\alpha \frac{d\alpha}{f(\alpha)}$
	Reaction Order	
F1. First Order	$(1-\alpha)$	$-\ln(1-\alpha)$
F2. Second Order	$(1-\alpha)^2$	$(1-\alpha)^{-1}-1$
F3. Third Order	$(1-\alpha)^3$	$(1/2)[(1-\alpha)^{-2}-1]$
Fn. N-Order	$(1-\alpha)^n$	$[1-(1-\alpha)^{1-n}]\,(1-n)^{-1}$
	Diffusional	
D1. One-dimensional diffusion	$1/2\alpha^{-1}$	α^2
D2. Two-dimensional diffusion	$[-\ln(1-\alpha)]^{-1}$	$(1-\alpha)\ln(1-\alpha)+\alpha$
D3. Three-dimensional diffusion (Jander)	$\frac{3}{2}(1-\alpha)^{\frac{2}{3}}\left[1-(1-\alpha)^{\frac{1}{3}}\right]^{-1}$	$\left[1-(1-\alpha)^{\frac{1}{3}}\right]^2$
D4. Ginstling-Brounshtein	$3[2((1-\alpha)^{-1/3}-1)]^{-1}$	$1-(2/3)\alpha-(1-\alpha)^{2/3}$
	Nucleation	
P2. Power Law	$2\alpha^{1/2}$	$\alpha^{1/2}$
P3. Power Law	$3\alpha^{2/3}$	$\alpha^{1/3}$
P4. Power Law	$4\alpha^{3/4}$	$\alpha^{1/4}$
P2/3. Power Law	$2/3\alpha^{-1/2}$	$\alpha^{3/2}$
A2. Avrami-Erofeev	$2(1-\alpha)\,[-\ln(1-\alpha)]^{1/2}$	$[-\ln(1-\alpha)]^{1/2}$
A3. Avrami-Erofeev	$3(1-\alpha)\,[-\ln(1-\alpha)]^{2/3}$	$[-\ln(1-\alpha)]^{1/3}$
A4. Avrami-Erofeev	$4(1-\alpha)\,[-\ln(1-\alpha)]^{3/4}$	$[-\ln(1-\alpha)]^{1/4}$
	Contracting Geometry	
R2. Contracting Area	$2(1-\alpha)^{1/2}$	$1-(1-\alpha)^{1/2}$
R3. Contracting Volume	$3(1-\alpha)^{2/3}$	$1-(1-\alpha)^{1/3}$

Table 3.4 presents the parameters a and b derived from the compensation effect method [Equation (14)] reported by Fernandez et al. (2022) and Torres-Sciancalepore et al. (2022). The R^2 values of the linear fit and the A values calculated using Equation (19) are reported for both samples analyzed.

Table 3.4 Compensation effect parameters and pre-exponential factor of biowaste pyrolysis

Biowaste	β (K/min)	a	b	R^2	A (1/s)
	5	0.27	–8.01	0.98	2.07×10^3
QW	10	0.27	–7.55	0.98	2.90×10^3
	15	0.29	–7.13	0.97	1.44×10^4
	5	0.26	–7.10	0.99	9.99×10^0
PFQW	10	0.30	–9.11	0.97	4.95×10^0
	15	0.29	–7.21	0.97	1.93×10^1
	10	0.34	–6.91	0.95	9.12×10^{10}
GS	15	0.26	–7.47	0.97	8.94×10^6
	20	0.26	–7.77	0.90	1.12×10^7
	10	0.38	–8.36	0.90	8.21×10^7
AP	15	0.31	–8.87	0.95	1.71×10^3
	20	0.35	–8.04	0.93	4.06×10^7
	10	0.34	–8.85	0.90	6.45×10^8
GM	15	0.29	–7.90	0.95	1.35×10^7
	20	0.29	–7.33	0.95	5.20×10^7

The A values for the pyrolysis of GS, GM, and AP are considerably higher than those corresponding to the pyrolysis of quince biowaste (QW and FPQW), resulting in elevated values of the Arrhenius specific rate constant, k(T). Nevertheless, E is also considerably higher in the quince biowaste reaction, such that the effect of A is offset by E. This is because an increase in A or a decrease in the value of E produces an increase in k(T).

3.6 REACTION MODEL

The conversion function, $f(\alpha)$, can be obtained by comparing experimental curves with those corresponding to theoretical models (Table 3.3). This is known as the *master-plots method* (Vyazovkin et al. 2011, Gotor et al. 2000).

Considering the integrated rate equation [Equation (10)] and Doyle's approximation for the temperature integral P(u):

$$P(u) = 0.0048e^{-1.0516u} \tag{20}$$

Evaluating $\alpha = 0.5$:

$$g(0.5) = \frac{AE}{\beta R} P(u_{0.5}) \tag{21}$$

where $u_{0.5}$ is $E/RT_{0.5}$, that is, at the temperature corresponding to $\alpha = 0.5$. Dividing Equation (10) between Equation (21):

$$\frac{g(\alpha)}{g(0.5)} = \frac{P(u)}{P(u_{0.5})} \tag{22}$$

The term on the left of Equation (22) represents the theoretical plot, while the term on the right represents the experimental plot. The function $g(\alpha)$ (and therefore, $f(\alpha)$) will be the adequate model if it approximates the experimental values, verifying as best as possible the exact form of Equation (22).

Figure 3.1 depicts the master plot, which encompasses the experimental data about QW pyrolysis and the theoretical models that have been reported in the bibliography. It is of significant importance to note that analogous behavior was also observed in the case of the FPQW, GS, GM, and AP.

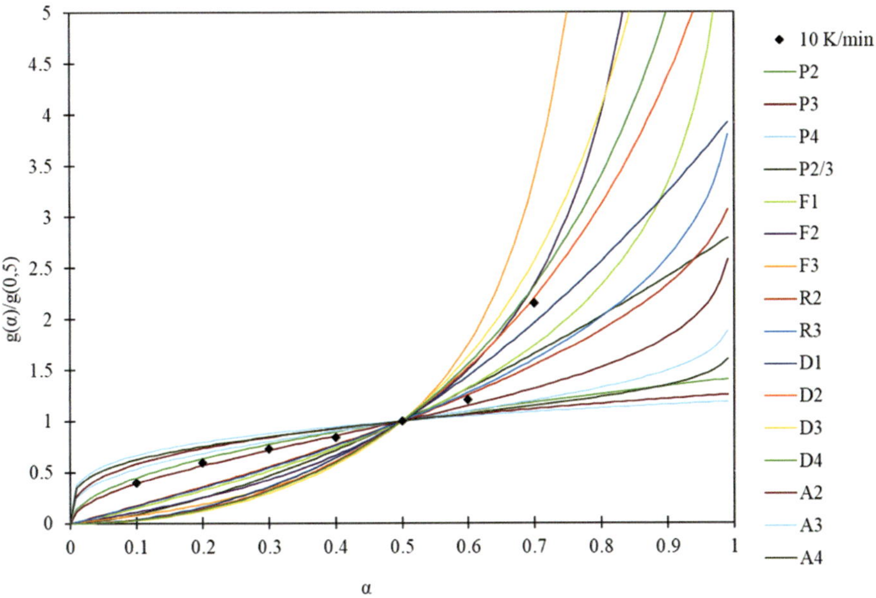

Figure 3.1 Master plots for reaction model (A2) determination of QW pyrolysis.

The experimental pyrolysis data exhibit a strong fit to the Avrami-Erofeev (A2) theoretical model of nucleation and growth. This fit is demonstrated in the A2 fit plot, which shows an R^2 value of 0.99. It should be highlighted that the fit of experimental data to theoretical curves is purely empirical and does not necessarily have a strict physical meaning.

3.7 ARTIFICIAL NEURAL NETWORKS

The Backpropagation Neural Network (BPNN) is a tool for predicting "experimental" results, such as those corresponding to the complex pyro-gasification processes.

This neural network model employs the Levenberg-Marquardt algorithm, which is well-suited for nonlinear systems. Temperature, heating rate, and the concentrations of carbon, hydrogen, and oxygen in agro-industrial organic wastes significantly influence pyrolysis and were therefore established as input parameters. All layers, except the output layer, which used the "purelin" function, were configured with the 'tansig' function as the transfer function (Muravyev et al. 2021). During training, the BPNN was used to fine-tune the network weights and biases to minimize the mean squared error (MSE). The net.trainParam.goal parameter was set to a desired performance goal of 0.001 using MATLAB ONLINE.

Figure 3.2 illustrates the grid topology of the BPNN used for the study of the three main components of MSW: plastic (PW), paper (PP), and organic matter (PO). The connections depicted represent the interconnected pathways through which information flows within the ANN.

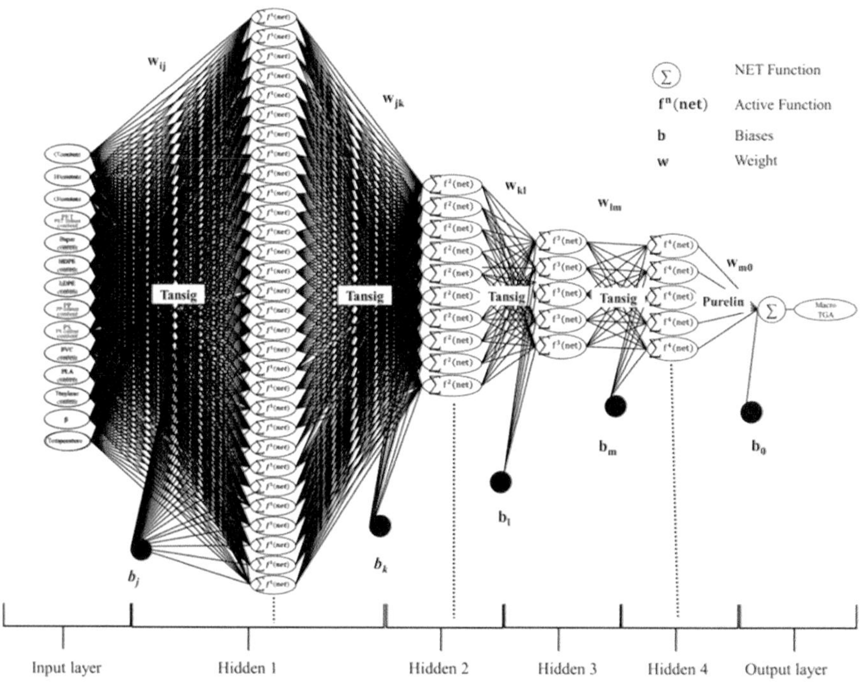

Figure 3.2 Grid topology of the BPNN.

These elements are essential for the network's functionality, determining how it processes and analyzes data. This intricate network of connections enables the ANN to acquire knowledge and represent complex relationships within the data. In Figure 3.2, each input variable node is connected to all nodes in the first hidden layer, with lines representing adjustable weights optimized during training. These connections extend through multiple hidden layers, where inputs are transformed by activation functions. The network culminates in an output layer, where the final ANN output is produced after all transformations, allowing the ANN to learn and model complex data relationships effectively.

The training matrix was created using experimental data from the TGA and macro-TGA studies described in Table 3.5. The input parameter matrix consisted of 14 columns representing the following variables: percent mass composition of carbon, hydrogen, oxygen, carbon from organic matter, PET (polyethylene terephthalate), paper, high-density polyethylene (HDPE), low-density polyethylene (LDPE), polystyrene (PS), polypropylene (PPR), polyvinyl chloride (PVC), polylactic acid (PLA), terylene, heating rate (β), and temperature. The output variable matrix consisted of a single column representing the remaining mass. A total of 21 data points were introduced from 47 TGA experiments with different solid samples and compositions. Therefore, we had an input matrix X of size 966 × 14 and an output matrix of 966 × 1. Of these data, 80% were used for training and 20% for validation.

Table 3.5 TGA data used for ANN training

Reference	TGA Experiment	Biomass	Plastic	Paper	Heating Rate (K/min)
Ansah et al. 2016	1	wood	–	–	10
	2	–	–	paper	10
	3	–	PET	–	10
	4	wood	PET	–	10
	5	wood	PET	–	10
	6	wood	PET	–	10
Burra and Gupta 2018	7	–	PPR	–	10
	8	–	PPR	–	20
	9	–	PPR	–	30
	10	–	PET	–	10
	11	–	PET	–	20
	12	–	PET	–	30
	13	pinewood	PPR		10
	14	pinewood	PPR		10
	15	pinewood	PPR		10
	16	pinewood	–	–	10
	17	pinewood	–	–	20
	18	pinewood	–	–	30
Gunasee et al. 2016	19	–	PVC	–	20
	20	–	PPR	–	20
	21	–	PS	–	20
	22	branches	–	–	20
	23	leaves	–	–	20
	24	grass	–	–	20
	25	–	–	cardboard	20
	26	–	–	hygienic paper	20
	27	branches	PPR	cardboard	20

(*Contd.*)

Reference	TGA Experiment	Biomass	Plastic	Paper	Heating Rate (K/min)
Gunasee et al. 2017	28	–	–	cellulose	10
	29	–	LDPE	–	10
	30	–	LDPE	cellulose	10
Hu et al. 2020	31	–	LDPE	–	10
	32	–	HDPE	–	10
	33	–	PPR	–	10
	34	–	PS	–	10
	35	–	terylene	–	10
	36	–	terylene	ldpe	10
	37	–	terylene	hdpe	10
	38	–	terylene	ppr	10
Torres-Sciancalepore et al. 2022	39	quince	–	–	5
	40	quince	–	–	10
	41	quince	–	–	15
	42	pectin–free quince	–	–	5
	43	pectin–free quince	–	–	10
	44	pectin–free quince	–	–	15
Sette et al. 2020	45	grape marc	–	–	10
	46	grape marc	–	–	15
	47	grape marc	–	–	20
	48	grape stalk	–		10
	49	grape stalk	–		15
	50	grape stalk	–		20
	51	apple pomace	–		10
	52	apple pomace	–		15

The prediction efficiency of different ANN architectures has been evaluated to determine the optimal number of neurons. Figure 3.3 illustrates the predicted and experimental mass left values of PW at 15 K/min using the neural network.

A notable consistency between theoretical and experimental data is observed, thereby demonstrating the effectiveness of the artificial neural network (ANN) model. The highest accuracy was achieved with the PW pyro-gasification process at 15 K/min, resulting in an R^2 value of 0.96, a mean absolute error (MAE) of 3.2, and a root mean square deviation (RMSE) of 3.99. These results indicate the high predictive ability of ANNs and indirectly validate the consistency of the experiments with the literature listed in Table 3.5. Furthermore, the study underscores the efficacy of macro-TGA as a valuable instrument for elucidating the kinetics of biomass decomposition in inert and low-oxygen atmospheres.

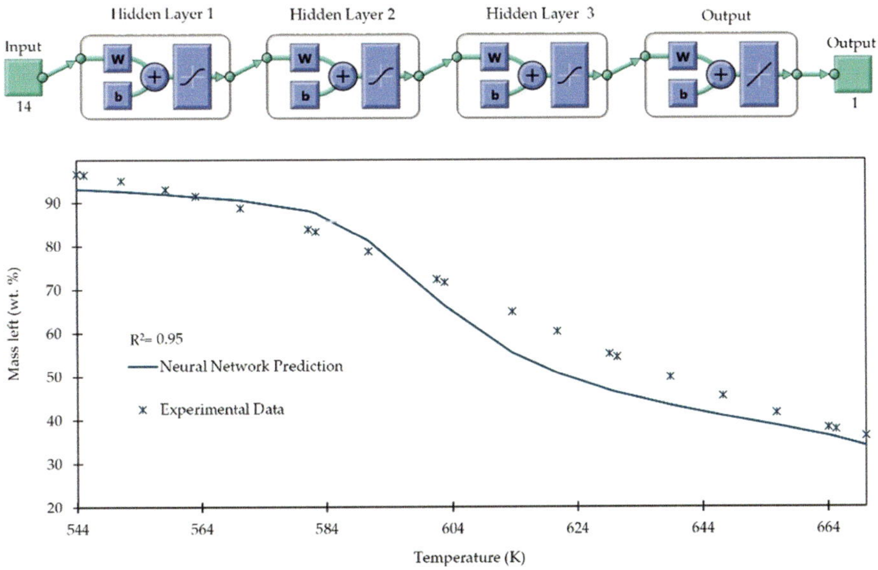

Figure 3.3 Neural network prediction and experiments mass left values of PW at 15 K/min.

3.8 CONCLUSIONS

This chapter presents the basics for the kinetic study of the slow pyro-gasification process of biowastes. It highlights the potential of adding value to biowastes by extracting bioactive compounds and subsequently pyro-gasifying them. Furthermore, the acid extraction treatment results in the production of a solid residue that is more reactive and more readily convertible through pyro-gasification.

The master-plots method demonstrated a suitable method to fit the experimental data to theoretical models to describe the conversion function. It is paramount to comprehend the fluctuations in activation energy throughout the pyro-gasification process. This understanding is essential for optimizing parameters and predicting feedstock behavior, thereby guiding future process design.

About artificial neural networks (ANN), a strong agreement between predicted and experimental data confirms the consistency of the results, thereby enhancing confidence in their reliability and significance. Macro-thermogravimetric analysis (Macro-TGA) effectively explores the decomposition kinetics in various waste materials, validated by ANN, underscoring their utility in characterizing pyro-gasification behaviors.

REFERENCES

Alves, J.L.F., J.C.G. da Silva, G.D. Mumbach, M. Di Domenico, R.F. de Sena, R.A.F. Machado, et al. 2020. Demonstrating the suitability of tamarind residues to bioenergy exploitation via combustion through physicochemical properties, performance indexes, and emission characteristics. Bioenergy Research. 13: 1308–1320.

Ansah, E., L. Wang and A. Shahbazi. 2016. Thermogravimetric and calorimetric characteristics during co-pyrolysis of municipal solid waste components. Waste Management. 56: 196–206.

ASTM, E.-82. 2013. Standard Test Method for Bulk Density of Densified Particulate Biomass Fuels. 2 pp.

Burra, K.G. and A.K. Gupta. 2018. Kinetics of synergistic effects in co-pyrolysis of biomass with plastic wastes. Applied Energy. 220: 408–418.

Coats, A.W. and J.P. Redfern. 1964. Kinetic parameters from thermogravimetric data. Nature. 201: 68–69.

Fernandez, A., P. Sette, M. Echegaray, J. Soria, D. Salvatori, G. Mazza, et al. 2022. Clean recovery of phenolic compounds, pyro-gasification thermokinetics, and bioenergy potential of spent agro-industrial bio-wastes. Biomass Convers Biorefin.

Fernandez, A., D. Zalazar-García, C. Lorenzo-Doncel, D.M. Yepes Maya, E.E. Silva Lora, R. Rodriguez, et al. 2024. Kinetic modeling of Co-Pyrogasification in Municipal Solid Waste (MSW) Management: Towards sustainable resource recovery and energy generation. Sustainability. 16: 4056.

Garvín, A., R. Ibarz and A. Ibarz. 2017. Kinetic and thermodynamic compensation. A current and practical review for foods. Food Research International. 96: 132–153.

Gotor, F.J., José, M. Criado, J. Malek and N. Koga. 2000. Kinetic analysis of solid-state reactions: The universality of master plots for analyzing isothermal and nonisothermal experiments. Journal of Physical Chemistry A. 104: 10777–10782.

Gunasee, S.D., M. Carrier, J.F. Gorgens and R. Mohee. 2016. Pyrolysis and combustion of municipal solid wastes: Evaluation of synergistic effects using TGA-MS. Journal of Analytical and Applied Pyrolysis. 121: 50–61.

Gunasee, S.D., B. Danon, J.F. Görgens and R. Mohee. 2017. Co-pyrolysis of LDPE and cellulose: Synergies during devolatilization and condensation. Journal of Analytical and Applied Pyrolysis. 126: 307–314.

Hu, Q., Z. Tang, D. Yao, H. Yang, J. Shao and H. Chen. 2020. Thermal behavior, kinetics, and gas evolution characteristics for the co-pyrolysis of real-world plastic and tyre wastes. Journal of Cleaner Production. 260: 121102.

Liu, L. and Q.X. Guo. 2001. Isokinetic relationship, isoequilibrium relationship, and enthalpy-entropy compensation. Chemical Reviews. 101: 673–695.

Muravyev, N.V., G. Luciano, H.L. Ornaghi, R. Svoboda and S. Vyazovkin. 2021. Artificial neural networks for pyrolysis, thermal analysis, and thermokinetic studies: The status quo. Molecules. 26: 1–25.

Protásio, T. de P., L. Bufalino, G.H.D. Tonoli, M.G. Guimarães Junior, P.F. Trugilho and L.M. Mendes. 2013. Brazilian lignocellulosic wastes for bioenergy production: Characterization and comparison with fossil fuels. Bioresources. 8: 1166–1185.

Rony, A.H., L. Kong, W. Lu, M. Dejam, H. Adidharma, K.A.M. Gasem, et al. 2019. Kinetics, thermodynamics, and physical characterization of corn stover (Zea mays) for solar biomass pyrolysis potential analysis. Bioresource Technology. 284: 466–473.

Sette, P., A. Fernandez, J. Soria, R. Rodriguez, D. Salvatori and G. Mazza. 2020. Integral valorization of fruit waste from wine and cider industries. Journal of Cleaner Production. 242.

da Silva, J.C.G., J.G. de Albuquerque, W.V. de A. Galdino, R.F. de Sena and S.L.F. Andersen. 2020. Single-step and multi-step thermokinetic study – Deconvolution method as a simple pathway for describe properly the biomass pyrolysis for energy conversion. Energy Conversion and Management. 209: 112653.

Torres-Sciancalepore, R., D. Asensio, D. Nassini and A. Fernandez. 2022. Assessment of the behavior of *Rosa rubiginosa* seed waste during slow pyrolysis process towards complete recovery: Kinetic modeling and product analysis. Energy Conversion and Management. 272: 116340.

Torres-Sciancalepore, R., A. Fernandez, D. Asensio, M. Riveros, M.P. Fabani, G. Fouga et al. 2022. Kinetic and thermodynamic comparative study of quince bio-waste slow pyrolysis before and after sustainable recovery of pectin compounds. Energy Conversion and Management. 252: 115076.

Vyazovkin, S. and C.A. Wight. 1999. Model-free and model-fitting approaches to kinetic analysis of isothermal and nonisothermal data. Thermochimica Acta. 340–341: 53–68.

Vyazovkin, S., A.K. Burnham, J.M. Criado, L.A. Pérez-Maqueda, C. Popescu and N. Sbirrazzuoli. 2011. ICTAC Kinetics Committee recommendations for performing kinetic computations on thermal analysis data. Thermochimica Acta. 520: 1–19.

Vyazovkin, S. 2015. Isoconversional Kinetics of Thermally Stimulated Processes. Springer Cham.

Yaro, N.S.A., M.H. Sutanto, N.Z. Habib, A. Usman, J.M. Kaura, A.A. Murana, et al. 2023. A comprehensive review of biochar utilization for low-carbon flexible asphalt pavements. Sustainability. 15: 6729.

Chapter 4

Using Biochar to Capture CO_2

Viviana Fernández-Maldonado*[a], Ana Laura Navas[a], Rosa Rodriguez[a], Anabel Fernandez[a] and Andrés Reyes-Urrutia[b]

[a]Instituto de Ingeniería Química, Fac. de Ingeniería, Grupo Vinculado al PROBIEN (Universidad Nacional de San Juan, CONICET, Universidad Nacional del Comahue), San Juan, Argentina

[b]Instituto de Investigación y Desarrollo en Ingeniería de Procesos, Biotecnología y Energías Alternativas, PROBIEN (CONICET, Universidad Nacional del Comahue), Neuquén, Argentina

4.1 INTRODUCTION

Climate change is one of the greatest challenges facing humanity in the 21st century. Anthropogenic activities, especially burning fossil fuels, deforestation, and intensive agriculture, have caused an unprecedented increase in greenhouse gas (GHG) emissions, with carbon dioxide (CO_2) being the main contributor. This increase in CO_2 in the atmosphere has altered the planet's energy balance, triggering devastating effects such as rising global temperatures, changes in weather patterns, aridification, and loss of biodiversity. In this context, it is imperative to develop and adopt sustainable technologies that allow the reduction of CO_2 emissions and the mitigation of their negative impacts.

One of the most promising solutions is using biochar, a material derived from organic biomass through a controlled pyrolysis process. This material has proven effective in capturing CO_2 from the atmosphere and improving soil quality, reducing emissions of other GHGs such as methane (CH_4) and nitrous oxide (N_2O),

*Corresponding author: V.N. Fernández Maldonado (vivifernandezm@unsj-cuim.edu.ar)

DOI: 10.1201/9781032726694-4

and contributing to agricultural fertility. Biochar stands out for its porous structure and its ability to be chemically modified, which allows for optimizing its performance as a carbon adsorbent.

This chapter comprehensively addresses the properties and applications of biochar in CO_2 capture, analyzing in detail the adsorption mechanisms involved, the physical and chemical activation techniques that enhance its adsorbent capacity, and the functionalization processes that allow its characteristics to be adapted to different needs. Innovative methods such as metallic impregnation are also explored, which expands the efficiency of biochar in environmental remediation and carbon storage.

Throughout the chapter, the importance of optimizing biochar production parameters, including biomass selection, pyrolysis conditions, and post-treatments, is highlighted to maximize its effectiveness in carbon capture and storage applications. Furthermore, the most widely used methodologies for modeling adsorption processes and assessing biochar performance are presented, providing a solid scientific basis for future research and industrial applications.

In summary, this chapter aims to report on the current state of knowledge around biochar and its role in climate change mitigation and inspire new lines of research that make the most of its potential as a tool to confront the environmental crisis. This multidimensional analysis highlights biochar as a bridge between technological innovation and sustainability, positioning it as a key resource in transitioning toward a greener and more resilient future.

4.2 GLOBAL WARMING AND CARBON CAPTURE

Global warming is defined as the increase in the average temperature on Earth. This is mainly due to the increase in the concentrations of different greenhouse gases (GHGs), such as water vapor, carbon dioxide (CO_2), nitrous oxide (N_2O), ozone (O_3), methane (CH_4), chlorofluorocarbons (CFCs) among others (Venkataramanan 2011). This effect occurs when greenhouse gases (GHGs) in the atmosphere trap heat from the sun, keeping the planet at a stable temperature. These gases act like the glass in a greenhouse, letting light through but preventing heat from entirely escaping into space (Kweku et al. 2018). However, human activity has contributed to climate change and global warming at an accelerated rate due to global industrialization since the early 20th century (Peter 2018). Industrialization and the advancement of the transportation sector are the primary sources of greenhouse gas emissions due to the burning of fossil fuels. Other sources, such as deforestation and intensive agriculture, especially meat production and waste management, also contribute to this phenomenon (Al-Ghussain 2019). Zhou et al. (2021a) demonstrated that crops have a significant impact on land surface temperature, with effects varying by location, season, and background climate. These findings have important implications for understanding climate change and underscore the need to consider the biological and biogeochemical effects of crops in future assessments of climate change and the formulation of mitigation and adaptation strategies. These human activities have significantly increased the concentrations of greenhouse gases in the atmosphere, leading to global warming and significant climate changes worldwide. However, arid lands are often considered the most fragile ecosystems and are very sensitive to climate change. Koutroulis (2019) examines the impact of climate change in arid areas and its influence on water availability and human habitability. It uses high-resolution climate projections where arid areas will increase by 7% of global land area by 2100.

Additionally, population increases are projected in dry areas, with significant implications for water availability and degradation of the ground. Park et al. (2018) estimated emerging aridification under different climate scenarios, identifying its emergence before certain levels of global warming, where they found that limiting warming to 1.5°C could reduce aridification and its negative consequences. Likewise, global warming and, consequently, climate change are significantly impacting the frequency of extreme precipitation worldwide, with important implications for the vulnerability and adaptation of communities to these changes. Papalexiou and Montanari (2019) highlight the importance of understanding and monitoring these changes to develop mitigation and adaptation strategies.

Another consequence of climate change is altering precipitation patterns worldwide, resulting in more prolonged droughts in some regions, intensifying aridification, and intense rainfall in others. This can cause serious problems for agriculture, food security, biodiversity, and the availability of fresh water. Papalexiou and Montanari (2019) assessed global changes in the frequency and magnitude of extreme precipitation during global warming acceleration from 1964 to 2013. The findings indicate that climate change is exerting a notable influence on extreme precipitation events and heavy rainfall globally, which holds substantial implications for the susceptibility and adaptation of communities to these transformations. The authors highlight the importance of understanding and monitoring these changes to develop mitigation and adaptation strategies. Likewise, the melting of polar caps and glaciers, due to the increase in temperature, contributes to the rising sea level. This has secondary effects on local ecosystems and the wildlife that depends on them, such as coral reefs (Hughes et al. 2018) and marine mammals (Albouy et al. 2020). Frölicher et al. (2018) project that marine heat waves (MHWs) will become more frequent, long-lasting, extensive, and intense in the future, with a significant increase in MHW days. These changes could push marine organisms and ecosystems to the limits of their resistance, which could cause irreversible changes.

Among the greenhouse gases, one of the most important in global warming is CO_2. The accumulation of carbon dioxide in the atmosphere due to human activity is mainly caused by burning fuel for different purposes, such as automobiles, cement production plants, power generation from coal-fired power plants, the transportation sector, and industrial manufacturing processes. Although routes have been proposed to decarbonize the use of fossil fuels, such as carbon capture and storage, which we will describe later, they remain dominated by the global energy system (Novotnik et al. 2020). However, forests act as important carbon sinks, storing large amounts of carbon in their biomass and the soil through capture, storage, and fixation. However, deforestation reduces the capacity of forest ecosystems to store carbon, further contributing to rising CO_2 levels in the atmosphere. Li et al. (2022a) compare carbon loss and its effect on climate change using Earth system model simulations and satellite observations, where they found that deforestation causes a further increase in carbon losses in aboveground biomass, particularly in the Amazon and Congo, due to warming and drought resulting from deforestation.

Carbon capture is a vital tool in the battle against climate change, but it comes with limitations and challenges. These include costs (Ozkan et al. 2022), the availability of suitable storage sites for captured carbon dioxide, and concerns regarding the safety and long-term effectiveness of underground CO_2 storage (Chen et al. 2022). However, it is considered an important part of a comprehensive strategy to reduce greenhouse gas emissions and limit global warming. Therefore, investigating carbon capture technologies is important as it is considered the only solution to mitigate CO_2 emissions. Currently, four main technologies are used for carbon capture. One is the pre-combustion route, another is post-combustion (being the most used) (Omoregbe et al. 2020), oxygen combustion (used to a lesser extent), and capture from industrial processes (e.g., oil refineries, biogas sweetening, and production of ammonia, cement, iron, and steel; see the work of Raza et al. 2019). Post-combustion involves capturing CO_2 after combustion, while pre-combustion involves separating CO_2 before combustion by converting the fuel into a mixture of H_2 and CO_2. Oxy-fueling uses pure oxygen in combustion to produce CO_2 and H_2O, which are then separated.

The captured CO_2 is compressed and transported through pipelines, ships, or tanks to its final destination for storage or reuse. This is used in oil and gas recovery (Ghiat and Al-Ansari 2021), agriculture, soda ash manufacturing, the food industry, and the production of value-added chemicals and fuels such as methanol and methane (Jones 2020), or stored in geological reservoirs or saline aquifers (Chen et al. 2022). The captured CO_2 is transported via pipelines, ships, or tanks to its final destination for storage or reuse (Chen et al. 2022). The chemical and petrochemical industry appears to be leading in integrating and implementing carbon capture and storage to reduce emissions. This sector is expected to contribute approximately 28% of the carbon emissions reductions planned for 2070, according to the International Energy Agency (IEA) scenario. Furthermore, the chemical and petrochemical industry has more operational and planned carbon capture facilities than any other industry, indicating a strong commitment to reducing carbon emissions (Olabi et al. 2022).

4.3 BIOCHAR FOR GREENHOUSE GAS FIXATION IN SOIL

Soils are an important source of GHG emissions such as carbon dioxide (CO_2), nitrous oxide (N_2O), and methane (CH_4) (Fernández-Maldonado et al. 2024). These gases are generated through natural processes and play a crucial role in the global carbon and nitrogen cycle. Microorganisms in the soil are an important component of ecosystems and play a key role in both cycles. These can participate in fundamental processes, such as carbon fixation and degradation, as well as in the nitrogen cycle, such as nitrification, denitrification, ammonization, and nitrogen fixation (Zhao et al. 2022). Finally, methane is generated mainly by microbial processes under anaerobic conditions, generally occurring in water-saturated environments, such as flooded soils or with high humidity levels, where oxygen availability is limited (Abagandura et al. 2019). Jiang et al. (2021) found that combining biochar and controlled irrigation reduced CH_4 emissions and modified the soil microbial community in rice fields under controlled irrigation. Biochar has been highlighted as a promising tool to mitigate soil-based greenhouse gas emissions as it contributes to carbon fixation and nitrogen fertilizer immobilization, reducing N availability and soil N_2O emissions (Abhishek et al. 2022). However, the effects of biochar on nitrogen mineralization and soil GHG emissions are ambiguous and may vary depending on soil conditions and agricultural practices (Craswell et al. 2021). Liang et al. (2019) found that the interaction of biochar and arbuscular mycorrhizal fungi (AMF) can improve plant growth and nutrient uptake and facilitate nutrient retention synergistically. However, the presence of AMF can increase N_2O emissions when biochar is added, influencing greenhouse gas emissions in wetland-soil plant systems. Also, it has been found that biochar can mitigate methane emissions in agricultural soils by immobilizing the gas and improving soil conditions that reduce methanogenic activity (Malyan et al. 2021).

The effectiveness of biochar in fixing greenhouse gases depends on several factors, such as the type of feedstock (Liu et al. 2021), pyrolysis conditions (such as temperature; see the work of Mosa et al. 2023), the properties of the soil (such as pH, porosity, and conductivity; see the work of Shakoor et al. 2021) and climate and application process (Xia et al. 2023). Hui (2021) summarized the general findings on the effects of biochar on soil physical, chemical, and microbial properties, nutrient availability, plant growth, biomass production and yield, GHG emissions, and carbon sequestration in the soil (see the work of Hui 2021). Due to variations in pyrolysis conditions, feedstock types, biochar application rates and methods, and possible interactions with other factors such as plant species and soil nutritional conditions, the results of the studies are inconclusive. However, most studies report positive effects of biochar on soil physical and chemical properties, microbial activities, plant biomass, and yield, and potential reductions in soil GHG emissions (Wang et al. 2018; Martí et al. 2021).

Additionally, biochar positively impacts on cropland restoration by reducing the toxicity of agricultural soils, improving plant nutrient availability, and promoting crop growth. Also, it contributes to the immobilization of metals in the soil, which benefits both soil health and crop yields (Xu et al. 2023). In their review, Ji et al. (2022) emphasize that biochars from crop residues and woody biomass are effective for organic pollution and greenhouse gas reduction, while those with high ash content are better for heavy metal pollution. Biochar applied to agricultural soils has potential benefits, although its impact remains debated and may vary depending on conditions. It is important to recognize the potential negative effects of biochar on soil so as not to underestimate its application (Kavitha et al. 2018).

In summary, biochar is revealed as a promising tool to reduce greenhouse gas emissions in the soil by immobilizing carbon and nitrogen, improving the physical properties of the soil, modifying the microbial activity related to the production of these gases, and increasing crop productivity, as noted by Zhang et al. (2021). In addition, it acts as an amendment that improves nutrient retention and supply, as well as microbial activity, reducing soil density. Also, biochar can be used to adsorb heavy metals, increase water retention capacity, and improve physical properties, as Abhishek et al. (2022) discussed. However, further research is required to verify the mechanisms of biochar-soil-microbial-plant interactions and understand how they can be better utilized in agriculture.

4.4 BIOCHAR AND CO_2 ADSORPTION

Biochar adsorption is a process by which it captures carbon dioxide (CO_2) molecules from the atmosphere or aqueous solutions. This is because it has a porous structure that allows it to retain CO_2 in its internal spaces and on the surface of its particles (Gong et al. 2019). The CO_2 adsorption process can help reduce the concentration of this greenhouse gas in the atmosphere, thus mitigating climate change. CO_2 adsorption by biochar is an active research area in the field of carbon capture and storage, and it is used to develop more efficient and economical technologies for the reduction of CO_2 emissions on a large scale. Promraksa and Rakmak (2020) found that palm residue biochar shows a high surface area and CO_2 absorption capacity, positioning it as a promising adsorbent for carbon dioxide capture. Furthermore, the varied origin of the biomass can influence the properties of the biochar and its adsorption capacity, as well as its production temperature, since higher temperatures generally increase both the surface area and the adsorption capacity of the biochar (Mamaghani et al. 2023). Figure 4.1 shows the possible morphology and the presence of surface functional groups on biochar.

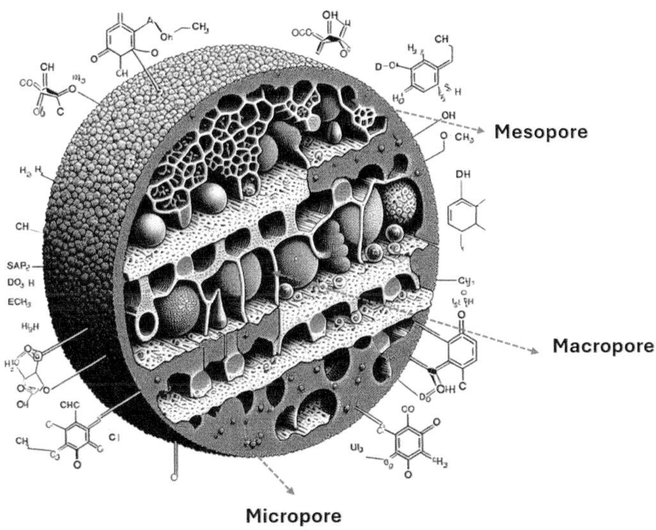

Figure 4.1 Morphology and the presence of surface functional groups on biochar.

The CO_2 adsorption capacity of biochar depends mainly on its physical-chemical properties, such as surface area, size and volume of pores, presence of basic functional groups and heteroatoms, the presence of functional groups of surfaces, presence of alkali and alkaline earth metals, hydrophobicity, non-polarity and aromaticity (Dissanayake et al. 2020b). These physical and chemical properties of biochar are closely related to the type of raw material used and the thermochemical conditions during its production (Bamdad et al. 2019). Chen et al. (2020) found that pyrolysis temperature influences the CO_2 adsorption capacity in algal biochar, where higher temperatures improve the adsorption capacity due to higher specific surface area and N doping groups.

The surface area is calculated as the ratio between the total surface area and the total mass of the biochar. An increase in the surface area provides more active sites for the physical adsorption of CO_2 (Guo et al. 2022). Therefore, a biochar with a larger surface area has a higher CO_2 absorption capacity. During the decomposition of volatile organic matter in the polymeric matrix of the feedstock, porous structures form within the biochar. Numerous studies have shown that a greater total volume of pores in biochar increases the availability of active sites for interaction with CO_2.

Furthermore, a high microporosity content has been highlighted as crucial for higher CO_2 adsorption capacity, especially under ambient temperature and low-pressure conditions (Shafawi et al. 2021). Oxygen-containing functional groups must be controlled because many tend to be acidic and are, therefore, unsuitable for CO_2 adsorption. Nitrogen-containing functional groups increase the surface basicity of the biochar, which is essential in its ability to adsorb CO_2 since they increase the affinity of the biochar for CO_2 (Cao et al. 2022). This indicates that as the N content in biochar increases, the amount of CO_2 that can be adsorbed also increases (Zhang et al. 2022a). These groups can be introduced to the biochar surface through reactions with different nitrogenous reagents or activation of the biochar using nitrogen-containing precursors. Zhang et al. (2022a) showed that the CO_2 adsorption capacity was improved by more than 40% compared to the original biochar due to the presence of more micropores, especially in the range of 0.8–1.5 nm, and the abundance of pyrrolic nitrogen in the MNBC3-400. These features improve the CO_2 capture capacity by increasing the van der Waals interaction between the adsorbent material and CO_2 molecules. Alkali and alkaline earth metals, such as Na, K, Ca, Mg, and Li, may increase the CO_2 adsorption capacity of biochar. Lahijani et al. (2018) showed that the CO_2 adsorption capacity of Mg-loaded biochar (80.0 mg/g) is higher than that of the original biochar (69.1 mg/g), which represents a 15.7% increase in sorption capacity. This increase is attributed to the synergistic effects of physical and chemical interactions and the presence of MgO on the surface that contributes to the chemisorption of CO_2 through the formation of carbonate. Biochar with hydrophobic properties and low polarity facilitates CO_2 adsorption by minimizing competition with water molecules (H_2O). Characteristics such as H/C and O/C ratios less than 0.2 indicate high aromaticity and chemical stability of the fixed carbon, resulting in higher hydrophobicity, low polarity, and improved ability to capture CO_2. Reducing the oxygen (O) and hydrogen (H) contents in biochar during carbonization is beneficial to increase its hydrophobicity and reduce its polarity, which is crucial for the selective adsorption of CO_2 compared to polar molecules such as water (Shafawi et al. 2021).

Methods used to improve the CO_2 adsorption capacity of biomass-derived materials include activation processes, surface treatments, incorporation of nitrogen-containing functional groups, and impregnation of alkali metal oxides (Jung et al. 2019). Treatments can be applied to the feedstock before or after pyrolysis to induce specific changes in the biochar. These treatments include chemical and physical modifications, as well as impregnation with elements or grafts. These methods aim to increase the affinity of CO_2 for adsorbents by modifying their surface properties. In their review, Qiao and Wu (2022) highlight that combining modification with nitrogen functional groups and the "mixed molten salt" synthesis method are effective strategies to improve the CO_2 adsorption capacity of activated carbons. In summary, according to the literature, chemical and physical activations are the most efficient activation methods to modify biochar and convert it into a potential catalyst. Chemical activation involves activating agents during carbonization to improve the adsorption capacity, focusing on improving the adsorption capacity of biochar.

On the other hand, physical activation involves increasing the porosity and surface area of the biochar by exposing it to gases such as CO_2 or steam at high temperatures, aiming to improve adsorption efficiency by eliminating highly reactive carbons during the process (Anto et al. 2021). Zhang et al. (2023) investigated the application of biochar for CO_2 capture in three types of biochar (bamboo charcoal, wood pellets, and coconut shell) modified by physical (H_2O, CO_2) and chemical ($ZnCl_2$, KOH, and H_3PO_4) activation to improve its performance in carbon capture. They found a positive correlation between surface area, micropore volume, and C/N ratio with CO_2 adsorption capacity, while the O/C ratio and pore width had a negative correlation. This indicates that biochars with high aromaticity and hydrophobicity are more suitable for CO_2 capture (Zhang et al. 2023). In conclusion, biochar, especially when properly activated and modified, can be an effective adsorbent for CO_2 capture, and its use could be a promising strategy to address the challenge of climate change.

4.5 PHYSICAL ACTIVATION: BY STEAM AND CO_2

The physical activation of biochar is a process in which biochar is subjected to high temperatures in the absence of oxygen to create a porous structure and increase its surface area. High-temperature carbonization typically performs this process, followed by activation with water vapor or CO_2. During the activation process, CO_2 and water vapor are assumed to react with different active sites on the carbon. The reaction with CO_2 increases the pore length, while the reaction with water vapor increases the pore radius (Maya et al. 2020). However, Korus et al. (2022) found that when both gases are combined, a significant increase in the reactivity of carbon derived from sewage sludge is observed. This suggests that combining CO_2 and H_2O may have a synergistic effect that improves the carbon's ability to react and form pores more efficiently. Zaini and Syed-Hassan (2022) discovered that activating palm shells with a sequential combination of CO_2 and water vapor resulted in the highest methane adsorption yield. Furthermore, this method yielded greater specific surface area and total pore volume than activation with either agent alone. The resulting carbon materials were mainly microporous, highlighting the importance of carefully choosing activation agents to optimize the porous and adsorption properties of the carbon materials.

The physical steam activation method increases the surface area of biochar by forcing high-temperature steam through its pores (Guo et al. 2022). Furthermore, steam activation increases the aromaticity and decreases the polarity of biochar, which is reflected in low values of the H/C and O/C molar ratios (Shafawi et al. 2021). Biochar undergoes partial gasification by steam, which improves devolatilization and promotes the formation of a crystalline structure. Oxygen from water molecules at the surface sites of carbon generates surface oxides and produces H_2. This generated H_2 then reacts with the surface sites of the carbon, forming hydrogen complexes on the surface and thus activating the biochar. Igalavithana et al. (2020) showed that steam activation on pine sawdust biochar increased its CO_2 adsorption capacity by enhancing the van der Waals forces between the biochar matrix and CO_2, thus facilitating its adsorption. Steam-activated biochar exhibits slightly higher CO_2 adsorption capacity and higher CO_2 selectivity over N_2 than non-steam-activated biochar.

Furthermore, the activation temperature and heating rate are crucial in biochar production as they directly influence the structure and properties of the material. Increasing the activation temperature promotes the decomposition of functional groups and the formation of pores in the biochar, thus increasing its surface area and adsorption capacity with a greater presence of mesoporous and microporous structures (Cao et al. 2022). On the other hand, a faster heating rate can accelerate the activation of adsorption sites, improving the process efficiency and contaminant removal capacity. Together, these factors contribute to improving the adsorbent properties of biochar, making it more effective in removing organic and inorganic contaminants (Zhang et al. 2023). De Castro et al. (2023) found that, overall, steam activation proved to be more efficient and cost-effective compared to CO_2 activation in the production of activated carbon from Brazil nut shells on a large scale.

CO_2 activation in biochar production involves exposing the carbonaceous material to carbon dioxide at high temperatures, typically between 200°C and 900°C. During this process, CO_2 decomposes into carbon monoxide (CO) and oxygen, which then react with the biochar, creating porosity and enhancing the material's surface area (Anto et al. 2021). CO_2 activation results in higher pore generation efficiency and specific surface area compared to other activation methods, which significantly improves the adsorption capacity of biochar for contaminant removal (Zhang et al. 2023). The process of CO_2 activation is a carbon-burning reaction that does not introduce oxygen-containing functional groups into the biochar. As temperature increases, CO_2 becomes reactive and binds to the carbon matrix via the reverse Boudouard reaction, forming a more porous structure and increasing the material's surface area (Maya et al. 2020). This increased porosity improves the affinity of CO_2 for the biochar, enhancing its adsorption capacity (Shafawi et al. 2021). While CO_2 activation is generally effective for improving microporosity, it has a smaller effect on the porous structure compared to steam activation (Zhang et al. 2023). One advantage of CO_2 activation is its ability to produce a carbon material with a narrow micropore structure, which is highly efficient in adsorbing specific compounds (Rashidi and Yusup 2020).

Additionally, CO_2 activation is favored for its ease of manipulation and the purity of CO_2, making it a preferred method in some cases (Zhang et al. 2022b). However, compared to steam activation, which generates a broader range of pore sizes and higher microporosity, CO_2 activation focuses on enhancing the material's microporosity (Zaini and Syed-Hassan 2022). Rashidi and Yusup (2020) also emphasize the sustainability of physical activation with CO_2, highlighting its potential to replace non-renewable precursors and improve biochar's market value. However, further research is needed to explore economic aspects, life cycle impacts, desorption and regeneration, and dynamic adsorption to ensure the viability and sustainability of this process in industrial applications.

Physical activation improves the properties of biochar by increasing its adsorption capacity for contaminants, nutrients, and gases, making it useful in applications such as the remediation of contaminated soils, water purification, and carbon capture. Furthermore, physical activation can modify the structure of biochar to adapt it to specific applications and improve its performance in various fields, such as agriculture, environmental engineering, and energy production. Sarkar et al. (2023) used activated carbon derived from petroleum coke to remove naphthenic acid from waters affected by synthetic oil sands processes. They found that CO_2 activation produced carbon with a higher mesopore volume fraction than water vapor activation. Prolonging the activation time with CO_2 increased the specific surface area and mesopore volume, which influenced the adsorption kinetics of naphthenic acids. Generally, a combined approach of technologies is suggested to purify tailings pond water.

In conclusion, the choice between water vapor activation and CO_2 activation depends on the specific applications and the desired properties of the biochar. Both methods are effective in improving the properties of biochar and making it suitable for various environmental and industrial applications. Continued research in this field is crucial to optimize these processes, improve the sustainability and efficiency of biochar, and expand its use in environmental remediation, water purification, and carbon capture.

4.6 CHEMICAL ACTIVATION (KOH)

Carbon capture and storage (CCS) is one of the most promising technologies to mitigate climate change, as it reduces CO_2 emissions to the atmosphere (Li et al. 2022b). Biochar, a carbonaceous material obtained from the pyrolysis of biomass, has emerged as an attractive adsorbent for CO_2 capture due to its high surface area, developed porosity, and abundance of functional groups on its surface (Ding and Liu 2020; Zhang et al. 2022c). However, its adsorption characteristics need further enhancement to efficiently use biochar in CCS applications (Ding and Liu 2020).

One of the most effective techniques to enhance the properties of biochar is chemical activation using potassium hydroxide (KOH) as an activating agent (Chen et al. 2020). This activation process induces a series of complex chemical reactions between the carbon in the biochar and KOH, forming a highly porous and developed carbon structure (Zhang et al. 2022b).

The typical interactions between the carbon in biochar and KOH during the activation process can be represented by the following chemical equations (Equations 1, 2, 3, and 4):

$$C + 2KOH \rightarrow K_2CO_3 + H_2 \quad (1)$$

$$K_2CO_3 \rightarrow K_2O + CO_2 \quad (2)$$

$$C + K_2O \rightarrow 2K + CO \quad (3)$$

$$6KOH + 2C \rightarrow 2K + 3H_2 + 2K_2CO_3 \quad (4)$$

These reactions generate various gaseous byproducts, such as CO, CO_2, and H_2, which escape through the carbon matrix, creating extensive porosity in the activated biochar. Additionally, the intercalation of metallic potassium into the carbon layers and its subsequent removal via acid washing contribute to pore formation.

Compared to physical activation, chemical activation with KOH significantly enhances the CO_2 adsorption capacity of biochar (Zhang et al. 2022c). This improvement is attributed to the alkaline treatment with KOH, which produces more condensed and ordered carbon structures, increasing both the size and volume of existing pores and generating new pores.

When KOH (in the form of species such as K_2CO_3 and K_2O) impregnates the highly ordered carbon layers of biochar, it can induce the formation of condensed carbon structures (Chen et al. 2020). This expands existing pores and develops new ones, enhancing the material's porosity. Additionally, the carbon in the biochar can react with adsorbed CO_2 to form CO, further contributing to micropore development (Huang et al. 2019).

Unlike steam-activated carbon, which is unsuitable for cyclic adsorption, CO_2 and KOH-activated carbons demonstrate excellent stability and can be regenerated and reused for at least 50 consecutive cycles. Moreover, at atmospheric pressure and low CO_2 temperature, KOH-activated carbon is considered the best alternative due to its high specific surface area and porosity.

Various studies have demonstrated that KOH-activated biochar exhibits a significantly higher maximum CO_2 adsorption capacity than unmodified biochar. For example, Ding and Liu (2020) found that porous biochar from algae (SCK-800–1 and ECK-800–1) prepared with a KOH/biomass weight ratio of 1:1 at 800°C exhibited the highest CO_2 adsorption capacity, reaching up to 5.8 mmol/g at 25°C and 1 bar CO_2 pressure. Sargassum and Enteromorpha, two types of macroalgae, have been used as raw materials for producing KOH-activated biochar (Ding and Liu 2020). The preparation method for these materials was similar: fresh sargassum was repeatedly washed with deionized water and dried to remove possible impurities. Then, the dried sargassum was ground and sieved to obtain powdered biomass materials of less than 50 mesh (Ding and Liu 2020). Subsequently, the powdered sargassum and solid KOH were thoroughly mixed in different KOH/biomass weight ratios (0, 1, 2, and 4). Finally, the mixtures were pyrolyzed at different temperatures (400°C, 600°C, and 800°C) for 2 hours in a tubular furnace under a continuous flow of N_2, which was referred to as a one-step carbonization and activation method. The resulting solid powders were then washed with dilute HCl to remove KOH residues and dried at 105°C (Ding and Liu 2020). The results showed that KOH-activated biochar had a higher specific surface area (SSA) and pore volume than inactivated biochar. For instance, sargassum biochar prepared at 800°C with a KOH/biomass ratio of 2:1 (SCK-800–2) exhibited a SSA surface area of 2,014 m^2/g and a pore volume of 1.06 cm^3/g, compared to inactivated sargassum biochar (SA-800) which had a surface area of only 686 m^2/g and a pore volume of 0.38 cm^3/g (Ding and Liu 2020). Moreover, KOH-activated biochar showed a higher CO_2 adsorption capacity than inactivated biochar. The SCK-800–2 biochar adsorbed up to 5.8 mmol/g of CO_2 at 25°C and 1 bar pressure, while inactivated SA-800 biochar only reached an adsorption capacity of 3.2 mmol/g under the same conditions (Ding and Liu 2020). This demonstrates that chemical activation with KOH is an effective strategy to significantly improve the CO_2 adsorption properties of biochar (Dissanayake et al. 2020a).

The mechanisms explaining the increased CO_2 adsorption capacity in KOH-activated biochar include (Li et al. 2022):

- Increased surface area and pore volume: KOH activation generates a highly porous and developed carbon structure, providing more surface area and pore volume for CO_2 adsorption.
- Formation of basic functional groups: The impregnation of potassium species such as K_2CO_3 and K_2O in the biochar creates basic sites on the surface, enhancing the affinity towards acidic CO_2 molecules.
- Improved porosity through intercalation and gasification reactions: The intercalation of metallic potassium into the carbon layers and its subsequent extraction, along with the release of gaseous byproducts, generate additional pores.
- Chemisorption of CO_2: Some basic functional groups in KOH-activated biochar can form stronger chemical bonds with CO_2 molecules.

In general, the chemical activation of biochar with KOH is an effective strategy to improve its performance in CO_2 capture (Jung et al. 2019). Generating a highly porous carbon structure and developing specific functional groups on the surface of activated biochar are the main factors contributing to a significant increase in CO_2 adsorption capacity (Zhang et al. 2023). In addition to the improved adsorption capacity, KOH activation also presents other advantages for biochar in CCS applications (Huang et al. 2019). For instance, KOH-activated biochar exhibits high stability and can be regenerated and reused over multiple adsorption-desorption cycles without significant loss of efficiency. This makes it an attractive and viable adsorbent for industrial-scale carbon capture and storage processes (Zhang et al. 2022c). However, it is important to consider that the chemical activation process with KOH also entails some disadvantages, such as higher energy costs and the generation of byproducts that must be adequately managed (Jung et al. 2019). Therefore, optimizing activation conditions, such as the KOH/biomass ratio, pyrolysis temperature, and retention times, is necessary to maximize the benefits of this technique while minimizing negative impacts (Zhang et al. 2022c).

Furthermore, ongoing research is crucial to develop new activation and modification strategies for biochar to achieve even higher CO_2 adsorption capacities and improve the material's stability and durability in long-term applications. These strategies may include combining chemical activation with KOH with other methods, such as physical activation, or incorporating metallic additives to enhance biochar properties further.

4.7 METAL IMPREGNATION

The adsorption of metal ions or heteroatoms onto the surface and into the pores of biochar is known as metal impregnation (Gopalan et al. 2022). This technique is employed to enhance the CO_2 and other pollutant absorption processes. Researchers use metal ions such as magnesium, iron, silver, zinc, etc., alongside increasing the specific surface area of the biochar to boost its adsorption capacity (Yang et al. 2019; Kumar et al. 2022). In recent years, several studies have focused on the metal impregnation of biochar. Magnesium, silver, zinc, copper, and iron are the most commonly incorporated metal ions. Metal-impregnated biochar significantly improves adsorption capacity compared to unmodified biochar (Francis et al. 2023). This is due to the presence of two solid phases: nano-crystals of metal oxides and the biochar matrix. Both phases contribute to adsorption through various mechanisms such as hydrogen bond formation, precipitation, electrostatic precipitation, and ligand exchange (Bushra and Remya 2024).

It has been found that CO_2 adsorption can be improved by increasing the alkalinity of the biochar surface (Francis et al. 2023). It has been found that CO_2 adsorption can be improved by increasing the alkalinity of the biochar surface (Lahijani et al. 2018). Therefore, introducing metal groups such as Na, Ca, Mg, Al, Ni, and Fe into biochar increases basic sites, presenting a promising option for enhancing its CO_2 adsorption capacity (Jung et al. 2019). Studies have shown that biochar incorporating Mg exhibits a higher CO_2 adsorption capacity (82.0 mg/g) compared to unmodified biochar (72.6 mg/g) at 25°C and 1 atm (Lahijani et al. 2018). Additionally, cyclic CO_2 capture experiments showed that Mg-loaded biochar has high-performance stability. Generally, metal oxyhydroxides are basic and tend to bind with acidic CO_2 molecules. Therefore, due to their ferromagnetic properties, biochar and metal oxyhydroxide composites, such as Fe_2O_3-biochar, can enhance CO_2 adsorption (Creamer et al. 2016). However, Creamer et al. (2016) found that a larger surface area does not always guarantee high CO_2 adsorption in aluminum oxide-modified biochar. Biochar impregnation is crucial for improving its physicochemical properties and forming composite materials that increase mass yield and adsorption capacity (Wen et al. 2023). Currently, biochar modified with metal oxides, such as CaO, ZnO, FeO, and MnO, and metal salts, such as $LaCl_3$, $MgCl_2$, $AlCl_3$, and $FeCl_3$, are available. These materials show new hydroxyl groups, larger pore sizes, and better surface areas, contributing to increased adsorption capacity (Zhou et al. 2021b).

There are three main approaches for synthesizing metal-activated biochar (MBAC): (1) wet impregnation with metal salt before carbonization; (2) impregnation of metal particles into activated biochar after carbonization; and (3) carbonization of biomass in the presence of metal oxide (Lahijani et al. 2018). Each method has advantages and disadvantages regarding cost, complexity, and control over the metal structure.

The presence of metal oxides such as Mg, Al, Cu, and Fe in biochar can increase active sites that attract CO_2 molecules (Jung et al. 2019; Qiao and Wu 2022). Studies have found that Ca-based MBAC generated at 700°C exhibits the highest CO_2 adsorption of 4.7 mmol/g under ambient conditions compared to non-impregnated activated carbon (Zhou et al. 2021b). Additionally, metal impregnation can improve the catalytic, magnetic, and adsorption properties of biochar (Creamer and Gao 2016). For example, the Fe_2O_3-biochar composite has ferromagnetic properties that facilitate its separation and recovery after adsorption (Creamer et al. 2016). Other studies have shown that biochar impregnated with Ni and Cu significantly improves its catalytic activity for reducing organic pollutants (Zhang et al. 2020).

In summary, the metal impregnation technique of biochar is an effective and promising strategy to enhance its CO_2 adsorption capacity. The controlled introduction of metal groups increases the basicity of the material and creates new properties that favor the capture and retention of carbon dioxide. This makes modified biochar a high-performance adsorbent for climate change mitigation applications. Additionally, impregnation with certain metals can provide additional benefits regarding catalytic and magnetic properties. Therefore, this technique presents an integral solution to enhance biochar efficiency in various environmental applications.

4.8 FUNCTIONALIZATION OF BIOCHAR SURFACE GROUPS TO ENHANCE CO_2 ABSORPTION

Functionalizing the surface groups of biochar is a crucial strategy to improve its CO_2 adsorption capacity (Sajjadi et al. 2019a). This process involves introducing specific functional groups onto the biochar surface, increasing its surface area, and providing additional active sites for adsorption. Surface functionalization of biochar is critical due to its ability to enhance CO_2 adsorption (Feng et al. 2021; Petrovic et al. 2022). Raw biochar, while containing some functional groups such as hydroxyls (–OH) and carbonyls (–C=O), generally lacks a sufficient quantity for efficient CO_2 adsorption applications. Adding specific functional groups can increase biochar hydrophilicity and improve its electrostatic interaction with CO_2 molecules (Sajjadi et al. 2019a; Igalavithana et al. 2020). Common treatments include amino acid modification, acid modification, sulfidation, and ultrasound-assisted amination. Fundamental concepts, techniques used, and the most important processes in functionalizing biochar for CO_2 capture are described below.

4.8.1 Fundamental Concepts

Biochar inherently possesses some functional groups like hydroxyls and carbonyls, but their quantity is relatively small compared to the total surface area of biochar (Cheng and Li 2018; Sajjadi et al. 2019a). Surface functionalization aims to increase these functional groups, thereby enhancing the adsorptive properties of biochar (Feng et al. 2021; Petrovic et al. 2022). The functional groups can include amines, carboxyls, phenols, and sulfonic groups, each contributing differently to CO_2 adsorption. In addition to its surface area and microporosity, the functional groups on the biochar surface are crucial for its CO_2 adsorption capacity (Cheng and Li 2018). These groups improve CO_2 capture at low pressures, independent of biochar porosity. High pressures, especially in pores larger than 1.0 nm, are essential to increase adsorption capacity (Amer et al. 2024).

4.8.2 Surface Basicity

CO_2 adsorption can be improved by increasing the surface alkalinity of biochar. Oxygen- and nitrogen-containing functional groups influence acidity and alkalinity. Groups like lactols, phenols, lactones, and carboxylic acids increase acidity, while chromenes, ketones, and pyrones increase basicity. Basic functional groups are more effective in adsorbing CO_2, an acidic gas, (Amer et al. 2024).

4.8.3 Aromaticity

The aromaticity of biochar, related to carbon stability and resistance to biodegradation, facilitates CO_2 adsorption (Amer et al. 2024). The van Krevelen diagram, based on H/C and O/C atomic ratios, measures this property. A low H/C and O/C ratio indicates high chemical stability. Higher aromaticity and hydrophobicity favor CO_2 sorption by reducing water presence on the biochar surface (Amer et al. 2024).

4.9 FUNCTIONALIZATION TECHNIQUES

Various techniques exist for functionalizing biochar surfaces, including acid modification, amination, sulfidation, and ultrasound modification (Sajjadi et al. 2019a; Petrovic et al. 2022). Depending on the specific modification objective, these techniques can be applied before or after pyrolysis.

4.9.1 Acid Modification

The acid modification involves treating biochar with strong acids like nitric acid. This process increases the amount of oxygen-containing functional groups on the biochar surface, enhancing its hydrophilicity and capacity to interact electrostatically with CO_2 molecules (Amer et al. 2024). However, this method can affect biochar morphology and pore structure, reducing its surface area. Additionally, high acid concentrations and long treatment times can be prohibitive in terms of cost and industrial safety.

4.9.2 Amination

Amination introduces amino groups onto the biochar surface (Tang et al. 2019). This can be achieved through various methods, such as nitrogen reduction and anchoring tetraethylenepentamine (TEPA) or polyethyleneimine reagents (Petrovic et al. 2022), or treating biochar with NH_3 at high temperatures. Amination improves CO_2 adsorption due to the presence of basic sites that strongly interact with CO_2 molecules (Sajjadi et al. 2019a). For instance, sawdust-derived biochar treated with TEPA showed significantly higher CO_2 adsorption capacity than non-functionalized biochar.

4.9.3 Nitro-Reduction Method

This method treats biochar with nitric and sulfuric acids to introduce nitro groups, which are then reduced to amino groups (Petrovic et al. 2022). This process involves the formation of nitronium ions that attack the aromatic carbon of biochar, creating temporary complexes that convert into nitro groups. These nitro groups are subsequently reduced to amines using a reducing agent (Sajjadi et al. 2019a). This method has been shown to increase the CO_2 adsorption capacity of biochar.

4.9.4 Surface Oxidation

Surface oxidation involves producing additional oxygen-containing functional groups (e.g., carboxyl, hydroxyl, phenolic groups) on the biochar surface. It is generally conducted by thermally treating biochar under aerobic conditions (Cheng and Li 2018). Some authors have notably increased the concentration of oxygen-containing functional groups, especially carboxyl groups (from 0.53 to 3.70 mmol/g), on hydrochar surfaces simply by heating the hydrochar to 300°C in air (Cheng and Li 2018). Besides thermal treatment in the air, oxidizing reagents like H_2O_2, O_3, HNO_3, and $KMnO_4$ can also achieve surface oxidation (Cheng and Li 2018).

4.9.5 Sulfidation

Sulfidation introduces sulfur-containing functional groups, such as sulfonyl groups, onto the biochar surface (Cheng and Li 2018). These groups can enhance the electrostatic interaction between biochar and CO_2 molecules, increasing adsorption capacity (Cheng and Li 2018). Although less common than acid modification or amination, sulfidation offers a viable alternative for biochar functionalization.

4.9.6 Ultrasound-Assisted Functionalization

Ultrasonic irradiation can assist in the functionalization of biochar by dispersing and activating the material's surface. This method has been combined with chemical amination to improve the CO_2 adsorption capacity of biochar. For example, pine wood biochar functionalized with TEPA in the presence of an activator showed enhanced CO_2 adsorption capacity after ultrasonic treatment.

4.10 IMPORTANT PROCEDURES

The typical procedure for biochar functionalization may vary depending on the specific method, but generally follows these steps:

1. Biochar Preparation: Biochar is produced from biomass through controlled pyrolysis.
2. Chemical Treatment: An appropriate chemical treatment (acid, amination, etc.) introduces the desired functional groups.
3. Activation: In some cases, biochar is thermally or ultrasonically activated to enhance the dispersion and reactivity of functional groups.
4. Characterization: Techniques like Fourier-transform infrared spectroscopy (FTIR) and scanning electron microscopy (SEM) are used to characterize surface changes and confirm the presence of new functional groups.

4.11 IMPACT OF FUNCTIONALIZATION ON BIOCHAR MORPHOLOGY AND STRUCTURE

Surface functionalization can significantly affect biochar morphology and pore structure (Igalavithana et al. 2020; Feng et al. 2021). For instance, acid modification can reduce biochar's surface area by altering its porous structure. However, this effect can be mitigated using techniques like autoclave functionalization, which combines high temperature and pressure to preserve or even enhance biochar's porous structure.

4.12 EVALUATION OF MORPHOLOGICAL CHANGES

Advanced techniques such as SEM and nitrogen adsorption-desorption can be used to evaluate the changes in morphology and structure of functionalized biochar (Cheng and Li 2018; Feng et al. 2021). These techniques allow for observing the porous structure and determining the specific surface area of biochar before and after functionalization (Igalavithana et al. 2020).

- Scanning Electron Microscopy (SEM): Provides detailed images of the biochar surface, showing changes in texture and porosity.
- Nitrogen Adsorption-Desorption: The Brunauer-Emmett-Teller method is used to measure the specific surface area and pore volume of functionalized biochar.

4.13 EXAMPLES OF SUCCESSFUL FUNCTIONALIZATION

A study on the amination of spruce sawdust-derived biochar showed that biochar functionalized with NH_2 through the nitro-reduction method had a CO_2 adsorption capacity of 2.79 mmol/g due to the presence of nitrogen-enriched functional groups that facilitated both physisorption and chemisorption of CO_2.

Another example includes functionalizing soybean straw biochar with KOH and urea, followed by thermal treatment. This biochar showed significant changes in amine functional groups during CO_2 adsorption and desorption reactions, providing deeper insights into the interactions between functional groups and CO_2 (Zhang et al. 2021).

Functionalizing the surface groups of biochar is an effective strategy to enhance its CO_2 adsorption capacity. Functionalization techniques such as acid modification, amination, and ultrasound use can introduce a variety of functional groups that increase the active surface area and improve interactions with CO_2 molecules. These advancements are crucial for developing efficient and sustainable adsorbents for CO_2 capture and storage in industrial applications.

4.14 ACID MODIFICATION OF BIOCHAR FOR ENHANCED CO_2 ADSORPTION

Acid modification of biochar is an effective technique for enhancing its CO_2 adsorption capacity. This process involves treating biochar with various acids, significantly altering its physical and chemical properties and thereby optimizing its efficiency as a gas adsorbent. Commonly used acids for this purpose include HCl, H_2SO_4, HNO_3, H_3PO_4, oxalic acid, and citric acid.

The primary objective of acid modification is to remove impurities such as metals and ash while introducing acidic functional groups to the surface of the biochar. These functional groups, such as carboxyls, phenols, and lactones, are crucial for increasing the affinity of biochar for CO_2, as they enhance electrostatic and chemical interactions between CO_2 and the biochar surface. Acid modification can significantly alter the surface area and porosity of biochar, depending on the type and concentration of the acid used.

4.14.1 HCl

Modifying sugarcane-derived biochar with 1M hydrochloric acid reduced the ash content from 29.5 to 11.8% and increased the surface area from 58.75 to 88.35 m^2/g. This provided hydrophobic adsorption sites for the adsorption of organic compounds such as pentachlorophenol (Peng et al. 2017).

4.14.2 HNO_3

Modification with 2M nitric acid has effectively introduced functional groups such as carboxyls, lactones, phenols, and carbonyls in bamboo-derived biochar, enhancing its gas adsorption capacity (Li et al. 2014). This treatment significantly increases the specific surface area and CO_2 adsorption capacity due to the introduction of oxygenated functional groups that strongly interact with CO_2.

4.14.3 H_3PO_4

Phosphoric acid is one of the most commonly used modification agents. Modification with H_3PO_4 generally slightly decreases the surface area and total pore volume of biochar but improves CO_2 adsorption capacity by incorporating phosphorus-containing functional groups (Taha et al. 2014; Zhou et al. 2017). These micropores and functional groups can play a crucial role in the physical and chemical adsorption of CO_2. For instance, activation with H_3PO_4 at 450°C significantly improved the CO_2 capture capacity of activated biochar compared to other acids. However, not all acids increase surface area. Treating rice straw biochar with 2% sulfuric acid reduced the surface area from 71.35 to 56.9 m^2/g (Yakout et al. 2015). Although this treatment may reduce surface area, it can introduce a higher amount of acidic functional groups that favor CO_2 adsorption.

4.14.4 Weak Acids (Oxalic Acid and Citric Acid)

These acids introduce carboxyl groups to the biochar surface through esterification. Although these treatments often decrease surface area, combining sulfuric acid with oxalic acid can significantly increase the surface area of biochar, thereby enhancing its CO_2 adsorption capacity (Sun et al. 2015; Vithanage et al. 2015).

Various studies have extensively evaluated the CO_2 adsorption capacity of modified biochar. Acid modification increases the surface area and porosity of biochar and introduces functional groups that enhance interaction with CO_2. Studies have shown that biochar activated with phosphoric acid and zinc chloride can have a high CO_2 capture capacity. Specifically, biochar activated with a combination of phosphoric acid and KOH showed an adsorption capacity of 18,040 mg/g at 30°C and 1 bar, which is significantly higher than the capacity of biochar activated with H_3PO_4 or $ZnCl_2$ alone (Heidari et al. 2019).

The effectiveness of acid-modified biochar in CO_2 capture has also been confirmed using advanced characterization techniques such as X-ray photoelectron spectroscopy (XPS). These analyses have revealed the presence of functional groups like –OH and –COOH on the biochar surface, indicating that acid modification effectively introduces active sites for CO_2 adsorption (Jing et al. 2014). Fourier-transform infrared spectroscopy (FTIR) has also been used to identify changes in the surface chemistry of biochar after acid modification. These studies have shown the appearance of bands corresponding to carboxyl and carbonyl groups, confirming the incorporation of these functional groups, which are crucial for CO_2 adsorption.

The acid modification of biochar generally involves immersing the biochar in an acid solution, followed by a drying and thermal activation stage. For example, in one study, eucalyptus wood was immersed in an H_3PO_4 or $ZnCl_2$ solution and dried overnight. Subsequently, it was carbonized at temperatures between 450°C and 500°C, depending on the acid used. This procedure created a highly porous and functionalized surface suitable for CO_2 adsorption.

Several factors can influence the effectiveness of biochar acid modification:

4.14.4.1 Acid Concentration

Different acid concentrations can have varying effects on the surface and porous structure of biochar. Higher acid concentrations can increase porosity but may also lead to excessive degradation of the biochar structure.

4.14.4.2 Type of Biomass

The raw material used to produce biochar can influence its response to acid treatment. Biochars derived from different biomasses may have different initial compositions and structures, affecting their subsequent modification.

4.14.4.3 Activation Conditions

The temperature and activation time are crucial in determining the final properties of the modified biochar. Higher activation temperatures generally improve porosity but may reduce the content of oxygenated functional groups (Sajjadi et al. 2019).

Using acid-modified biochar for CO_2 capture has potential applications in various areas, including climate change mitigation, industrial gas purification, and air quality improvement. The CO_2 adsorption capacity of modified biochar can be further optimized by combining chemical and physical activation techniques and adjusting acid modification conditions (Sajjadi et al. 2019).

Acid modification of biochar is a promising technique for enhancing its CO_2 adsorption properties. Proper selection of acid type and concentration is crucial for maximizing adsorption capacity, as different acids and treatment conditions can have varying effects on surface area and functional group introduction. Continued development in this field may lead to the production of highly efficient biochar for environmental applications, particularly in CO_2 capture.

4.15 SURFACE AMINATION

Gas adsorption is significantly influenced by the physical properties of biochar, particularly its surface area and porosity. During biomass pyrolysis, high-temperature dehydration leads to the formation of pores of various categories: nano (<0.9 nm), micro (<2 nm), meso (2–50 nm), and macro (>50 nm) pores. According to an analysis conducted by Leng et al. (2021), the surface area, total pore volume, micropore volume, and mesopore volume of pristine biochar obtained through pyrolysis typically fall within the ranges of 8–132 m^2/g, 0.016–0.083 cm^3/g, 0.012–0.060 cm^3/g (with a median value of 0.024 cm^3/g), and 0.007–0.020 cm^3/g (with a median value of 0.009 cm^3/g), respectively.

The primary factors that may affect the surface area and porosity of biochar include the type of biomass or feedstock, pyrolysis conditions, and biochar modification techniques (physical, chemical, or physicochemical) (Leng et al. 2021). Depending on the type of biomass, cellulose-rich biomass materials are known to produce biochar with more micropores during pyrolysis, while pyrolysis of lignin-rich biomass feedstocks results in biochar with high macroporosity (Sajjadi et al. 2019b). Additionally, high pyrolysis and activation temperatures lead to a larger surface area and increased pore size (Feng et al. 2017a, b). While micropores are highly desired for gas adsorption, excessive widening of pores at high pyrolysis or activation temperatures can adversely affect the gas adsorption capacity of the biochar (Lahijani et al. 2018; Zubbri et al. 2020). Specific surface area and microporosity are interrelated properties; generating a high volume of small micropores results in a large specific surface area, thus contributing to high CO_2 adsorption. Multiple studies have shown that a highly microporous adsorbent possesses superior CO_2 adsorption capacity (Durá et al. 2016). It has even been reported that micropore volume is far more important than total pore volume for CO_2 adsorption, especially at ambient temperature and low pressure. The presence of micropores is essential for gas adsorption because adsorption occurs in micropores due to the diffusional activation energy of these pores (Ghanbari and Kamath 2019). CO_2 adsorption in micropores depends on the adsorption temperature and pressure. Ultra-micropores (<0.7 nm) and pores smaller than 1 nm are the dominant pores responsible for CO_2 adsorption at 25°C and 0°C, respectively. At pressures up to 1 bar, CO_2 adsorption is governed by micropores smaller than 1 nm. At pressures below 0.1 and 1 bar, CO_2 adsorption is mostly dominated by micropores of 0.5 nm and smaller than 1 nm, respectively. Therefore, the porosity of biochar should be carefully controlled to achieve high CO_2 adsorption capacity. In this context, the effects of various modification techniques on the porosity properties of biochar, specifically used for CO_2 capture, are addressed.

There are several methods for functionalizing carbon surfaces with nitrogen groups. Ammonia is commonly used, where adsorbent particles are placed in a tube furnace. The adsorbent is gradually heated to the specified temperature, typically with N_2 purging. Once the set point temperature is reached, the N_2 is replaced with NH_3 or an NH_3 mixture. Zhang et al. (2016) modified soybean straw biochar with NH_3 over a temperature range of 500–900°C. This process enhanced the surface area (from 1.5 to as high as 496 m^2/g) and increased the CO_2 adsorption capacity to 1.8 mmol/g.

Other methods of introducing nitrogen include adding nitrogen-rich proteins and amino acids. Thote et al. (2010) synthesized nitrogen-enriched carbon using local soybean as the nitrogen source (soy protein), followed by chemical activation with zinc chloride and physical activation with CO_2. The surface area of the synthesized nitrogen-enriched carbon increased to 811 m^2/g, and the breakthrough adsorption capacity increased to 0.5 mmol/g at 120°C. Plaza et al. (2007) applied different alkylamines to activated carbon (Norit CGP) through a wet impregnation method to increase the basicity and nitrogen content. The impregnation decreased the surface area (from 1,762 to 90 m^2/g) and did not increase adsorption capacity. The raw activated carbon showed the highest CO_2 adsorption capacity. It was proposed that the amine might block a fraction of the pores, reducing the surface area available for adsorption.

A subset of the biochars was thermally activated using a diluted air-nitrogen mixture at a moderate temperature (560°C) and compared with non-activated chars. The relationship between the impact of porous structures and nitrogen-containing groups on enhancing the CO_2 adsorption capacity of biochar was assessed. Further investigation into the functionalization of these biochars can enhance adsorption and allow the chars to be tailored to target gases (such as other acid gases like H_2S).

4.16 CO_2 ADSORPTION MECHANISM

Greenhouse gases are the main causes of global warming, with CO_2 being the most predominant, accounting for around 75% of total emissions worldwide (Marchal et al. 2011). It is necessary to reduce CO_2 emissions into the atmosphere to prevent an increase in the average temperature of the planet. Thus, developing porous materials for storing and capturing gas molecules (particularly CO_2 capture and storage) has attracted great interest in academic and industrial communities.

Adsorption is key for separating CO_2 from mixtures, such as flue gases (CO_2/N_2) or biogas (CO_2/CH_4). It is a process in which the molecules of a gas, liquid, or dissolved solids, known as adsorbates, adhere to the surface of another material, the adsorbent.

Adsorption mechanisms can be mainly classified into two categories: physisorption (physical adsorption) and chemisorption (chemical adsorption). The International Union of Pure and Applied Chemistry (IUPAC) defines chemisorption as an adsorption process in which strong chemical bonds are formed between the surface of an adsorbent and the molecules of an adsorbate, resulting in a monolayer of these molecules adhered to the adsorbent surface. This type of adsorption is highly specific due to the nature of the chemical bonds involved and is generally stronger than physical adsorption (physisorption), where the interactions are weaker (such as Van der Waals forces), with the possibility of successive layers of adsorbate on the adsorption surface. Temperature has an opposing effect in each case. When physisorption predominates, raising the temperature increases the Brownian motion of the adsorbate molecules, decreasing adsorption and favoring desorption (Cao et al. 2022).

From another perspective, physisorption is an exothermic phenomenon, which, as such, is not favored by increases in temperature. On the other hand, when chemisorption is present alongside physisorption, increases in temperature do not cause such significant effects as in the case of pure physisorption.

An adsorbent must meet a set of characteristics such as high chemical and mechanical stability, high availability, low cost, high selectivity toward the adsorbate (in this case, CO_2), and sufficient surface interactions. The latter implies that these interactions should neither be too strong nor too weak; if they are too strong, they hinder the regeneration of the adsorbent, and if they are too weak, the adsorption efficiency is low (Yu et al. 2012). There are various solid adsorbents for CO_2 capture, including activated carbons, ion exchange resins, zeolites, porous silicates, metal oxides, organic-inorganic hybrid adsorbents (e.g., amine-modified silicas, metal-organic frameworks [MOFs]), covalent organic frameworks (COFs), and composite materials (Pereira et al. 2023).

For biochar as an adsorbent and CO_2 as an adsorbate, physisorption primarily depends on the pore structure and specific surface area (SSA), whereas chemical adsorption depends on the functional groups on the surface through Lewis acid-base interactions (Guo et al. 2022). Regarding pore structure and SSA, various studies have shown that biochar characteristics depend primarily on preparation and modification methods, as well as the nature of the biomass used as raw material. Slow pyrolysis tends to produce more porous biochar than hydrothermal carbonization. Additionally, higher final pyrolysis temperatures tend to enhance SSA (Cao et al. 2022).

Concerning the effect of biomass type as a raw material, the main components of biomass are lignin, cellulose, and hemicellulose, along with extractable compounds (proteins, carbohydrates, and lipids) and inorganic compounds that form part of the ash after biomass combustion. Generally, woody biomass has a higher content of cellulose, hemicellulose, and lignin and a lower content of inorganic compounds than grasses, resulting in higher carbonized material content after pyrolysis in woody biomass.

4.17 CHEMISTRY OF CO_2 SORPTION

Purely physical adsorption can be enhanced with chemical adsorption, involving the interaction of functional groups on the exposed surface of biochar with CO_2 to form chemical bonds. The presence of these functional groups primarily depends on the biomass type, dictated by its elemental composition, which directly influences the biochar composition and defines its chemical properties, such as surface functional groups, acidity, and alkalinity. Nitrogen in biomass contributes to the generation of basic groups, like amino groups; thus, biomass with higher nitrogen content will yield biochar that is more alkaline compared to biomass with lower nitrogen content.

It is also possible to add functional groups to biochar through different treatments. Among the options available are the incorporation of Lewis basic functional groups, including basic metal sites (Lahijani et al. 2018; Cao et al. 2022), and others. Concerning the addition of basic functional groups, CO_2 is a weak Lewis acid gas, which allows it to easily bond with a basic adsorbent (Igalavithana et al. 2020). Xu et al. (2019) enhanced the adsorption capacity of biochar by doping it with nitrogen. The authors characterized the content and species of nitrogen in the doped biochars, studied the mechanisms of nitrogen-related group formation, and evaluated the CO_2 adsorption capacity, among other activities. The organic functional groups on the surface of the biochar samples were identified using Fourier-transform infrared spectroscopy (FTIR). The authors concluded that most of the nitrogen introduced into the biochar was present in the form of amino groups ($-NH_2$) and cyanide groups ($-C{\equiv}N$) and that, due to their basic properties, the CO_2 adsorption performance was improved. The incorporation of amino functional groups can lead to a significant increase in CO_2 capture. CO_2 and amines can react through various pathways. For example, primary amines ($-NH_2$) and secondary amines (–NH) can react directly with CO_2 to form carbamates (Cao et al. 2022; Pereira et al. 2023). In a similar vein, Wang et al. (2020) doped nitrogen into porous carbon for CO_2 capture from wet combustion gases. FTIR spectra evidenced the presence of C–N bonds, demonstrating the effectiveness of the doping. Ultimately, the authors concluded that nitrogen-doped porous carbon exhibits a good capacity for capturing CO_2 from high-humidity combustion gases.

Concerning the incorporation of metallic sites into the carbon matrix, Lahijani et al. (2018) incorporated metal salts into biochar and subsequently subjected it to heat treatment under an inert atmosphere of nitrogen (N_2). X-ray diffraction analysis revealed the presence of metal oxides primarily, including aluminum (Al_2O_3), calcium in the form of calcite ($CaCO_3$), iron in the form of magnetite ($Fe_{2.932}O_4$), magnesium oxide (MgO), sodium nitrite ($NaNO_2$), and nickel oxide (NiO). The authors concluded that biochar loaded with Mg achieved the highest CO_2 adsorption capacity (80.0 mg/g). The enhanced adsorption capacity can be attributed to the formation of special functional groups on the biochar surface during thermal treatment with incorporated metals. Equation (5) shows the reaction of carbon dioxide (CO_2) with the adsorbent deposited on the biochar surface, resulting in the formation of carbonate ions (CO_3^{2-}).

$$\text{Mg–O} + CO_2 \rightarrow \text{Mg–O}- - -CO_2 \tag{5}$$

4.18 METHODS OF MODELING CO_2 SORPTION

Over the years, numerous isotherm models and various approaches have been developed for mathematical modeling. One of the most valuable methods for identifying the most accurate correlation, measuring the distribution of the adsorbate, and enabling the theoretical analysis of the adsorbate-adsorbent complex is linear regression, which has been widely used. Its utility stems from the simplicity of the linear equation and its adaptability across a broad range of process parameters, such as temperature and pressure. However, it has become clear that non-linear regression is more effective for selecting the best-fitting isotherm model. This approach avoids the need to transform non-linear isotherm equations into their linear forms, thus eliminating biases that can lead to incorrect interpretations of isotherm model parameters and their application in adsorption system design. Non-linear regression minimizes the error distribution between experimental equilibrium data and the predicted curves from the isotherm models. This optimization process requires specifying error functions to assess the fit of a specific isothermal curve (Kumar 2007).

Different adsorption isotherm models can be employed to model CO_2 adsorption equilibrium data on biochar. The non-linear equations of the examined isotherm models, their theoretical assumptions, and a detailed explanation of their parameters are provided and discussed below (Kumar and Sivanesan 2006).

4.19 LANGMUIR ISOTHERM MODEL

The Langmuir isotherm applies to monolayer adsorption at homogeneous sites on the surface of a solid adsorbent. The model assumes a finite number of active sites on the solid surface, each capable of adsorbing a single adsorbate molecule. Equilibrium is achieved when all active sites are occupied by adsorbed molecules, forming a monomolecular layer (Langmuir 1918).

In essence, the Langmuir isotherm is characterized by a gradual approach to maximum adsorption, corresponding to the complete coverage of the solid surface by a monomolecular layer of adsorbate. Additionally, it is assumed that there is no interaction between the adsorbed molecules on the surface and that the heat of adsorption is constant, regardless of the surface coverage degree. The nonlinear equation of the Langmuir isotherm model is expressed as follows (Equation 6):

$$q_e = q_m \frac{K_L P}{1 + K_L P} \tag{6}$$

Here, q_e is the amount of adsorbed adsorbate at the equilibrium state (mol/g), q_m is the maximum monolayer adsorption capacity (mol/g), K_L is the equilibrium constant between the gas and adsorbed phases (1/Pa), and P is the relative pressure of the adsorbate (Pa).

4.20 FREUNDLICH ISOTHERM MODEL

The Freundlich isotherm model is an empirical equation without a theoretical basis and is not limited to monolayer formation (Freundlich 1906). It describes the non-ideal, reversible process of monolayer or multilayer adsorption on a heterogeneous surface. According to the theory of Freundlich, the number of adsorbed molecules at complete surface coverage cannot exceed the number of active sites. Different from the Langmuir model, which assumes that the heat of adsorption remains constant regardless of surface coverage, the Freundlich model assumes that the adsorption energy decreases logarithmically as the number of available adsorption sites decreases, i.e., as the adsorbate coverage on the solid surface increases. The non-linear equation of the Freundlich model is expressed mathematically as follows (Equation 7):

$$q_e = K_F^{Pn_F} \tag{7}$$

K_F is the Freundlich constant (1/Pa), and n_F is the heterogeneity factor (–). Physisorption is indicated when the n_F value is more than one, whereas chemisorption is shown by a number less than unity (Haghseresht and Lu 1998).

4.21 FUTURE WORKS

Although biochar is emerging as a promising tool for CO_2 capture and storage, several research areas need to be explored to maximize its effectiveness and economic viability. It is essential to optimize the processes of biochar production and activation, determining the ideal pyrolysis conditions, such as temperature, time, and atmosphere, that allow maximizing its surface area and porosity. It is also essential to evaluate the effectiveness of new combinations of physical and chemical activation methods, as well as the pretreatments applied to biomass.

The development of advanced techniques to incorporate functional groups, whether basic or metallic, on the surface of biochar is another priority, along with the exploration of combinations with metals and the addition of nitrogen groups that improve their capacity for selective adsorption of CO_2. In addition, exhaustive analyses must be carried out on the emissions associated with its production, as well as studies that determine the economic and environmental sustainability of its implementation on an industrial scale.

In parallel, it is crucial to investigate strategies to regenerate biochar efficiently and with low energy consumption after multiple adsorption cycles. It is also necessary to refine existing adsorption models to predict biochar behavior under different operating conditions and develop simulation tools to optimize the design of biochar-based CO_2 capture systems.

Finally, interdisciplinary collaboration between areas such as chemistry, materials engineering, and environmental sciences is essential to overcome the challenges associated with implementing biochar. These joint efforts will not only increase its effectiveness in capturing CO_2 but will also strengthen its role as a sustainable and viable technology to mitigate the effects of climate change at a global level.

4.22 CONCLUSIONS

Biochar emerges as a promising solution to the global challenge of climate change, particularly in capturing and storing carbon dioxide (CO_2). This chapter explores the fundamental aspects that make biochar an effective material for CO_2 adsorption, highlighting its physicochemical properties, activation techniques, and adsorption mechanisms.

It was shown that the capacity of biochar to fix CO_2 depends largely on its porous structure, specific surface area, and the presence of suitable functional groups on its surface, factors that are determined by production conditions and subsequent modifications. Techniques such as physical and chemical activation, metal impregnation, and functionalization of surface groups have shown significant results in improving its performance as an adsorbent.

Furthermore, biochar contributes to mitigating atmospheric CO_2 and offers collateral benefits in agricultural applications, remediation of contaminated soils, and water purification. However, developing technologies to optimize biochar production and functionalization and a sustainable and economically viable approach is essential for its large-scale adoption.

In summary, biochar represents a key tool in the transition towards a more sustainable future, offering a bridge between technological innovation and climate action. Its strategic implementation, supported by continuous research, can significantly mitigate greenhouse gas emissions and build climate-resilient systems.

NOMENCLATURE AND ACRONYMS

Al	:	Aluminum
Al_2O_3	:	Aluminum oxide
$AlCl_3$	:	Aluminum chloride
AMF	:	Arbuscular Mycorrhizal Fungi
C	:	Carbon
$-C{\equiv}N$	:	Cyanide groups
Ca	:	Calcium
$CaCO_3$	:	Calcium carbonate
CaO	:	Calcium Oxide
CCS	:	Carbon Capture and Storage
CFCs	:	Chlorofluorocarbons
CH_4	:	Methane
CO	:	Carbon monoxide
CO_2/CH_4	:	Carbon dioxide-to-methane ratio
CO_2/N_2	:	Carbon dioxide-to-nitrogen ratio
CO_2	:	Carbon dioxide
CO_3^{2-}	:	Carbonate ions
COF/COFs	:	Covalent Organic Frameworks

–COOH	:	Carboxyl group
Cu	:	Copper
Fe	:	Iron
$Fe_{2.932}O_4$	:	Iron in the form of magnetite
Fe_2O_3	:	Iron oxide
$FeCl_3$	:	Ferric chloride
FeO	:	Ferrous oxide
FTIR	:	Fourier-Transform Infrared Spectroscopy
GHG/s	:	Greenhouse Gases
H/C	:	Hydrogen-to-carbon ratio
H	:	Hydrogen
H_2	:	Dihydrogen
H_2O	:	Water
H_2O_2	:	Hydrogen peroxide
H_2S	:	Hydrogen sulfide
H_2SO_4	:	Sulfuric acid
H_3PO_4	:	Phosphoric acid
HCl	:	Hydrochloric acid
HNO_3	:	Nitric acid
IEA	:	International Energy Agency
IUPAC	:	International Union of Pure and Applied Chemistry
K	:	Potassium
K_2CO_3	:	Potassium carbonate
K_2O	:	Potassium oxide
$KMnO_4$	:	Potassium Permanganate
KOH	:	Potassium Hydroxide
$LaCl_3$	:	Lanthanum chloride
MBAC	:	Metal Activated Biochar
Mg	:	Magnesium
$MgCl_2$	:	Magnesium chloride
MgO	:	Magnesium Oxide
MHW/s	:	Marine heat waves
MnO	:	Manganous oxide
MOF/s	:	Metal-Organic Frameworks
N_2	:	Nitrogen
N_2O	:	Nitrous Oxide
Na	:	Sodium
$NaNO_2$	:	Sodium nitrite
–NH	:	Secondary amine
$–NH_2$	:	Amino group
NH_3	:	Ammonia
Ni	:	Nickel
NiO	:	Nickel oxide
Norit CGP	:	Commercial activated carbon
NO_x	:	Nitrogen Oxides
O/C	:	Oxygen-to-carbon ratio

O	:	Oxygen
O_3	:	Ozone
–OH	:	Hydroxide
pH	:	Hydrogen potential
SEM	:	Scanning Electron Microscopy
SSA	:	Specific Surface Area
TEPA	:	Tetraethylenepentamine
XPS	:	X-ray photoelectron spectroscopy
$ZnCl_2$	:	Zinc chloride
ZnO	:	Zinc oxide

REFERENCES

Abagandura, G.O., R. Chintala, S.S. Sandhu, S. Kumar and T.E. Schumacher. 2019. Effects of biochar and manure applications on soil carbon dioxide, methane, and nitrous oxide fluxes from two different soils. Journal of Environmental Quality. 48: 1664–1674.

Abhishek, K., A. Shrivastava, V. Vimal, A.K. Gupta, S.K. Bhujbal, J.K. Biswas, et al. 2022. Biochar application for greenhouse gas mitigation, contaminants immobilization and soil fertility enhancement: A state-of-the-art review. Science of the Total Environment. 853: 158562.

Albouy, C., V. Delattre, G. Donati, T.L. Frölicher, S. Albouy-Boyer, M. Rufino, et al. 2020. Global vulnerability of marine mammals to global warming. Scientific Reports. 10: 548.

Al-Ghussain, L. 2019. Global warming: review on driving forces and mitigation. Environmental Progress & Sustainable Energy. 38: 13–21.

Amer, N.M., P. Lahijani, M. Mohammadi and A.R. Mohamed. 2024. Modification of biomass-derived biochar: A practical approach towards development of sustainable CO_2 adsorbent. Biomass Conversion and Biorefinery. 14: 7401–7448.

Anto, S., M.P. Sudhakar, T.S. Ahamed, M.S. Samuel, T. Mathimani, K. Brindhadevi, et al. 2021. Activation strategies for biochar to use as an efficient catalyst in various applications. Fuel. 285: 119205.

Bamdad, H., K. Hawboldt, S. MacQuarrie and S. Papari. 2019. Application of biochar for acid gas removal: Experimental and statistical analysis using CO_2. Environmental Science and Pollution Research. 26: 10902–10915.

Bushra, B. and N. Remya. 2024. Biochar from pyrolysis of rice husk biomass—characteristics, modification, and environmental application. Biomass Conversion and Biorefinery. 14: 5759–5770.

Cao, L., X. Zhang, Y. Xu, W. Xiang, R. Wang, F. Ding, et al. 2022. Straw and wood based biochar for CO_2 capture: Adsorption performance and governing mechanisms. Separation and Purification Technology. 287: 120592.

Chen, Y.D., F. Liu, N.Q. Ren and S.H. Ho. 2020. Revolutions in algal biochar for different applications: State-of-the-art techniques and future scenarios. Chinese Chemical Letters. 31: 2591–2602.

Chen, S., J. Liu, Q. Zhang, F. Teng and B.C. McLellan. 2022. A critical review on deployment planning and risk analysis of carbon capture, utilization, and storage (CCUS) toward carbon neutrality. Renewable and Sustainable Energy Reviews. 167: 112537.

Cheng, F. and X. Li. 2018. Preparation and application of biochar-based catalysts for biofuel production. Catalysts. 8: 346.

Craswell, E.T., P.M. Chalk and B.B. Kaudal. 2021. Role of 15N in tracing biologically driven nitrogen dynamics in soils amended with biochar: A review. Soil Biology and Biochemistry. 162: 108416.

Creamer, A.E. and B. Gao. 2016. Carbon-based adsorbents for postcombustion CO_2 capture: A critical review. Environmental Science & Technology. 50: 7276–7289.

Creamer, A.E., B. Gao and S. Wang. 2016. Carbon dioxide capture using various metal oxyhydroxide-biochar composites. Chemical Engineering Journal. 283: 826–832.

De Castro, P.M., S.R. Ferrarini, M.D. Rimoli, A.A. Merlo, R.M. Nogueira and E.M. Pires. 2023. Preparation and characterization of steam and CO_2 activated carbon from Brazil nut shell. Bioscience Journal. 39: e39054.

Ding, S. and Y. Liu. 2020. Adsorption of CO_2 from flue gas by novel seaweed-based KOH-activated porous biochars. Fuel. 260: 116382.

Dissanayake, P.D., S.W. Choi, A.D. Igalavithana, X. Yang, D.C. Tsang, C.H. Wang, et al. 2020a. Sustainable gasification biochar as a high efficiency adsorbent for CO_2 capture: a facile method to designer biochar fabrication. Renewable and Sustainable Energy Reviews. 124: 109785.

Dissanayake, P.D., S. You, A.D. Igalavithana, Y. Xia, A. Bhatnagar, S. Gupta, et al. 2020b. Biochar-based adsorbents for carbon dioxide capture: A critical review. Renewable and Sustainable Energy Reviews. 119: 109582.

Durá, G., V.L. Budarin, J.A. Castro-Osma, P.S. Shuttleworth, S.C. Quek, J.H. Clark, et al. 2016. Importance of micropore–mesopore interfaces in carbon dioxide capture by carbon-based materials. Angewandte Chemie. 128: 9319–9323.

Feng, D., Y. Zhao, Y. Zhang, Z. Zhang, H. Che and S. Sun. 2017a. Experimental comparison of biochar species on in-situ biomass tar H_2O reforming over biochar. International Journal of Hydrogen Energy. 42: 24035–24046.

Feng, D., Y. Zhao, Y. Zhang, Z. Zhang, L. Zhang, J. Gao, et al. 2017b. Synergetic effects of biochar structure and AAEM species on reactivity of H_2O-activated biochar from cyclone air gasification. International Journal of Hydrogen Energy. 42: 16045–16053.

Feng, D., D. Guo, Y. Zhang, S. Sun, Y. Zhao, Q. Shang, et al. 2021. Functionalized construction of biochar with hierarchical pore structures and surface O-/N-containing groups for phenol adsorption. Chemical Engineering Journal. 410: 127707.

Fernández-Maldonado, V. and A.L. Navas, M.P. Fabani, G. Mazza and R. Rodríguez. 2024. A multi-temporal analysis on the dynamics of the impact of land use and land cover on NO_2 and CO emissions in Argentina for sustainable environmental management. Sustainability 16: 4400.

Francis, J.C., A. Nighojkar and B. Kandasubramanian. 2023. Relevance of wood biochar on CO_2 adsorption: A review. Hybrid Advances. 100056.

Freundlich, H.M.F. 1906. Over the adsorption in solution. Journal of Physical Chemistry. 57: 1100–1107.

Frölicher, T.L., E.M. Fischer and N. Gruber. 2018. Marine heatwaves under global warming. Nature. 560: 360–364.

Ghanbari, S. and G. Kamath. 2019. Dynamic simulation and mass transfer study of carbon dioxide capture using biochar and MgO-impregnated activated carbon in a swing adsorption process. Energy Fuels. 33: 5452–5463.

Ghiat, I. and T. Al-Ansari. 2021. A review of carbon capture and utilisation as a CO_2 abatement opportunity within the EWF nexus. Journal of CO_2 Utilization. 45: 101432.

Gong, H., Z. Tan, L. Zhang and Q. Huang. 2019. Preparation of biochar with high absorbability and its nutrient adsorption–desorption behaviour. Science of the Total Environment. 694: 133728.

Gopalan, J., A. Buthiyappan and A.A.A. Raman. 2022. Insight into metal-impregnated biomass-based activated carbon for enhanced carbon dioxide adsorption: A review. Journal of Industrial and Engineering Chemistry. 113: 72–95.

Guo, S., Y. Li, Y. Wang, L. Wang, Y. Sun and L. Liu. 2022. Recent advances in biochar-based adsorbents for CO_2 capture. Carbon Capture Science Technology. 4: 100059.

Haghseresht, F. and G.Q. Lu. 1998. Adsorption characteristics of phenolic compounds onto coal-reject-derived adsorbents. Energy Fuels. 12: 1100–1107.

Heidari, A., E. Khaki, H. Younesi and H.R. Lu. 2019. Evaluation of fast and slow pyrolysis methods for bio-oil and activated carbon production from eucalyptus wastes using a life cycle assessment approach. Journal of Cleaner Production. 241: 118394.

Huang, G.G., Y.F. Liu, X.X. Wu and J.J. Cai. 2019. Activated carbons prepared by the KOH activation of a hydrochar from garlic peel and their CO_2 adsorption performance. New Carbon Materials. 34: 247–257.

Hughes, T.P., J.T. Kerry, A.H. Baird, S.R. Connolly, A. Dietzel, C.M. Eakin, et al. 2018. Global warming transforms coral reef assemblages. Nature. 556: 492–496.

Hui, D. 2021. Effects of biochar application on soil properties, plant biomass production, and soil greenhouse gas emissions: A mini-review. Agricultural Sciences. 12: 213–236.

Igalavithana, A.D., S.W. Choi, J. Shang, A. Hanif, P.D. Dissanayake, D.C. Tsang, et al. 2020. Carbon dioxide capture in biochar produced from pine sawdust and paper mill sludge: Effect of porous structure and surface chemistry. Science of the Total Environment. 739: 139845.

Ji, M., X. Wang, M. Usman, F. Liu, Y. Dan, L. Zhou, et al. 2022. Effects of different feedstocks-based biochar on soil remediation: A review. Environmental Pollution. 294: 118655.

Jiang, Z. and S. Yang. 2021. Biochar improved soil health and mitigated greenhouse gas emission from controlled irrigation paddy field: Insights into microbial diversity. Journal of Cleaner Production. 318: 128595.

Jing, X.R., Y.Y. Wang, W.J. Liu, Y.K. Wang and H. Jiang. 2014. Enhanced adsorption performance of tetracycline in aqueous solutions by methanol-modified biochar. Chemical Engineering Journal. 248: 168–174.

Jones, W.D. 2020. Carbon capture and conversion. Journal of the American Chemical Society. 142: 4955−4957.

Jung, S., Y.K. Park and E.E. Kwon. 2019. Strategic use of biochar for CO_2 capture and sequestration. Journal of CO_2 Utilization. 32: 128–139.

Kavitha, B., P.V.L. Reddy, B. Kim, S.S. Lee, S.K. Pandey and K.H. Kim. 2018. Benefits and limitations of biochar amendment in agricultural soils: A review. Journal of Environmental Management. 227: 146–154.

Korus, A., J.P. Gutierrez, A. Szlęk, J. Jagiello and A.P. Hornung. 2022. Pore development during CO_2 and H_2O activation associated with the catalytic role of inherent inorganics in sewage sludge char and its performance during the reforming of volatiles. Chemical Engineering Journal. 446: 137298.

Koutroulis, A.G. 2019. Dryland changes under different levels of global warming. Science of the Total Environment. 655: 482–511.

Kumar, K.V. and S. Sivanesan. 2006. Pseudo second order kinetics and pseudo isotherms for malachite green onto activated carbon: comparison of linear and non-linear regression methods. Journal of Hazardous Materials. 136: 721–726.

Kumar, K.V. 2007. Optimum sorption isotherm by linear and non-linear methods for malachite green onto lemon peel. Dyes and Pigments. 74: 595–597.

Kumar, A., E. Singh, R. Mishra, S.L. Lo and S. Kumar. 2022. A green approach towards sorption of CO_2 on waste-derived biochar. Environmental Research. 214: 113954.

Kweku, D.W., O. Bismark, A. Maxwell, K.-A. Desmond, K.B. Danso, E.A. Oti-Mensah, et al. 2018. Greenhouse effect: greenhouse gases and their impact on global warming. Journal of Scientific Research and Reports. 17: 1–9.

Lahijani, P., M. Mohammadi and A.R. Mohamed. 2018. Metal incorporated biochar as a potential adsorbent for high capacity CO_2 capture at ambient conditions. Journal of CO_2 Utilization. 26: 281–293.

Langmuir, I. 1918. The adsorption of gases on plane surfaces of glass, mica and platinum. Journal of the American Chemical Society. 40: 1361–1403.

Leng, L., Q. Xiong, L. Yang, H. Li, Y. Zhou, W. Zhang, et al. 2021. An overview on engineering the surface area and porosity of biochar. Science of the Total Environment. 763: 144204.

Li, Y., J. Shao, X. Wang, Y. Deng, H. Yang and H. Chen. 2014. Characterization of modified biochars derived from bamboo pyrolysis and their utilization for target component (furfural) adsorption. Energy Fuels. 28: 5119–5127.

Li, Y., P.M. Brando, D.C. Morton, D.M. Lawrence, H. Yang and J.T. Randerson. 2022a. Deforestation-induced climate change reduces carbon storage in remaining tropical forests. Nature Communications. 13: 1964.

Li, K., D. Zhang, X. Niu, H. Guo, Y. Yu, Z. Tang, et al. 2022b. Insights into CO_2 adsorption on KOH-activated biochars derived from the mixed sewage sludge and pine sawdust. Science of the Total Environment. 826: 154133.

Liang, J.F. 2019. Interactive effects of biochar and AMF on plant growth and greenhouse gas emissions from wetland microcosms. Geoderma. 346: 11–17.

Liu, Z., J. Tang, X. Ren and S.M. Schaeffer. 2021. Effects of phosphorus modified nZVI-biochar composite on emission of greenhouse gases and changes of microbial community in soil. Environmental Pollution. 274: 116483.

Malyan, S.K., S.S. Kumar, R.K. Fagodiya, P. Ghosh, A. Kumar, R. Singh, et al. 2021. Biochar for environmental sustainability in the energy-water-agroecosystem nexus. Renewable and Sustainable Energy Reviews. 149: 111379.

Mamaghani, Z.G., K.A. Hawboldt and S. MacQuarrie. 2023. Adsorption of CO_2 using biochar—review of the impact of gas mixtures and water on adsorption. Journal of Environmental Chemical Engineering. 109643.

Marchal, V., R. Dellink, D. Van Vuuren, C. Clapp, J. Chateau, B. Magné, et al. 2011. OECD environmental outlook to 2050. Organization for Economic Cooperation and Development. 8: 397–413.

Martí, E., J. Sierra, X. Domene, M. Mumbrú, R. Cruañas and M.A. Garau. 2021. One-year monitoring of nitrogen forms after the application of various types of biochar on different soils. Geoderma. 402: 115178.

Maya, J.C., R. Macías, C.A. Gómez and F. Chejne. 2020. On the evolution of pore microstructure during coal char activation with steam/CO_2 mixtures. Carbon. 158: 121–130.

Mosa, A., M.M. Mansour, E. Soliman, A. El-Ghamry, M. El Alfy and A.M. El Kenawy. 2023. Biochar as a soil amendment for restraining greenhouse gases emission and improving soil carbon sink: Current situation and ways forward. Sustainability. 15: 1206.

Novotnik, B., A. Nandy, S.V. Venkatesan, J.R. Radović, J.D.L. Fuente, S. Nejadi, et al. 2020. Can fossil fuel energy be recovered and used without any CO_2 emissions to the atmosphere? Reviews in Environmental Science and Biotechnology. 19: 217–240.

Olabi, A.G., T. Wilberforce, K. Elsaid, E.T. Sayed, H.M. Maghrabie and M.A. Abdelkareem. 2022. Large scale application of carbon capture to process industries–A review. Journal of Cleaner Production. 362: 132300.

Omoregbe, O., A.N. Mustapha, R. Steinberger-Wilckens, A. El-Kharouf and H. Onyeaka. 2020. Carbon capture technologies for climate change mitigation: A bibliometric analysis of the scientific discourse during 1998–2018. Energy Reports. 6: 1200–1212.

Ozkan, M., S.P. Nayak, A.D. Ruiz and W. Jiang. 2022. Current status and pillars of direct air capture technologies. iScience. 25: 4.

Papalexiou, S.M. and A. Montanari. 2019. Global and regional increase of precipitation extremes under global warming. Water Resources Research. 55: 4901–4914.

Park, C.E., S.J. Jeong, M. Joshi, T.J. Osborn, C.H. Ho, S. Piao, et al. 2018. Keeping global warming within 1.5°C constrains emergence of aridification. Nature Climate Change. 8: 70–74.

Peng, H., P. Gao, G. Chu, B. Pan, J. Peng and B. Xing. 2017. Enhanced adsorption of Cu (II) and Cd (II) by phosphoric acid-modified biochars. Environmental Pollution. 229: 846–853.

Pereira, D., R. Fonseca, I. Marin-Montesinos, M. Sardo and L. Mafra. 2023. Understanding CO_2 adsorption mechanisms in porous adsorbents: A solid-state NMR survey. Current Opinion in Colloid and Interface Science. 64: 101690.

Peter, S.C. 2018. Reduction of CO_2 to chemicals and fuels: A solution to global warming and energy crisis. ACS Energy Letters. 3: 1557–1561.

Petrovic, B., M. Gorbounov and S.M. Soltani. 2022. Impact of surface functional groups and their introduction methods on the mechanisms of CO_2 adsorption on porous carbonaceous adsorbents. Carbon Capture Science & Technology. 3: 100045.

Plaza, M.G., C. Pevida, A. Arenillas, F. Rubiera and J.J. Pis. 2007. CO_2 capture by adsorption with nitrogen enriched carbons. Fuel. 86: 2204–2212.

Promraksa, A. and N. Rakmak. 2020. Biochar production from palm oil mill residues and application of the biochar to adsorb carbon dioxide. Heliyon. 6: e04019.

Qiao, Y. and C. Wu. 2022. Nitrogen enriched biochar used as CO_2 adsorbents: A brief review. Carbon Capture Science Technology. 2: 100018.

Rashidi, N.A. and S. Yusup. 2020. Biochar as potential precursors for activated carbon production: parametric analysis and multi-response optimization. Environmental Science and Pollution Research. 27: 27480–27490.

Raza, A., R. Gholami, R. Rezaee, V. Rasouli and M. Rabiei. 2019. Significant aspects of carbon capture and storage–A review. Petroleum. 5: 335–340.

Sajjadi, B., T. Zubatiuk, D. Leszczynska, J. Leszczynski and W.Y. Chen. 2019a. Chemical activation of biochar for energy and environmental applications: A comprehensive review. Reviews in Chemical Engineering. 35: 777–815.

Sajjadi, B., W.Y. Chen and N.O. Egiebor. 2019b. A comprehensive review on physical activation of biochar for energy and environmental applications. Reviews in Chemical Engineering. 35: 735–776.

Sarkar, B., H. Fricska, Q. Gao, S. Tong and C.Q. Jia. 2023. Adsorption of single-ring model naphthenic acid from oil sands tailings pond water using physically activated petroleum coke. Canadian Journal of Chemical Engineering. 101: 4374–4384.

Shafawi, A. N., A. R. Mohamed, P. Lahijani, and M. Mohammadi. 2021. Recent advances in developing engineered biochar for CO_2 capture: An insight into the biochar modification approaches. Journal of Environmental Chemical Engineering. 9: 106869.

Shakoor, A., M.S. Arif, S.M. Shahzad, T.H. Farooq, F. Ashraf, M.M. Altaf, et al. 2021. Does biochar accelerate the mitigation of greenhouse gaseous emissions from agricultural soil? - A global meta-analysis. Environmental Research. 202: 111789.

Sun, L., D. Chen, S. Wan and Z. Yu. 2015. Performance, kinetics, and equilibrium of methylene blue adsorption on biochar derived from eucalyptus sawdust modified with citric, tartaric, and acetic acids. Bioresource Technology. 198: 300–308.

Taha, S.M., M.E. Amer, A.E. Elmarsafy and M.Y. Elkady. 2014. Adsorption of 15 different pesticides on untreated and phosphoric acid treated biochar and charcoal from water. Journal of Environmental Chemical Engineering. 2: 2013–2025.

Tang, S., N. Shao, C. Zheng, F. Yan and Z. Zhang. 2019. Amino-functionalized sewage sludge-derived biochar as sustainable efficient adsorbent for Cu(II) removal. Waste Management. 90: 17–28.

Thote, J.A., K.S. Iyer, R. Chatti, N.K. Labhsetwar, R.B. Biniwale and S.S. Rayalu. 2010. In situ nitrogen enriched carbon for carbon dioxide capture. Carbon. 48: 396–402.

Venkataramanan, M. 2011. Causes and effects of global warming. Indian Journal of Science and Technology. 4: 226–229.

Vithanage, M., A.U. Rajapaksha, M. Zhang, S. Thiele-Bruhn, S.S. Lee and Y.S. Ok. 2015. Acid-activated biochar increased sulfamethazine retention in soils. Environmental Science and Pollution Research. 22: 2175–2186.

Wang, C., J. Liu, J. Shen, D. Chen, Y. Li, B. Jiang, et al. 2018. Effects of biochar amendment on net greenhouse gas emissions and soil fertility in a double rice cropping system: A 4-year field experiment. Agriculture, Ecosystems & Environment. 262: 83–96.

Wang, Z., N. Goyal, L. Liu, D.C. Tsang, J. Shang, W. Liu, et al. 2020. N-doped porous carbon derived from polypyrrole for CO_2 capture from humid flue gases. Chemical Engineering Journal. 396: 125376.

Wen, C., T. Liu, D. Wang, Y. Wang, H. Chen, G. Luo, et al. 2023. Biochar as the effective adsorbent to combustion gaseous pollutants: preparation, activation, functionalization, and the adsorption mechanisms. Progress in Energy and Combustion Science. 99: 101098.

Xia, F., Z. Zhang, Q. Zhang, H. Huang and X. Zhao. 2023. Life cycle assessment of greenhouse gas emissions for various feedstocks-based biochars as soil amendment. Science of The Total Environment. 168734.

Xu, X., Y. Zheng, B. Gao and X. Cao. 2019. N-doped biochar synthesized by a facile ball-milling method for enhanced sorption of CO_2 and reactive red. Chemical Engineering Journal. 368: 564–572.

Xu, P., Y. Gao, Z. Cui, B. Wu, B. Yan, Y. Wang, et al. 2023. Research progress on effects of biochar on soil environment and crop nutrient absorption and utilization. Sustainability. 15: 4861.

Yakout, S.M., A.E.H.M. Daifullah and S.A. El-Reefy. 2015. Pore structure characterization of chemically modified biochar derived from rice straw. Environmental Engineering Management Journal (EEMJ). 14: 473–480.

Yang, X., S. Zhang, M. Ju and L. Liu. 2019. Preparation and modification of biochar materials and their application in soil remediation. Applied Sciences. 9: 1365.

Yu, C.H., C.H. Huang and C.S. Tan. 2012. A review of CO_2 capture by absorption and adsorption. Aerosol and Air Quality Research. 12: 745–769.

Zaini, M.S. and S.S.A. Syed-Hassan. 2022. Effects of different physical activation agents on adsorbent pore development and methane uptake. Recent Innovations in Chemical Engineering (Formerly Recent Patents on Chemical Engineering). 15: 127–137.

Zhang, X., J. Wu, H. Yang, J. Shao, X. Wang, Y. Chen, et al. 2016. Preparation of nitrogen-doped microporous modified biochar by high temperature CO_2 –NH_3 treatment for CO_2 adsorption: Effects of temperature. RSC Advances. 6: 98157–98166.

Zhang, H., J. Shao, S. Zhang, X. Zhang and H. Chen. 2020. Effect of phosphorus-modified biochars on immobilization of Cu (II), Cd (II), and As (V) in paddy soil. Journal of Hazardous Materials. 390: 121349.

Zhang, Y., J. Wang and Y. Feng. 2021. The effects of biochar addition on soil physicochemical properties: A review. Catena. 202: 105284.

Zhang, J., D. Huang, J. Shao, X. Zhang, S. Zhang, H. Yang, et al. 2022a. A new nitrogen-enriched biochar modified by ZIF-8 grafting and annealing for enhancing CO_2 adsorption. Fuel Processing Technology 231: 107250.

Zhang, C., S. Sun, S. He and C. Wu. 2022b. Direct air capture of CO_2 by KOH-activated bamboo biochar. Journal of the Energy Institute. 105: 399–405.

Zhang, C., S. Sun, S. Xu and C. Wu. 2022c. CO_2 capture over steam and KOH activated biochar: Effect of relative humidity. Biomass and Bioenergy. 166: 106608.

Zhang, C., Y. Ji, C. Li, Y. Zhang, S. Sun, Y. Xu, et al. 2023. The application of biochar for CO_2 capture: influence of biochar preparation and CO_2 capture reactors. Industrial and Engineering Chemistry Research. 62: 17168−17181.

Zhao, Y., X. Li, Y. Li, H. Bao, J. Xing, Y. Zhu, et al. 2022. Biochar acts as an emerging soil amendment and its potential ecological risks: a review. Energies. 16(1): 410.

Zhou, N., H. Chen, Q. Feng, D. Yao, H. Chen, H. Wang, et al. 2017. Effect of phosphoric acid on the surface properties and Pb (II) adsorption mechanisms of hydrochars prepared from fresh banana peels. Journal of Cleaner Production. 165: 221–230.

Zhou, D., J. Xiao, S. Frolking, S. Liu, L. Zhang, Y. Cui, et al. 2021a. Croplands intensify regional and global warming according to satellite observations. Remote Sensing of Environment. 264: 112585.

Zhou, Y., S. Qin, S. Verma, T. Sar, S. Sarsaiya, B. Ravindran, et al. 2021b. Production and beneficial impact of biochar for environmental application: a comprehensive review. Bioresource Technology 337: 125451.

Zubbri, N.A., A.R. Mohamed, N. Kamiuchi and M. Mohammadi. 2020. Enhancement of CO_2 adsorption on biochar sorbent modified by metal incorporation. Environmental Science and Pollution Research. 27: 11809–11829.

Chapter 5

Life Cycle Assessment of Biochar: Methodology, Technologies, and Environmental Sustainability

Daniela Zalazar-García*[a], Jose Soria[b], Angie Espinosa Sarmiento[c], Diego Yepes Maya[d], Celina Podetti[a] and Mathias Riveros-Gomez[a]

[a]Instituto de Ingeniería Química, Fac. de Ingeniería, Grupo Vinculado al PROBIEN (Universidad Nacional de San Juan, CONICET, Universidad Nacional del Comahue), San Juan, Argentina

[b]Instituto de Investigación y Desarrollo en Ingeniería de Procesos, Biotecnología y Energías Alternativas, PROBIEN (CONICET, Universidad Nacional del Comahue), Neuquén, Argentina

[c]Mechanical Engineering Institute, Federal University of Itajuba, Av. BPS 1303, Itajuba, MG, Brazil

[d]Excellence Group in Thermal Power and Distributed Generation (NEST), Institute of Mechanical Engineering, Federal University of Itajubá, MG, Brazil

5.1 INTRODUCTION TO BIOCHAR AND ITS ENVIRONMENTAL RELEVANCE

In the last decade, the biochar production from the thermochemical treatment of lignocellulosic waste has attracted broad interest due to its potential as an environmentally friendly approach to waste management and soil upgrading. Biochar derived from biomass has many advantages, including improved soil

*Corresponding author: Daniela Zalazar-García (dzalazar@unsj.edu.ar)

DOI: 10.1201/9781032726694-5

conditions, reduced greenhouse gas emissions, and effective waste management strategies. Moreover, biochar production through pyrolysis or gasification presents a low-emission alternative to traditional waste disposal methods, further emphasizing its role in environmental conservation and energy recovery. This sustainable methodology facilitates effective waste management while simultaneously enhancing the value of agricultural waste by converting it into valuable resources and reintegrating it into the circular economy. To weigh the sustainability of biochar production fairly, the environmental implications of its very process must be well perceived and quantified. Among many such tools, the Life Cycle Assessment (LCA) is one of the most effective in evaluating the impacts (Bergman et al. 2016).

Life Cycle Assessment is an internationally standardized methodology and systematic procedure widely adopted for estimating environmental impacts emanating from all stages of a product life cycle. The primary goal of LCA is to offer a detailed assessment of the environmental aspects and potential impacts linked to a product, thereby supporting more sustainable decision-making (ISO 14040 2006). LCA encompasses an array of many impact categories, including global warming potential, acidification, eutrophication, and resource depletion, to provide a complete view of environmental performance. Analyzing the diverse set of categories in LCA enables the identification of trade-offs and synergies between the different impacts. Therefore, this makes it very useful for providing a comprehensive analysis for industries and policymakers towards reducing environmental footprints and enhancing resource efficiency (Hamedani et al. 2019).

LCA is especially useful for biochar production. A thorough assessment of environmental impact includes direct emissions, energy use, and indirect effects from the supply chain. The indirect implications cover raw material extraction, transportation, and waste management, presenting a complete picture of the environmental footprint biochar production causes. As such, both the potential environmental impacts of biochar production can be assessed in terms of its effect on GHG emissions, energy use, and resource depletion in quantitative values, therefore enabling pinpointing of opportunities and critical improvements in the sustainability of biochar production. Since action to reach global climate target levels is urgent, efficient reduction of greenhouse gas emissions will need to become an unavoidable strategy, given the preeminent role that biochar produced from biomass, particularly agro-industrial wastes, can play (Zhu et al. 2022). Figure 5.1 provides a visual representation of the stages and elements involved in a life cycle assessment.

5.2 FUNDAMENTALS OF LIFE CYCLE ASSESSMENT (LCA)

5.2.1 Life Cycle Analysis Methodology

The environmental analysis typically follows the Life Cycle Assessment (LCA) methodology as outlined in the ISO 14040 standard (Nemecek et al. 2024), which

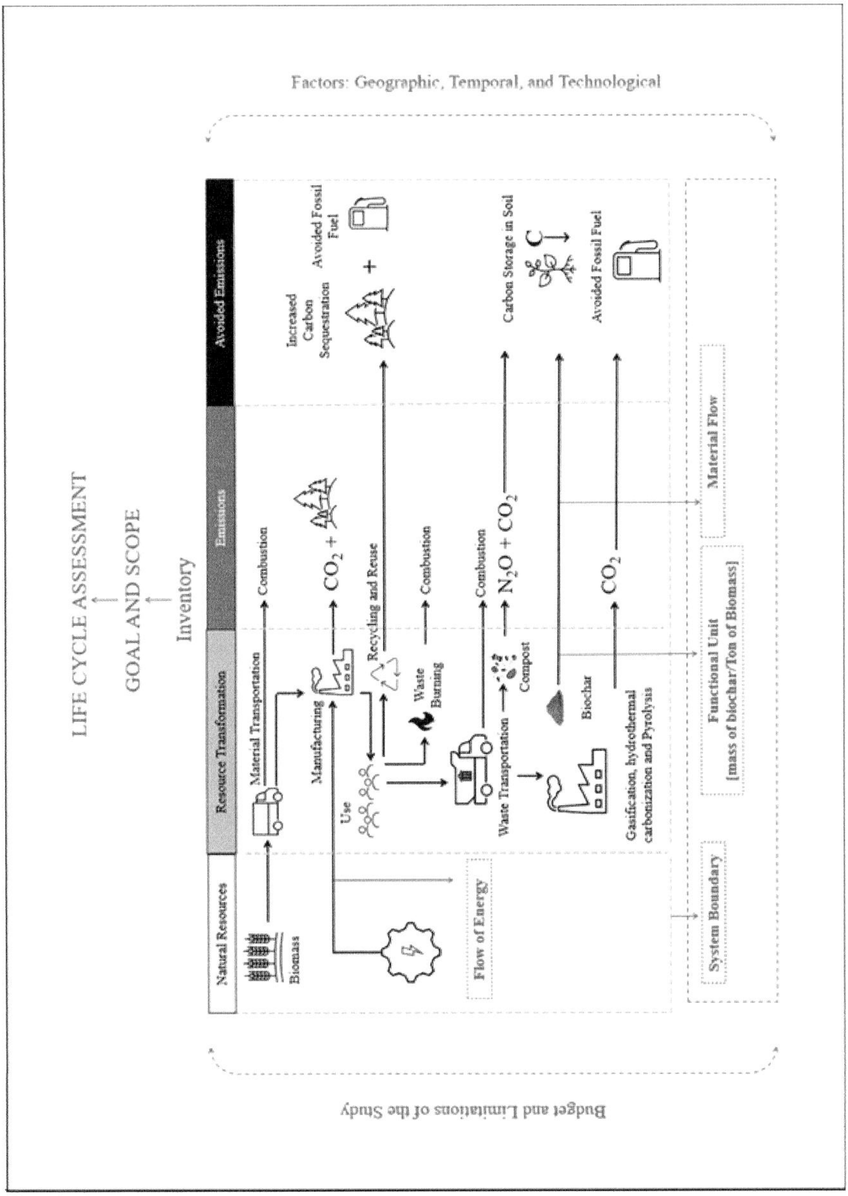

Figure 5.1 Stages and elements involved in a life cycle assessment.

systematically considers the inputs and outputs of the process under evaluation (Figure 5.2). This methodology generally includes four stages: (1) definition of objectives and scope, (2) life cycle inventory, (3) impact assessment, and (4) interpretation of results. Various software tools, such as SimaPro, are commonly employed to conduct LCA. Environmental analyses often involve quantitative estimates using Midpoint indicators, with widely applied methods like ReCiPe Midpoint (I). LCA is usually carried out through a multifunctional approach, involving system subdivision, expansion, or allocation. The International Reference Life Cycle Data System (ILCD) provides guidelines for these processes, ensuring consistency and transparency in LCA practices.

Numerous studies have focused on assessing the environmental impacts of different biomass feedstocks in producing electricity, biogas, biofuels, and biochar. Bio-oil is recognized for its potential in energy production, while biochar is highly valued for its role in soil remediation, reducing pollution, enriching biotic populations, and contributing to carbon sequestration and climate change mitigation (Abhishek et al. 2022; Zhang et al. 2019).

In the case of biochar, its performance and associated environmental impacts depend on the physical and chemical properties of the material, which are closely related to the conditions of the pyrolysis process and the composition of the raw materials used. It has the potential to sequester carbon and has been used in various applications due to important characteristics such as surface area, pore volume, calorific value, surface functional groups, cation exchange capacity, and structural stability (Gupta et al. 2021).

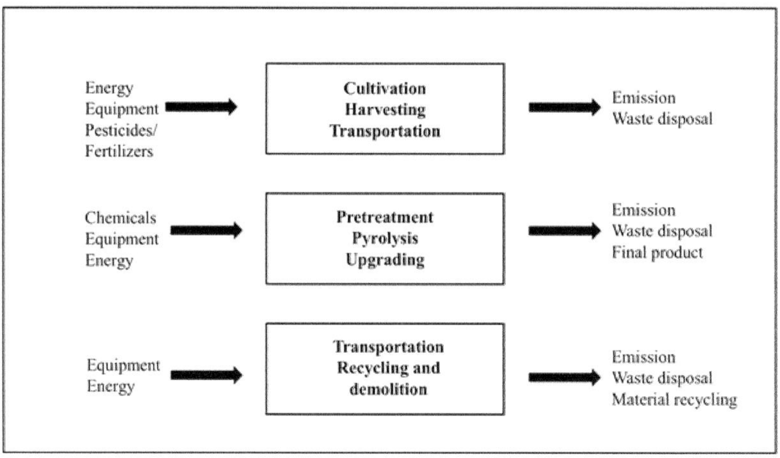

Figure 5.2 Generalized system for an LCA analysis.

The typical biomass utilization system using fast pyrolysis usually includes biomass pretreatment, fast pyrolysis, cyclone separation, condensation, and hydrotreatment, so the whole process can have a significant environmental impact (Yang et al. 2020). For this reason, it is necessary to analyze in detail the impacts associated with the process and, in this way, it is possible to make the most appropriate decisions to favor the use of waste and reduce environmental impacts.

5.2.2 LCA Types

Two primary methodologies characterize the two types of LCA: attributional and consequential LCA. Attributional LCA (ALCA) aims to describe the environmentally relevant physical flows to and from the life cycle and its subsystems. It is often used to assess impacts associated with product production and use at a certain point in time. ALCA is particularly handy for pointing out the environmental performance of current biochar production practices and comparing them across different production methods (Bergman et al. 2016). On the other hand, Consequential LCA (CLCA) considers changes in the life cycle that affect the environment, even to the extent of increasing or decreasing volume, along with new policies for implementation. Thus, CLCA is a more dynamic tool that can help to acquire insights into the broader, system-wide impact of decisions on biochar production. Examples are CLCA, which can evaluate the impacts of scaling up biochar production or implementation of new regulations that impact the source of biomass or pyrolysis technologies (Bergman et al. 2016). As a result, hybrid LCA approaches can be applied to improve the relevance and applicability of LCA results, which combine aspects of both ALCA and CLCA. These features enable hybrid LCA to capture both static and dynamic characteristics of biochar production systems, thus achieving a comprehensive understanding of their environmental impacts. Incorporating these approaches helps to correct the limitations of each methodology taken alone; therefore, it gives a balanced perspective for sustainable decision-making (Hamedani et al. 2019).

5.2.3 LCA Computational Tools

Life Cycle Assessment computational tools are the backbone of detailed and practical environmental impact studies. Popular LCA software includes SimaPro, GaBi, and OpenLCA, which provide automation in calculations to take off the complexity of the LCA process (Arena et al. 2016). These tools rely on vast databases that ensure the inventory of unit processes across industries is inclusive and, therefore, drastically reduce the time taken to collect data to account for environmental impact. Life Cycle Inventory (LCI) and Life Cycle Impact Assessment (LCIA) are essential phases in the broader framework of Life Cycle Assessment (LCA). LCI involves collecting and quantifying data on the inputs (such as energy, water, and materials) and outputs (like emissions, waste, and pollutants) throughout the entire life cycle of a product or process. This comprehensive data forms the foundation for subsequent analysis. In the LCIA phase, the data from LCI is analyzed to assess potential environmental impacts, including climate change, air and water pollution, and resource depletion. This phase converts raw inventory data into meaningful environmental indicators, providing a clearer understanding of the ecological footprint associated with the system under study. SimaPro is a typical example of this software, which has been used in academic and industrial research with its embedding of multiple LCIA methods, scenario modeling, and Monte Carlo simulations for uncertainty analysis. GaBi offers advanced modeling with the possibility to visualize material and energy flow, optimize processes, and evaluate environmental and economic performance. OpenLCA, as open-source software, shows remarkable success in life-cycle assessment evaluations of products and services; however, results from the same LCI data could be different for different, non-identical software applications (Herrmann and Moltesen 2015). Microsoft Excel or R software may also be used for simple calculations on impact category assessment using the manual calculation method. For example, global warming potential is a must-integrated indicator in this kind of assessment evaluation (Lefebvre et al. 2021). On the other hand, software packages such as CHEMCAD, COCO, and Aspen further enhance LCA studies by allowing the simulation of chemical processes, hence broadening the applications to complex systems.

5.3 BIOCHAR PRODUCTION: TECHNOLOGIES AND ENVIRONMENTAL CONSIDERATIONS

5.3.1 Technologies for Biochar Production

Biochar can be considered a solid residue derived from the thermochemical decomposition of biomass, mainly in a controlled oxygen atmosphere, with a wide range of environmental benefits (Sharma et al. 2024). Biochar characteristics and performance will be influenced by the type of biomass processed, the thermochemical route utilized, and the operating parameters (Malyan et al. 2021; Wang and Wang 2019). Some of the most used processes for biomass conversion are pyrolysis, gasification, torrefaction, and carbonization.

5.3.1.1 Pyrolysis

Pyrolysis has been used since ancient times as a thermochemical treatment for biomass and is conducted in a range of temperatures from 250–600°C (Hoang et al. 2021). During the process, the hydrocarbon chains are decomposed, resulting in liquid (bio-oil), solid (biochar), and gaseous (biogas) phases, which are collected throughout the different process stages (Jeyasubramanian et al. 2021). Pyrolysis consists of three main stages. The first stage involves removing moisture and volatiles and breaking bonds between compounds at temperatures ranging from room temperature to 200°C (Tomczyk et al. 2020). Once the moisture is removed, the organic components begin to disintegrate. In the second stage, hemicellulose and cellulose decompose, and finally, lignin is broken down at temperatures above 500°C (Yaashikaa et al. 2019).

Pyrolysis product yields (bio-oil, biogas, and biochar) highly depend on operational conditions, type of biomass used, reactor design, residence time, temperature, etc. (Tomczyk et al. 2020). Regarding char formation, enhanced performance has been found in pyrolysis processes with low heating rates at temperatures around 450°C (Uddin et al. 2018).

Depending on certain parameters, pyrolysis can be categorized as fast, flash, and slow pyrolysis. Fast pyrolysis is mainly focused on obtaining the liquid fraction (bio-oil), under temperature conditions within 450–650°C, heating rates between 103°C/s and 104°C/s, and short residence times below 2 s (Uddin et al. 2018). Yields in this type of pyrolysis usually reach 60–70 wt.% bio-oil, 13–25% of the gaseous portion, and 12–15% of biochar (Sharma et al. 2024). The flash pyrolysis and fast pyrolysis, pursue bio-oil production with yields around 75 wt.% and biochar 12/13 wt.%. (Kan et al. 2016) Conditions in this process are close to 1,000°C with heating rates greater than 700°C/s and a residence time below 0.5 s (Kan et al. 2016). Moreover, slow pyrolysis is more focused on char production, reaching efficiencies up to 20–50 wt.%, at temperatures from 300–500°C that enhance the process, heating rates of 0.1–1°C/s, and residence times from min to hours (Sharma et al. 2024; Pecha and Garcia-Perez 2020). Consequently, low heating rates and low temperatures should be prioritized to improve biochar production.

5.3.1.2 Gasification

Gasification is a thermochemical process mainly oriented to the conversion of biomass into syngas (gaseous mix of H_2, CO, and CH_4) at temperatures between 700–1,200°C (Sharma et al. 2023), under a partially oxidizing atmosphere; using a mixture of gases, CO_2, or steam to improve the conversion rate (Sharma et al. 2020), it is a multi-step process composed mainly of drying, pyrolysis, partial oxidation, and reduction. An advantage of gasification is that moist biomasses (up to 50%) can be used, resulting in yields higher than 50% and a small portion (approx. 10%) in biochar (Sharma et al. 2024). Additionally, various reactor configurations can be used depending on the requirements (fixed bed, entrained, moving bed, and bubbling fluidized bed) (Jeyasubramanian et al. 2021). The biochar resulting from biomass gasification has shown to be stabler and more resistant than that obtained from pyrolysis or torrefaction, due to its larger pore area (Sharma et al. 2023) and it also contains a greater proportion of functional groups such as –OH and –COOH, which makes it an excellent choice to be used in different areas such as soil treatment, water, and CO_2 absorption (Igalavithana et al. 2020). However, it is essential to consider that the amount of biochar produced in the gasification process can be quite insignificant, so the feasibility of the process must be evaluated for this purpose (Amalina et al. 2022).

5.3.1.3 Torrefaction

Torrefaction is a thermochemical process that releases organic volatiles produced by the cellulose, hemicellulose, and lignin degradation from biomass, as well as removing excess water, making it used as a pre-treatment stage, increasing the calorific value of materials, hydrophobicity, reactivity, and flammability (Amalina et al. 2022). The process is operated at temperatures between 200–300°C, under an inert environment, with low heating rates from 0.5 to 3 h (Iwuozor et al. 2022; Jeyasubramanian et al. 2021). Carbonized material obtained from torrefaction facilitates further processes such as handling, transport, and storage. It is considered an environmentally friendly and cost-effective process compared to other processes such as pyrolysis and gasification. Studies have shown that increasing the temperature and residence time improves the results obtained, reaching 70–80% yields (Enaime et al. 2020; Kanwal et al. 2019). However, it is important to highlight that the torrefied material cannot be considered biochar due to its chemical and physical characteristics (Daful and R Chandraratne 2020).

5.3.1.4 Carbonization

Carbonization is a specific type of pyrolysis that differs primarily in its operating temperature range and its focus on producing solid carbon materials. While pyrolysis is a broader thermal decomposition process carried out in the absence of oxygen to produce gases, liquids, and solids, carbonization specifically targets high carbon-content solids (Pahnila et al. 2023). This process is a thermal process performed in the absence of oxygen, removing all the non-carbon elements such as hydrogen, oxygen, and nitrogen, resulting in biochar and volatile material (Seow et al. 2022). The temperature setting ranges from 300–800°C, and the volatiles emitted may be more concentrated, depending on the heating rate. Unlike other pyrolysis methods, carbonization is a direct method, which makes it more cost-effective due to its high yields of high-quality biochar (Zakaria et al. 2023). Also, it is a low-emission process since the resulting gases can be burned and used to generate energy.

5.3.1.5 Hydrothermal Carbonization (HTC)

It is a thermochemical process to obtain liquid, solid, and gaseous products, using temperature ranges from 180–300°C and pressure up to 2–6 MPa (Sharma et al. 2024; Qin et al. 2022), with water as the solvent and reagent, this is why it is known as wet pyrolysis (Zhang et al. 2018). This characteristic confers greater performance to the process because biomass can be used without the previous drying stage, providing an economic advantage that makes it an alternative to slow pyrolysis. Residence times for this process are around 1–16 h (Amalina et al. 2022), producing biochar with high carbon content, greater porosity, and better properties compared to that from conventional pyrolysis (Seow et al. 2022; Wu et al. 2022) with yields between 36–75%, and can be used for environmental remediation, soil improvement, and as an energy resource (Qin et al. 2022; Jeyasubramanian et al. 2021). HTC is considered an environmentally friendly technique due to its operating conditions, costs, and the fewer chemical reactions involved in the process (Falco et al. 2011; Seow et al. 2022).

5.3.2 Life Cycle Analysis of Biochar Production Technologies

Production conditions, such as temperature, heating rate, and residence time, are crucial. For instance, slow and intermediate pyrolysis offers distinct benefits depending on the feedstock used, directly impacting the amount of carbon retained in the biochar and, consequently, its efficiency as a carbon sink. Equipment like pyrolytic kilns, pyrolytic stoves, and fluidized bed reactors have varying yields and environmental benefits. For example, pyrolytic stoves have demonstrated the greatest environmental benefits in terms of greenhouse gas emissions.

To better understand the impact of these pyrolysis processes on biochar production and their sustainability implications, Table 5.1 provides a comparative analysis of various biochar production technologies. This comparison highlights the differences in temperature ranges, biochar yield, and their corresponding effects on sustainability metrics, offering a comprehensive overview of each method's strengths and limitations.

Table 5.1 Comparative analysis of biochar production technologies

Process	Temperature (°C)	Biochar (Wt.%)	Impact on Sustainability
Slow pyrolysis	300–400	~35	—Carbon stabilization —Soil improvement —Requires significant energy —Maximizes biochar yield
Fast pyrolysis Flash pyrolysis	~500–700 ~900–1,200	12–25 12–20	—Produces syngas (main product) —Lower biochar yield —High energy efficiency —Requires greater energy consumption —Less focus on biochar production
Gasification	~800	~10	—Produces syngas as the main product —Lower biochar yield —High energy efficiency —Greater energy consumption —Less focus on biochar production
Torrefaction	200–300	~75	—Moisture content reducing —Stability improving —Lower biochar yield —More energy-efficient and cost-effective than other technologies
Carbonization	~300–600	~85	—High efficiency in producing biochar —Low environmental impact due to controlled emissions —Generates high-quality biochar suitable for agricultural applications
Hydrothermal carbonization	200–300	30–75	—Efficient process for wet biomass —Hydrochar with high carbon content —Fewer emissions and greater process efficiency

(Sharma et al. 2024; Iwuozor et al. 2022; Hoang et al. 2021; Qin et al. 2022; Wu et al. 2022; Al-Rumaihi et al. 2022)

Regarding slow pyrolysis, high residence time and low energy efficiency are the major challenges; however, biochar is the main product with high conversion rates and no limitations in terms of biomass that can be used with this process (Wang et al. 2020; Al-Rumaihi et al. 2022). In contrast, fast pyrolysis offers short residence times; however, it is necessary to have the proper particle size to achieve high efficiency. The biochar obtained from this process usually contains aromatics and is more stable than from slow pyrolysis (Sharma et al. 2024). The gasification process offers flexibility and efficiency; however, the initial capital is usually high, considering the amount of biochar obtained (Qin et al. 2022). Torrefaction yields a material with excellent characteristics, such as hydrophobicity; however, it is considered a pretreatment process due to the carbonaceous material's properties (Hoang et al. 2021). Finally, hydrothermal carbonization provides a unique advantage by allowing high-moisture content biomass to be processed, reducing drying costs, and enabling multiple feedstocks to be used. It may be considered that the residence times involved are high.

5.4 LIFE CYCLE INVENTORY (LCI) ANALYSIS

This phase involves collecting data on all inputs and outputs of the biochar production process. It is done as a list, table, or record, categorizing raw material, energy use, emissions, and waste. The system boundary requires a clear list of the type and quantity of materials and energy flowing through it. These could include data related to biomass collection and transportation, conditions for pyrolysis or gasification (temperature, residence time, and reactor type), post-treatment technology, and biochar application. LCI analysis can be considered reliable if there is a compelling collection of data and such data are as complete as possible. For example, in a biochar production system using agricultural residues, the LCI would involve collecting data on the energy and fuel used for harvesting and transporting the agro-industrial waste to the pyrolysis facility. The pyrolysis process conditions, such as operating temperature and residence time of the reactor, would be meticulously documented. The measurements will be emissions, such as CO_2, CO, and particulate matter. Other data to be captured includes the energy demand of the pyrolysis reactor. Data captured under LCI would be transportation and field application of the produced biochar and its impact on carbon sequestration and soil fertility.

Integrating the LCA studies considers the primary data capture, including the actual production system, and secondary data captured from available databases and literature. Such a combination gives a robust and realistic dataset. Additionally, it is often possible to undertake sensitivity analyses to learn what the impacts of changes in both data quality and assumptions will have on the outcome of an LCI. Critical uncertainties can be identified, allowing focused efforts to enhance data collection and analysis (Lee et al. 2020; Puettmann et al. 2020).

5.4.1 Life Cycle Impact Assessment (LCIA)

In this stage, LCI data are processed to obtain their potential environmental impacts. This step includes categorizing and characterizing inventory flows into many different categories of effects (e.g., global warming potential, acidification, eutrophication, and human toxicity). Tools developed by the US Environmental Protection Agency for the Reduction and Assessment of Chemical and Other

Environmental Impacts (TRACI) of operation are often drawn upon during this stage to create uniform results for impact assessment (Bergman et al. 2016). The LCIA stage involves selecting impacts that fall under a category, classification, and characterization. These are used to translate inventory data into the relevant indicators of environmental performance.

Also, LCIA includes normalization and weighting to compare and rank several environmental impacts. Normalization compares impact results with a reference value, such as a regional or global average impact. Weighing provides relative importance to different impact categories based on stakeholders' preferences or policy goals. Such methods result in LCIA output that is more understandable and relevant, and the results also provide a concise representation associated with vital environmental concerns and measures that potentially avoid them.

For example, in a biochar production system using pyrolysis, the LCI might record significant CO_2 and NO_x emissions from the process. In LCIA, these releases are characterized by global warming potential and acidification impact categories. CO_2 emissions would be quantified regarding their contribution to climate change, while NO_x emissions would be assessed for their role in acid rain formation. Normalization allows for the comparison of the impacts of biochar production with average emissions from typical practices in conventional agricultural waste management and brings out the relative advantages or disadvantages that such a production process may have. In this case, the weighting would indicate that global warming potential is of high concern and quite relevant to climate policy. Such analysis would give ground for identifying the significant impact issues on the environment for directions towards enhancing pyrolysis conditions and increasing the sustainability of biochar systems (Zhu et al. 2022).

5.5 ASSESSMENT AND COMPARISON OF ENVIRONMENTAL IMPACTS

In assessing and comparing the environmental impacts of biochar production technologies, it is crucial to evaluate the sustainability and ecological footprint of different processes. This assessment analyzes various factors, including greenhouse gas emissions, energy consumption, and overall environmental benefits. Comparison helps identify which technologies offer the most significant environmental advantages and which may require improvements to reduce their negative impacts.

5.5.1 LCA of Biochar from Biomss Pyrolysis or Gasification

The life cycle assessment (LCA) of biochar production from biomass pyrolysis and gasification provides valuable insights into the environmental performance of these technologies. This assessment evaluates the entire lifecycle of biochar production, from feedstock acquisition to final product use and disposal.

5.5.1.1 Pyrolysis

This process shows more favorable environmental outcomes. It is associated with lower impacts on greenhouse gas emissions and other environmental indicators. Pyrolysis is more efficient in producing biochar with a higher carbon content, which enhances its role as a carbon sink. The results from the LCA suggest that pyrolysis is generally more beneficial for the environment, particularly when considering the potential for global warming and reduced energy consumption compared to gasification.

5.5.1.2 Gasification

The LCA results for gasification indicate that this process is generally less environmentally friendly than fast pyrolysis. Gasification tends to have higher impacts on various environmental indicators, such as the potential for global warming and human toxicity potential, due to its energy-intensive nature and lower biochar yield. The gasification process produces syngas as the main product and typically requires more energy, leading to greater emissions and environmental impacts.

Both technologies have their specific advantages and limitations. For example, while fast pyrolysis demonstrates greater environmental benefits, it also has a lower biochar yield than slow pyrolysis. Therefore, choosing between these technologies should be informed by a comprehensive understanding of their environmental impacts and suitability for the intended biochar applications.

5.6 ENVIRONMENTAL IMPACT INDICATORS

The main environmental impact indicators considered in the LCAs of biochar production technologies include:

- Greenhouse Gas Emissions: Evaluate the global warming potential of the gases emitted during the biochar production process. Technologies that capture and reuse these gases can significantly reduce their net emissions.
- Energy Use: Measures energy consumption throughout the biochar life cycle, including direct energy used in production and energy incorporated into inputs such as fertilizers and water.
- Water Consumption: Evaluate the amount of water consumed during biochar production, considering the direct water used in the process and the water incorporated into the inputs.
- Impacts on Soil Quality: Considers the positive effects of biochar on soil quality, such as improved water retention and increased fertility, as well as potential negative consequences, such as the release of contaminants.

Life cycle analysis is key to evaluating the different biochar production technologies from a holistic environmental perspective. Comparative studies indicate that the choice of biochar production technology can significantly influence its ecological impacts. For an effective and sustainable implementation of biochar, it is essential to consider the technical and economic aspects and the environmental impacts throughout its life cycle.

5.7 INTERPRETING THE RESULTS

Conclusively, the results obtained from both the LCI and the LCIA must be construed to give the conclusions drawn from the results and the recommendations made. This process involves making conclusions relating to determining the significant environmental impacts, how robust and reliable these results are, and suggesting ways of improvement or alternatives to lower the ecological footprint of biochar. In interpreting the results, this last phase should also incorporate sensitivity and uncertainty analyses for robust conclusions. This is important in determining "hot spots" or areas that significantly impact the environment and devising methods to mitigate the consequences.

In addition to identifying the environmental hot spots and mitigation, an interpretation phase should be conducted, which allows a critical review of the LCA study from outside expertise to enforce the credibility and transparency of the LCA study. This review process can find potential biases, validate the assumptions, and ensure that the applied methodology follows known standards. By including feedback from outside, the LCA results become more dependable and credible, so that based on these, decisions can be made confidently with greater acceptance by stakeholders (Puettmann et al. 2020).

For example, in the interpretation phase, it might be found that the most significant cause of environmental impact comes from very high energy use in the thermochemical process. A sensitivity analysis could reveal that the adjustments in operating temperature and residence time have more significant effects on overall energy efficiency and emission levels. Their critical review by external experts may also lead to further reactor design optimization aimed at maximum heat recovery while reducing the application of energy inputs. The experts could recommend alternative feed drying methods with further energy use reduction. These insights help refine the biochar production process to make it more sustainable, precisely by reducing the carbon footprint and improving resource use efficiency in the entire biochar system.

5.8 FINAL THOUGHTS ON TECHNOLOGY

- Slow pyrolysis: Optimized to produce solid residues like biochar and some liquid and gaseous byproducts. Its versatility makes it suitable for energy recovery, waste management, and material production applications.
- Carbonization: Ideal for producing high-carbon solids like biochar or charcoal, focusing on maximizing carbon content and stability. It is well-suited for applications such as soil enhancement, filtration, energy storage, and carbon sequestration strategies.

- Fast Pyrolysis: Suitable for producing bio-oil and gases, with a lower biochar yield.
- Flash pyrolysis: Focused on producing bio-oil and gases with highly porous biochar, used in carbon capture applications.
- Gasification: Generates mainly syngas, with biochar as a by-product; suitable for catalytic and carbon capture processes.
- Hydrothermal Carbonization (HTC): Effective for wet biomass, it produces hydrochar with a high energy density and is suitable for various energy and environmental applications.

Adjusting operational conditions, selecting appropriate technologies, and implementing energy-saving measures can enhance the sustainability of biochar production. Refining processes based on LCA results and external feedback will make achieving a more efficient and environmentally friendly biochar production system possible.

5.9 CONCLUSIONS

Life cycle assessment (LCA) is essential for evaluating the environmental impact of biochar production. Various LCA types are employed based on specific study objectives. Attributional LCA (ALCA) provides a static evaluation of physical flows and environmental impact at a given time, which is ideal for comparing current biochar production practices. Conversely, Consequential LCA (CLCA) analyzes dynamic impacts on the entire system, considering factors such as production scaling and new regulatory policies, which are useful for long-term and systemic evaluations. A hybrid approach combines both analyses for a more comprehensive and valuable perspective in sustainable decision-making.

Computational tools facilitate these analyses. Software like SimaPro, GaBi, and OpenLCA automates complex calculations using extensive databases. Tools such as CHEMCAD, COCO, and Aspen also extend the analysis by incorporating chemical process simulations. Manual Excel or R methods can also be applied for simple calculations, albeit with less precision and scope.

Regarding biochar production, various technologies like pyrolysis, gasification, torrefaction, carbonization, and hydrothermal carbonization (HTC) have different environmental and efficiency considerations. Pyrolysis is the primary and most efficient method for biochar production, while gasification and torrefaction offer high-quality biochar for specific applications. HTC is notable for its economic and environmental advantages, especially for wet biomass.

Thus, selecting the appropriate LCA methodology and biochar production technology should be based on specific objectives and sustainability considerations, maximizing yield and minimizing environmental impacts. This ensures comprehensive evaluation and informed decisions towards sustainable practices.

REFERENCES

Abhishek, K., A. Shrivastava, V. Vimal, A.K. Gupta, S.K. Bhujbal, J.K. Biswas, et al. 2022. Biochar application for greenhouse gas mitigation, contaminants immobilization, and soil fertility enhancement: A state-of-the-art review. Science of the Total Environment. 853: 158562.

Al-Rumaihi, A., M. Shahbaz, G. Mckay, H. Mackey and T. Al-Ansari. 2022. A review of pyrolysis technologies and feedstock: A blending approach for plastic and biomass towards optimum biochar yield. Renewable and Sustainable Energy Reviews. 167: 112715.

Amalina, F., A.S.A. Razak, S. Krishnan, H. Sulaiman, A.W. Zularisam and M. Nasrullah. 2022. Biochar production techniques utilizing biomass waste-derived materials and environmental applications – A review. Journal of Hazardous Materials Advances. 7.

Arena, N., J. Lee and R. Clift. 2016. Life cycle assessment of activated carbon production from coconut shells. Journal of Cleaner Production. 125: 68–77.

Bergman, R.D., H. Gu, D.S. Page-Dumroese and N.M. Anderson. 2016. Life cycle analysis of biochar. *In*: Biochar: A regional supply chain approach in view of climate change mitigation. Cambridge University Press. Cambridge, United Kingdom. pp. 46–69.

Daful, A.G. and M. R Chandraratne. 2020. Biochar production from biomass waste-derived material. pp. 370–378 *In*: Saleem Hashmi and Imtiaz Ahmed Choudhury (eds). Encyclopedia of Renewable and Sustainable Materials: Volume 1–5. Elsevier.

Enaime, G., A. Baçaoui, A. Yaacoubi and M. Lübken. 2020. Biochar for wastewater treatment-conversion technologies and applications. Applied Sciences. 10(10): 3492.

Falco, C., N. Baccile and M.M. Titirici. 2011. Morphological and structural differences between glucose, cellulose and lignocellulosic biomass derived hydrothermal carbons. Green Chemistry. 13: 3273–3281.

Gupta, S., S. Muthukrishnan and H.W. Kua. 2021. Comparing influence of inert biochar and silica rich biochar on cement mortar – Hydration kinetics and durability under chloride and sulfate environment. Construction and Building Materials. 268(12): 121142.

Hamedani, S.R., T. Kuppens, R. Malina, E. Bocci, A. Colantoni and M. Villarini. 2019. Life cycle assessment and environmental valuation of biochar production: Two case studies in Belgium. Energies (Basel). 12: 1–21.

Herrmann, I.T. and A. Moltesen. 2015. Does it matter which Life Cycle Assessment (LCA) tool you choose? - A comparative assessment of SimaPro and GaBi. Journal of Cleaner Production. 86: 163–169.

Hoang, A.T., H.C. Ong, I.M.R. Fattah, C.T. Chong, C.K. Cheng, R. Sakthivel, et al. 2021. Progress on the lignocellulosic biomass pyrolysis for biofuel production toward environmental sustainability. Fuel Processing Technology. 223(7): 1066997.

Igalavithana, A.D., S.W. Choi, P.D. Dissanayake, J. Shang, C.H. Wang, X. Yang, et al. 2020. Gasification biochar from biowaste (food waste and wood waste) for effective CO_2 adsorption. Journal of Hazardous Materials. 391: 121147.

ISO 14040. 2006. Environmental assessment - Life cycle assessment - Principles and framework. International Standard Organisation. 1997: 1–20.

Iwuozor, K.O., E. Chizitere Emenike, J.O. Ighalo, F.O. Omoarukhe, P.E. Omuku and A. George Adeniyi. 2022. A review on the thermochemical conversion of sugarcane bagasse into biochar. Cleaner Materials. 6: 100162.

Jeyasubramanian, K., B. Thangagiri, A. Sakthivel, J. Dhaveethu Raja, S. Seenivasan, P. Vallinayagam, et al. 2021. A complete review on biochar: Production, property, multifaceted applications, interaction mechanism and computational approach. Fuel. 292: 120243.

Kan, T., V. Strezov and T.J. Evans. 2016. Lignocellulosic biomass pyrolysis: A review of product properties and effects of pyrolysis parameters. Renewable and Sustainable Energy Reviews. 57: 1126–1140.

Kanwal, S., N. Chaudhry, S. Munir and H. Sana. 2019. Effect of torrefaction conditions on the physicochemical characterization of agricultural waste (sugarcane bagasse). Waste Management. 88: 280–290.

Lee, M., Y.L. Lin, P. Te Chiueh and W. Den. 2020. Environmental and energy assessment of biomass residues to biochar as fuel: A brief review with recommendations for future bioenergy systems. Journal of Cleaner Production. 251: 119714.

Lefebvre, D., A. Williams, G.J.D. Kirk, J. Meersmans, S. Sohi, P. Goglio, et al. 2021. An anticipatory life cycle assessment of the use of biochar from sugarcane residues as a greenhouse gas removal technology. Journal of Cleaner Production. 312: 127764.

Malyan, S.K., S.S. Kumar, R.K. Fagodiya, P. Ghosh, A. Kumar, R. Singh, et al. 2021. Biochar for environmental sustainability in the energy-water-agroecosystem nexus. Renewable and Sustainable Energy Reviews. 149: 111379.

Nemecek, T., A. Roesch, M. Bystricky, P. Jeanneret, J. Lansche, M. Stüssi, et al. 2024. Swiss agricultural life cycle assessment: A method to assess the emissions and environmental impacts of agricultural systems and products. International Journal of Life Cycle Assessment. 29: 433–455.

Pahnila, M., A. Koskela, P. Sulasalmi and T. Fabritius. 2023. A review of pyrolysis technologies and the effect of process parameters on biocarbon properties. Energies (Basel). 16(19): 6936.

Pecha, M.B. and M. Garcia-Perez. 2020. Pyrolysis of lignocellulosic biomass: oil, char, and gas. pp. 581–619. *In*: Anju Dahiya (ed.). Bioenergy (2nd Ed.). Biomass to Biofuels and Waste to Energy. Academic Press.

Puettmann, M., K. Sahoo, K. Wilson and E. Oneil. 2020. Life cycle assessment of biochar produced from forest residues using portable systems. Journal of Cleaner Production. 250: 119564.

Qin, F., C. Zhang, G. Zeng, D. Huang, X. Tan and A. Duan. 2022. Lignocellulosic biomass carbonization for biochar production and characterization of biochar reactivity. Renewable and Sustainable Energy Reviews. 157: 112056.

Seow, Y.X., Y.H. Tan, N.M. Mubarak, J. Kansedo, M. Khalid, M.L. Ibrahim, et al. 2022. A review on biochar production from different biomass wastes by recent carbonization technologies and its sustainable applications. Journal of Environmental Chemical Engineering. 10: 107017.

Sharma, P.K., A.K. Sharma, R.H. Pulla and P.K. Sahoo. 2020. Performance analysis of a medium-scale downdraft gasifier using Lantana camera biomass as feeding material. Energy Sources, Part A: Recovery, Utilization and Environmental Effects. 46(1): 15087–15101.

Sharma, A.K., P.K. Ghodke, N. Goyal, P. Bobde, E.E. Kwon, K.Y.A. Lin, et al. 2023. A critical review on biochar production from pine wastes, upgradation techniques, environmental sustainability, and challenges. Bioresource Technology. 387: 129632.

Sharma, T., I.G. Hakeem, A.B. Gupta, J. Joshi, K. Shah, A.K. Vuppaladadiyam, et al. 2024. Parametric influence of process conditions on thermochemical techniques for biochar production: A state-of-the-art review. Journal of the Energy Institute. 113: 101559.

Tomczyk, A., Z. Sokołowska and P. Boguta. 2020. Biochar physicochemical properties: pyrolysis temperature and feedstock kind effects. Reviews in Environmental Science and Bio/Technology. 19: 191–215.

Uddin, M.N., K. Techato, J. Taweekun, M.M. Rahman, M.G. Rasul, T.M.I. Mahlia, et al. 2018. An overview of recent developments in biomass pyrolysis technologies. Energies. 11(11): 3115.

Wang, J. and S. Wang. 2019. Preparation, modification and environmental application of biochar: A review. Journal of Cleaner Production. 227: 1002–1022.

Wang, L., Y.S. Ok, D.C.W. Tsang, D.S. Alessi, J. Rinklebe, H. Wang, et al. 2020. New trends in biochar pyrolysis and modification strategies: feedstock, pyrolysis conditions, sustainability concerns and implications for soil amendment. Soil Use and Management. 36: 358–386.

Wu, Y., H. Wang, H. Li, X. Han, M. Zhang, Y. Sun, et al. 2022. Applications of catalysts in thermochemical conversion of biomass (pyrolysis, hydrothermal liquefaction and gasification): A critical review. Renewable Energy. 196: 462–481.

Yaashikaa, P.R., P. Senthil Kumar, S.J. Varjani and A. Saravanan. 2019. Advances in production and application of biochar from lignocellulosic feedstocks for remediation of environmental pollutants. Bioresource Technology. 292: 122030.

Yang, X., D. Han, Y. Zhao, R. Li and Y. Wu. 2020. Environmental evaluation of a distributed-centralized biomass pyrolysis system: A case study in Shandong, China. Science of the Total Environment. 716: 136915.

Zakaria, M.R., M.A. Ahmad Farid, Y. Andou, I. Ramli and M.A. Hassan. 2023. Production of biochar and activated carbon from oil palm biomass: Current status, prospects, and challenges. Industrial Crops and Products. 199: 116767.

Zhang, B., M. Heidari, B. Regmi, S. Salaudeen, P. Arku, M. Thimmannagari, et al. 2018. Hydrothermal carbonization of fruit wastes: A promising technique for generating hydrochar. Energies (Basel). 11(8): 2022.

Zhang, Z., Z. Zhu, B. Shen and L. Liu. 2019. Insights into biochar and hydrochar production and applications: A review. Energy. 171: 581–598.

Zhu, X., C. Labianca, M. He, Z. Luo, C. Wu, S. You, et al. 2022. Life-cycle assessment of pyrolysis processes for sustainable production of biochar from agro-residues. Bioresource Technology. 360: 127601.

Chapter 6

Biochar as an Amendment Soil

Romina Zabaleta[a], Paula Fabani[a, b], Rosa Rodriguez[a] and Ana Laura Navas* [a]

[a]Instituto de Ingeniería Química, Fac. de Ingeniería, Grupo Vinculado al PROBIEN (Universidad Nacional de San Juan, CONICET, Universidad Nacional del Comahue), San Juan, Argentina

[b]Instituto de Biotecnología, Facultad de Ingeniería, Universidad Nacional de San Juan, San Juan, Argentina

6.1 PROBLEM SOILS AND THEIR EFFECTS ON PLANTS: EFFECTS OF PHYSICAL AND CHEMICAL CONSTRAINTS

Soil is the Earth's surface layer, formed by the interaction of parent rock, climate, living organisms, and time; its composition is based on minerals, organic matter, water, and air. It is the habitat for a great diversity of organisms interacting dynamically. In addition, the soil serves as a medium that provides physical support and nutrients to plants, stores carbon, and performs other essential functions, making it fundamental to the development of life. This composition can now also be influenced by human activity, either positively or negatively.

Plant growth in the soil is a complex process involving the dynamic interaction of physical, chemical, and biological factors that influence the availability of nutrients, water, and environmental conditions necessary for plant development. Roots, for example, are the anchoring medium and mechanical support for plants, allowing for the absorption of water and various nutrients, as well as the process of biosynthesis and storage of chemical compounds, which in turn influence the

*Corresponding author: Ana Laura Navas (analauranavas@gmail.com)

DOI: 10.1201/9781032726694-6

interactions with abiotic and biotic factors in the soil environment (Richardson et al. 2009).

Depending on the agricultural area being studied, a wide range of soils is possible, from very healthy to poor. In healthy soils, roots can penetrate the soil particles, absorb water and nutrients, and consume oxygen (O_2), allowing plants to grow healthily and produce maximum yields. In contrast, poor soils do not allow roots to grow, preventing the absorption of water and nutrients and reducing the normal development of plants. The characteristics of soils that are naturally poor or have been polluted by human activity and their effects on plant growth are listed below.

6.1.1 Plant Growth in Naturally Problem Soils

Soil formation is a slow and protracted process in which many factors, including climate, plants, animals, and microorganisms, influence both physical and chemical properties. However, studies suggest that only 5% of soils are residual soils formed from pre-existing parent materials, and about 95% of the world's soils have been moved or transported to their present location (Yu et al. 2019).

a) The Impact of the Physical Constraints

The physical properties of soil depend on its formation, history, the influence of weathering, and human activity. They refer to its texture, bulk density, structure, pores, depth of organic matter, and temperature, and are related to a key factor influencing plant growth and development, soil compaction. This phenomenon affects a large percentage of soils intended for cultivation, at a depth of about 30 cm. It also affects bulk density, reducing pore space and water absorption, which implies poor aeration of the soil, affecting the growth of roots, as they cannot expand in search of the water and nutrients necessary for their proper growth. This affects the permeability and diffusivity of gases, creating anaerobic conditions in the soil, which is an advantageous condition for the process of soil denitrification, as well as reducing the activities of beneficial microorganisms in the soil (Yu et al. 2019). As a consequence of these phenomena, plants cannot develop under optimal conditions and do not have adequate size and productivity.

Several studies indicate that a significant proportion of the world's arable land is subject to compaction stress, a major cause of which is vehicle traffic during tillage operations. For example, Kroulík et al. (2009) used Global Positioning System tracking devices to find that conventional (uncontrolled) farming practices for wheat production covered between 56% and 88% of the field with at least one wheel pass during tillage. Another study found that 39% of the cost of restoring land affected by soil degradation was due to soil compaction, while 12% of the total cost was due to soil erosion, which is indirectly influenced by soil compaction (Graves et al. 2015). Work is therefore underway to investigate the effects of soil compaction and its management on sustainable crop production (Shaheb et al. 2021), as well as alternative crop traffic management systems (Godwin et al. 2022).

Subsurface compaction can also be caused by inappropriate agricultural practices such as monoculture, poor crop nutrition, and soil erosion, which also limit the productive, environmental, and economic potential of the soil. To maintain soil fertility several researchers and experts in the field have proposed solutions such as "break crops," which aim to break the life cycle of pests while providing biological benefits to the soil, and "crop diversification," which improves the use of natural resources and increases ecological resilience (Yang et al. 2021c).

b) The Impact of the Chemical Constraints

For plant growth and development, soil chemical limitations are related to pH, acidity, salinity, sodicity, and nutrient deficiencies. Soils with a pH <5.5 are considered acidic and account for approximately 30% of the world's frozen land and 50% of the world's arable land. At this pH, manganese (Mn) and aluminum (Al) become more soluble, while the bioavailability of calcium (Ca), magnesium (Mg), and phosphorus (P) decreases. This has many consequences for crops. For example, the increased availability of Al in the soil initially causes the roots to elongate, but as a consequence, the cellular structure of the root tip is destroyed, which then affects the uptake of water and nutrients by the roots. On the other hand, in acidic soils, P is easily fixed by clay minerals, making it unavailable for root uptake. Therefore, Al toxicity and P deficiency are two important constraints that hinder crop development in acid soils (Zheng 2010).

Conversely, the abundance of carbonates and bicarbonates in the soil means that soils with a pH greater than 8 are considered alkaline. Alkaline and calcareous soils are found in 30% of soils. This results in reduced bioavailability of micronutrients such as iron (Fe), copper (Cu), manganese (Mn), and zinc (Zn) in plants. In particular, low Fe availability causes plant leaves to become chlorotic and subsequently necrotic, which affects plant development and therefore productivity. In extreme cases, iron deficiency can lead to total crop failure (Chen and Barak 1982).

In terms of fertilization, the most common nutrients added to soils are potassium (K), nitrogen (N), and phosphorus (P), as these are the most important for crop development and are often depleted or unavailable due to intensive land use.

In the case of N-rich fertilizer application, previous studies have shown that it is not used efficiently, as expected crop yields are not achieved, e.g. through leaching, which also leads to potential groundwater contamination (Wang et al. 2021). The same phenomenon is observed with P, leading to the eutrophication of water bodies. The loss of P in dissolved and particulate form due to topography, soil type, and soil P concentration has generated the need to apply this element. However, in some cases, the application of P at concentrations higher than required by crops has generated an increase of P in the surface soil, and consequently, the need to develop nutrient management strategies based on this compound (McDowell and Sharpley 2001).

K is one of the essential elements for healthy plant growth and productivity as it is involved in the transport of water, nutrients, and sugars, enzyme activity, photosynthesis, and starch and protein synthesis. Deficiency is indicated by a change in leaf color from dark green to light green, then to yellow, followed by wilting and finally senescence, which also affects crop growth and yield. Due to its importance within the plant system, where it plays a vital role as a regulator, K is considered the second most important element after N for overall plant growth and is involved in combating abiotic stress caused by the environment (Johnson et al. 2022).

The micronutrients considered essential for plant development are Fe, Zn, Mn, Cu, nickel (Ni), boron (B), molybdenum (Mo), and chloride (Cl), whose presence or bioavailability in the soil is a function of soil pH and anthropogenic activities.

Salinity (a high concentration of dissolved salts) and sodicity (a high proportion of sodium (Na)) in soils occur mainly in arid and semi-arid regions. These processes can be attributed to natural causes such as climate, lithology, topography, and pedology, as well as anthropogenic activities, including human agricultural land use, particularly irrigated agriculture (Stavi et al. 2021). A soil is considered saline if certain characteristics of pH (8), exchangeable sodium percentage (ESP) (<8.5), and electrical conductivity (EC) (<4 dS/m) are met, while in sodic soils these values must be >8.5, less than 4 dS/m and greater than 15, respectively (Ismayilov et al. 2021). In some cases, salinity and sodicity are found together in soils due to their dependence on the parent rock, and both can alter plant growth, reducing development and hence their size and productivity.

6.1.2 Plant Growth in Anthropogenic Problem Soils

The use of various chemicals in agriculture, or the improper disposal of residues, results in the presence of artificial chemicals in the soil. These chemicals, if present in large quantities, lead to soil contamination and pose a risk to the health of people and the ecosystem living in and around the site. These include phthalates, polycyclic aromatic hydrocarbons (PAH), pesticides, chlorophenols, and heavy metals. These have various effects on crop growth and production, as well as associated risks in the environment (Lü et al. 2018; Angon et al. 2024).

a) Organic Pollutants

The polycyclic aromatic hydrocarbons (PAH) family is a class of hazardous organic pollutants commonly found in problematic soils, formed by the incomplete combustion of organic materials such as coal, oil, petrol, and wood. These compounds are important to identify and quantify because of their recognized toxicity and, in particular, their carcinogenic potential. They reach the soil through the air as the wind carries them and become deposited on the soil. Studies have shown that the concentration of these pollutants in the soil increases over time and with the increase of industrialized regions adjacent to agricultural fields (Keyte et al. 2013). Several reports indicate that PAHs are relatively stable and recalcitrant in soils, making them difficult to degrade and remove from contaminated soil by conventional treatment methods.

One of the most common families of toxic pollutants generated by industries such as pharmaceuticals, petrochemicals, plastics, iron and steel, textiles, paper, and paints is the chlorophenols (CPs), including monochlorophenols (MCP), dichlorophenols (DCP), trichlorophenols (TCP), among others. The toxicity of chlorophenol derivatives is proportional to the number and position of the chlorine atoms in the benzene ring. Moreover, precursors of CPs are used in the formulation of certain agricultural products, such as pesticides, herbicides, and fungicides, which have a negative impact on soil fertility and affect the microbial communities inhabiting these habitats (Yadav et al. 2023).

There are also reports indicating that PAHs and CPs pose a potential risk to suburban agriculture, as some of these compounds are carcinogenic, mutagenic, and persistent, resulting in high toxicity to both plants and humans (Keyte et al. 2013; Cui et al. 2022; Yadav et al. 2023).

Additionally, the coloring substances and synthetic pigments used mainly in the industry (cosmetics, textiles, pharmaceuticals, and paper, among others) form a group of chemical pollutants that have raised concern and interest due to their high toxicity to the environment when not properly managed and disposed of. Their negative effect on the environment is because their presence in water has a negative impact on O_2 availability (DO), O_2 diffusion, and the passage of sunlight, affecting the aquatic ecosystem and human health (Pagalan et al. 2020; Tang et al. 2021). These include methylene blue, methylene orange, and G-orange, organic dyes that are resistant to biodegradation and stable to light, which tend to destroy DNA structures and threaten the health of the environment (Oladoye et al. 2022).

b) Inorganic Pollutants

Inorganic pollutants can originate from anthropogenic sources such as mining, metallurgical activities, pesticides, and fertilizers containing heavy metals (HV) such as arsenic (As), cadmium (Cd), lead (Pb), copper (Cu), nickel (Ni), chromium (Cr), mercury (Hg) and zinc (Zn). These metals are present in the environment to varying degrees (food, soil, water, and air), but when present at dangerous levels (depending on the element), they affect soil quality and biological functions due to their toxicity and persistence in soil, as they are not biodegradable (Lu et al. 2017; Uchimiya et al. 2020; Angon et al. 2024). A brief description of their effects on soil is given below:

As: It is a carcinogenic trace element that is toxic to living organisms, affects plant growth, and is therefore of greater concern than other heavy metals. Its presence in soils is mainly due to the use of pesticides on crops and its use in extractive mining. Concentrations in contaminated soils vary widely, ranging from 1 mg/kg to 116,000 mg/kg (Nguyen et al. 2021).

Cd: Soil contains this element naturally (average concentration of 0.27 mg/kg), although levels of more than 360 mg/kg can be found in contaminated soils as a result of the same factors mentioned for As. High concentrations of Cd in soil cause phytotoxicity in plants, thereby affecting their metabolism. Cd in soils can be present in different physicochemical forms, and its ecological impact depends on this, as its mobility, bioavailability, uptake, and accumulation by plants and living organisms are affected. In addition, as a mobile element, it can enter the food chain through soil and water, accumulating in rice grains, for example. Its intake causes health problems in humans, affecting the pulmonary, cardiovascular, and musculoskeletal systems (Rahim et al. 2022).

Pb: It shares about 10% of the total pollution produced by heavy metals. The soil contains Pb at an average level of 32 mg/kg, so higher levels indicate contamination. Pb forms complexes with phosphates and carbonates in the soil and has adverse effects on plants (metabolic functions, growth, and photosynthetic activity). Like Cd, it can be transferred through the food chain to animals and humans, and due to its carcinogenic properties, its presence in soils is a serious concern for human health (Collin et al. 2022).

Cu, Ni, and Zn: These metals are essential elements for animals, plants, and humans. All three are involved in various metabolic functions of plants, allowing them to develop properly, bear fruit, and produce good yields. Cu, Ni, and Zn are naturally present in soils at average concentrations of 2–60 mg/kg, 9–98 mg/kg, and 33–88.88 mg/kg, respectively. Higher concentrations in cultivated soils are due to contamination from mining and combustion sources (Yu et al. 2019).

Hg: It is a persistent toxic pollutant; it does not degrade in the environment and becomes mobile due to the volatility of the element and several of its compounds (Gworek et al. 2020). The transformation of mercury (Hg) into the more toxic and bioaccumulative form methylmercury (MeHg) in soils and sediments can lead to the biomagnification of MeHg through the food chain, rendering it into a toxic element that poses ecological and health risks (Yang et al. 2021b).

6.2 BIOLOGICAL ACTIVITIES IN PROBLEMS SOILS

Soil biodiversity reflects the diversity of living organisms, including microorganisms (e.g., bacteria, fungi, protozoa, and nematodes), mesofauna (e.g., mites and springtails), as well as the more familiar macrofauna (e.g., earthworms and termites). Plant roots can also be considered soil organisms because they interact symbiotically with other soil components. All these organisms interact with one another and with the different plants and animals in the ecosystem, forming a complex biological framework.

Soil organisms are essential for the sustainable functioning of all ecosystems, acting as drivers of nutrient cycling, regulating soil organic matter dynamics, and carbon sequestration, as well as mitigating greenhouse gas emissions. They also modify soil physical structure and water regimes, increase the amount and efficiency of nutrient acquisition by vegetation, and improve plant health.

Similarly, the distribution of microorganisms in the soil is an important factor in determining the optimal conditions for crop development and productivity. The enzymes and hormones produced by soil microorganisms also contribute to soil properties, as they are important indicators of soil quality and health. The problem, however, is that soil microbial communities are particularly sensitive

6.5.1 Physical Property Improvements

The impact of biochar on soil physical properties is both profound and multifaceted. One of the most critical improvements is the reduction in bulk density (Ndede et al. 2022), which alleviates soil compaction, a common issue in agricultural soils that impedes root growth and limits the movement of air and water through the soil profile. By decreasing bulk density, biochar facilitates deeper root penetration, allowing plants to access a greater volume of soil resources, including water and nutrients. Moreover, the enhanced porosity introduced by the inherent porous structure of biochar is vital for improving soil aeration and drainage. This increase in porosity ensures that roots and beneficial soil microorganisms have access to the O_2 they need, while also preventing waterlogging, which can lead to root diseases and reduced plant vigor. Furthermore, the role of biochar in increasing aggregate stability is crucial for protecting soil against erosion (Baiamonte et al. 2019). Stable soil aggregates resist the disintegration caused by water and wind, thus preserving the soil structure and preventing the loss of valuable topsoil. These physical improvements are integral to optimizing the soil environment, supporting plant growth, and enhancing overall agricultural productivity.

6.5.2 Chemical Property Improvements

The chemical enhancements generated by biochar amendment are equally significant, particularly in the context of nutrient management and soil pH regulation (Kannan et al. 2021; Thi et al. 2021; Yang et al. 2022). The alkaline nature of biochar is particularly advantageous for soils with high acidity, a condition that can severely limit plant growth by reducing nutrient availability and impairing root function. By neutralizing soil acidity, biochar creates a more hospitable environment for plants, promoting healthier and more vigorous growth. In addition to its pH-modifying effects, biochar presents a high cation exchange capacity (CEC), which plays a fundamental role in nutrient retention. CEC is a measure of a soil's ability to hold and exchange cations of positively charged ions, which include essential nutrients like K, Ca, and Mg. The high CEC of biochar allows it to effectively capture these nutrients, preventing them from leaching away with rainfall or irrigation. This retention ensures a steady supply of nutrients to plants over time, reducing the need for frequent applications of chemical fertilizers. The presence of various functional groups on the biochar surface, such as carboxyl, hydroxyl, and carbonyl groups, further enhances its ability to form stable complexes with soil minerals, which improves the long-term storage and availability of nutrients in the soil. These chemical improvements are crucial for maintaining soil fertility, reducing environmental impacts from fertilizer runoff, and promoting sustainable agricultural practices.

6.6 ENHANCING PLANT GROWTH IN PROBLEMATIC SOILS THROUGH BIOCHAR AMENDMENT

6.6.1 Introduction to Biochar and its Role in Plant Growth

Biochar is a carbon-rich material obtained through pyrolysis, where organic materials like agricultural waste, wood chips, or manure are thermally decomposed in the absence of O_2 (Haider et al. 2022a; Qian et al. 2023). While the process yields a highly stable form of carbon, it also produces a material that can be used to enhance soil properties and promote plant growth (Guo et al. 2020).

Several studies have documented the ability of biochar to improve soil structure, increase water-holding capacity, and enhance nutrient retention (Palansooriya et al. 2019; Qian et al. 2023). These soil improvements directly impact plant growth by providing a more favorable environment for root development, nutrient uptake, and microbial interactions (El-Naggar et al. 2019; Qian et al. 2023).

Key mechanisms by which biochar influences plant growth include (El-Naggar et al. 2019):

Increased Nutrient Availability: By adsorbing nutrients and reducing leaching losses, biochar provides a more sustained release of essential nutrients like N, P, and K for plant uptake (Palansooriya et al. 2019; Qian et al. 2023).

Improved Soil Aeration and Structure: The biochar porous nature improves soil texture, enhancing root penetration and overall plant health (Palansooriya et al. 2019; Qian et al. 2023).

Enhanced Microbial Activity: The complex surface area of biochar creates a habitat for beneficial microbes that can promote plant growth by fixing N or enhancing P availability (Palansooriya et al. 2019; Qian et al. 2023).

6.6.2 Biochar Effect on Soil Nutrient Dynamics and Crop Productivity

The use of biochar has been found to increase nutrient retention in soils, thus reducing fertilizer needs and promoting sustainable agriculture (Alkharabsheh et al. 2021). Its ability to hold nutrients within its structure can be attributed to its high CEC (Agegnehu et al. 2017; Zhang et al. 2020a). By preventing nutrient leaching, biochar improves nutrient-use efficiency, especially in sandy soils or highly weathered tropical soils where nutrient losses are common (El-Naggar et al. 2019; Palansooriya et al. 2019). Key impacts on soil nutrient dynamics include:

Increased N Retention: Biochar can immobilize excess N, reducing its loss through volatilization or leaching (Alkharabsheh et al. 2021). This is particularly important in agricultural systems reliant on synthetic fertilizers, as biochar enhances N-use efficiency and minimizes environmental pollution (Palansooriya et al. 2019; Dai et al. 2020).

P Availability: The biochar interaction with soil P can vary based on soil pH and the pyrolysis temperature of the biochar (Agegnehu et al. 2017; Qian et al. 2023). In acidic soils, biochar can make P more available to plants by raising the pH and reducing P fixation. This has a direct influence on root development and crop yield (Jones et al. 2012; Zhang et al. 2020a).

Several studies show that biochar amendments have led to substantial increases in crop yields:

Maize and Rice: Enhanced N and P availability in biochar-amended soils has been shown to increase maize and rice yields by 10–40%, depending on soil type and climate conditions.

Wheat and Soybean: Biochar has also shown promise in increasing wheat and soybean yields by improving water and nutrient management in arid and semi-arid regions.

However, these yield improvements are highly context-specific, with factors like biochar type, application rate, and soil conditions playing a crucial role in determining the outcome (Zhang et al. 2020a).

6.6.3 Biochar Role in Enhancing Water Retention and Plant Growth in Arid Regions

The high porosity of biochar and surface area make it an excellent amendment for improving soil water retention, particularly in arid and semi-arid regions (Brtnicky et al. 2021). By increasing the water-holding capacity of soils, biochar helps mitigate drought stress in plants, which is increasingly important in the face of climate change (Schmidt et al. 2014; El-Naggar et al. 2019). Key mechanisms for improved water retention include:

Soil Porosity Enhancement: Biochar particles create channels within the soil matrix, allowing for better water infiltration and reducing surface runoff (Jones et al. 2012; Palansooriya et al. 2019).

Water-Holding Capacity: Biochar can retain water within its porous structure, making it available for plant uptake during dry periods (Qian et al. 2023). This can improve drought resistance in crops and contribute to more stable yields (Schmidt et al. 2014; Dai et al. 2020). Studies have shown that crops grown in biochar-amended soils have a higher water-use efficiency:

Drought Tolerance: Plants like maize, wheat, and tomatoes grown with biochar under water-limited conditions show greater root development, improved water uptake, and enhanced growth compared to controls. These benefits are particularly noticeable in soils with low organic matter content (Schmidt et al. 2014; Palansooriya et al. 2019).

Reduced Irrigation Needs: In semi-arid regions, biochar reduces the need for frequent irrigation, offering a potential solution to water scarcity in agricultural systems (Yu et al. 2019; Qian et al. 2023). This is particularly beneficial for crops like wheat and barley, which are often subject to seasonal water stress (Jones et al. 2012).

6.6.4 Biochar as a Soil Amendment for Improving Soil pH and Plant Nutrient Uptake

Soil pH significantly influences nutrient availability for plants, and biochar can act as a liming agent, particularly in acidic soils (Qian et al. 2023). By raising the pH, biochar can enhance the availability of nutrients like P, Ca, and Mg, all of which are crucial for plant growth (El-Naggar et al. 2019; Brtnicky et al. 2021).

Influence of biochar on soil pH and nutrient uptake:

Neutralizing Soil Acidity: In acidic soils, biochar can neutralize excess hydrogen ions, reducing Al toxicity and improving root growth and nutrient uptake (Palansooriya et al. 2019). This is especially important for acid-sensitive crops like legumes and certain grains (Yu et al. 2019).

Enhanced Micronutrient Availability: Biochar can also help in making micronutrients like iron, Mn, and Zn more available in soils with suboptimal pH conditions (Agegnehu et al. 2017; Haider et al. 2022a). This improvement is critical for achieving balanced plant nutrition and maximizing crop yields (El-Naggar et al. 2019; Zhang et al. 2020; Haider et al. 2022a). Research has demonstrated the biochar potential to restore productivity in acidic soils (Yu et al. 2019; Adhikari et al. 2022). For example, studies in tropical regions with acidic soils have shown that applying biochar can increase crop productivity by 20–50%, particularly in crops such as coffee and cocoa, which are sensitive to soil pH (Schmidt et al. 2014; Dai et al. 2020).

6.6.5 Interaction of Biochar with Soil Microbial Communities and Plant Health

One of the most promising aspects of the biochar impact on plant growth is its interaction with soil microbial communities (Palansooriya et al. 2019; Haider et al. 2022a). Biochar creates a stable habitat for microorganisms, resulting in increased microbial biomass and diversity (Zhang et al. 2020a). These microorganisms, in turn, play vital roles in nutrient cycling and plant health (El-Naggar et al. 2019). Effects of biochar on soil microbiota and plant-microbe interactions:

N-Fixing Bacteria: Biochar can enhance the colonization of N-fixing bacteria in the rhizosphere, leading to increased N availability for leguminous crops (Zhang et al. 2020a; Haider et al. 2022a). This symbiotic relationship is crucial for crops like beans, peas, and alfalfa, which rely on biological N fixation for growth (Jones et al. 2012).

Mycorrhizal Fungi: Biochar can also promote the growth of mycorrhizal fungi, which enhance P uptake and improve plant stress tolerance. This is particularly beneficial for crops like wheat, maize, and sorghum, which form symbiotic relationships with these fungi to access otherwise unavailable soil P (Agegnehu et al. 2017).

Disease Suppression: The biochar ability to alter the soil microbiome can suppress certain soil-borne pathogens, reducing disease incidence in plants (Agegnehu et al. 2017). Studies have shown reductions in diseases like root rot and damping-off in biochar-amended soils, improving plant health and yield stability (Brtnicky et al. 2021; Adhikari et al. 2022).

6.6.6 Challenges and Limitations of Biochar Use in Agriculture

While biochar has shown great promise as a soil amendment, its effectiveness is highly variable and context-dependent (Yu et al. 2019; Adhikari et al. 2022). Several factors can influence the outcomes of biochar application, including the feedstock used, pyrolysis conditions, and soil type (Jones et al. 2012; Kavitha et al. 2018; Brtnicky et al. 2021). Furthermore, there are some challenges associated with its widespread adoption:

Cost and Accessibility: Producing high-quality biochar can be costly, and its availability may be limited in certain regions (Kavitha et al. 2018).

Potential for Negative Interactions: In some cases, biochar may immobilize nutrients, particularly if applied in excess, leading to reduced nutrient availability for plants (Kavitha et al. 2018; Adhikari et al. 2022). This can result in lower yields, particularly in soils already rich in organic matter (Schmidt et al. 2014; Agegnehu et al. 2017). Future research should address these challenges, focusing on optimizing biochar production, application methods, and understanding the long-term effects of biochar on soil health and plant growth (Kavitha et al. 2018; Adhikari et al. 2022).

6.7 CO-APPLICATION OF BIOCHAR AND MICROORGANISMS AS SOIL AMENDMENT

6.7.1 General Concept and Definitions

The co-application of biochar refers to the synergistic use of biochar in combination with other remediation techniques for contaminated soils, such as bioaugmentation, phytoremediation, and biostimulation, to optimize bioremediation processes (Wu et al. 2021; Yang et al. 2023). This technique is considered a key tool within hybrid remediation strategies, which combine various biotechnological approaches to achieve efficient contaminant removal and soil restoration (Kuppusamy et al. 2016; Haider et al. 2022b). Biochar, produced through the pyrolysis process of biomass, can act as a contaminant adsorbent while also serving as a support for microbial proliferation, enhancing both the stability and efficiency of the bioremediation system (Zhou et al. 2021; Dike et al. 2022).

The porous and chemically reactive nature of biochar allows it to adsorb and retain organic and inorganic contaminants in the soil, reducing their mobility and bioavailability (Dike et al. 2022; Zheng et al. 2022). However, the synergistic action of specialized microorganisms is essential for the complete biodegradation of persistent contaminants such as PAHs and heavy metals (Couto et al. 2010; Zhou et al. 2021; Yang et al. 2023). The co-application of biochar and microorganisms not only has the potential to improve the degradation of these contaminants but also to restore soil health by promoting an increase in microbial biomass, enhancing soil enzymatic activities, and fostering ecological resilience (Haider et al. 2022b; Yang et al. 2023).

6.7.2 Classification of Biochar Co-Application

There are different ways to classify biochar co-application based on the remediation methods used in conjunction and the implementation techniques (Ren et al. 2020; Dike et al. 2022). The main categories are described below:

6.7.2.1 Co-Application Based on Remediation Techniques

Bioaugmentation: This technique involves introducing exogenous microorganisms or specialized microbial consortia for the degradation of specific contaminants, such as hydrocarbons, heavy metals, or persistent organic compounds (Dike et al. 2022). Biochar acts as a matrix to improve the survival and activity of the introduced microorganisms, protecting them from adverse conditions in the soil, such as toxicity or nutrient competition (Wei et al. 2021).

Bio-Stimulation: Bio-stimulation involves the addition of nutrients or compounds that stimulate the growth and activity of indigenous contaminant-degrading microorganisms (Lü et al. 2016). Biochar acts in this process as a vehicle that enhances the availability of nutrients and water, and by adsorbing contaminants, reduces their toxicity to microorganisms (Dike et al. 2022). In studies using wood chip-derived biochar, the combined bio-stimulation and biochar application showed greater efficiency in hydrocarbon degradation than either technique alone (García-Delgado et al. 2015; Wu et al. 2021).

In another study, the addition of biochar and nutrients to soil contaminated with hydrocarbons favored the proliferation of indigenous PAH-degrading microorganisms, increasing the degradation rate by over 50% compared to using biochar alone (Ren et al. 2020).

Phytoremediation: Phytoremediation is a technique that uses plants to remove, degrade, or stabilize contaminants in the soil (Dike et al. 2022). Biochar can complement this process by improving soil structure, increasing water and nutrient retention, and reducing contaminant toxicity to plants through adsorption (Xu et al. 2018). Recent studies have shown that the co-application of biochar with phytoremediator plants such as *Medicago sativa* significantly enhances the stability of heavy metals in the soil, reducing their bioavailability and improving plant growth (Zhang et al. 2019; Ren et al. 2020; Jan et al. 2023).

Biochar can also enhance microbial activity in the rhizosphere, supporting the joint action of plants and microorganisms in contaminant remediation (Ren et al. 2020; Dike et al. 2022). This interaction is crucial, as root exudates from plants can facilitate the release of dissolved organic compounds (DOC) from biochar, which in turn can serve as a carbon source for degrading microorganisms (Tang et al. 2016; Haider et al. 2022b).

Combined or Hybrid Techniques: This approach involves combining two or more remediation techniques (bioaugmentation, biostimulation, or phytoremediation) with biochar application, generating a synergistic effect that maximizes contaminant degradation (Dike et al. 2022; Yang et al. 2023). For example, it has been demonstrated that the co-application of biochar, bioaugmentation, and phytoremediation significantly improves the removal of contaminants such as hydrocarbons and heavy metals from contaminated soils (Guo et al. 2021).

6.7.2.2 Classification Based on Co-Application Method

Separate Application: In this method, biochar and microorganisms (or plants) are introduced into the soil independently, allowing interactions to occur naturally within the soil environment (Dike et al. 2022). This allows biochar to adsorb contaminants, while microorganisms establish themselves within biochar pores, forming a biofilm that promotes microbial activity (Frankel et al. 2016; Yang et al. 2023).

Microbial Immobilization on Biochar: This approach involves the pre-adsorption or immobilization of microorganisms on the biochar surface before their introduction into the soil. Immobilization ensures that microorganisms are in close contact with the adsorbed contaminants, increasing the efficiency of persistent organic compound degradation (Wei et al. 2021). This method also protects microorganisms from unfavorable environmental conditions, such as dryness or contaminant toxicity.

Microbial immobilization on biochar is an emerging technique that aims to protect microbes from adverse soil conditions, such as contaminant toxicity, and ensure their proximity to target compounds (Wei et al. 2021). This process involves the adsorption of microorganisms into biochar pores, where they form biofilms that enhance their stability and degradation efficiency (Dike et al. 2022).

Immobilization has proven particularly effective in soils contaminated with hydrocarbons. In a study using Sphingomonas sp. immobilized on biochar, a 59% removal of PAHs was observed, compared to 21% in bioaugmentation treatments without immobilization (Song et al. 2021). Furthermore, immobilization can improve enzymatic activity and microbial respiration, leading to higher contaminant removal rates (Dike et al. 2022).

Co-Application with Enzyme Immobilization: Instead of using complete microbial cells, microbial enzymes can be immobilized on biochar, offering additional advantages such as faster contaminant degradation and the ability to operate in environments with low nutrient levels (Imam et al. 2021; Dike et al. 2022). Enzymes, such as laccases and peroxidases, can degrade complex contaminants more efficiently, overcoming bioavailability limitations that affect microbial cells (Saravanan et al. 2021).

6.7.3 Types of Biochar Co-Application

6.7.3.1 Co-Application of Biochar with Bioaugmentation

Bioaugmentation involves the addition of exogenous microorganisms with specialized capacities to degrade specific contaminants. When combined with biochar, the latter acts as a physical support that facilitates microbial colonization and provides a protected microhabitat for the microorganisms (Guo et al. 2021). Recent studies have shown that the co-application of biochar with bioaugmentation significantly enhances hydrocarbon removal in contaminated soils (Dike et al. 2022). For instance, the co-application of wheat straw-derived biochar with free-living microorganisms achieved hydrocarbon removal rates ranging from 35 to 89%, compared to bioaugmentation alone (24%) or biochar alone (8–41%) (Guo et al. 2021). Additionally, the use of biochar in bioaugmentation helps mitigate the negative effects of toxic byproduct accumulation during microbial degradation, allowing for more sustained microbial activity over time (Dike et al. 2022). The results of enzymatic activity, microbial biomass, and soil respiration also show that biochar provides a favorable environment for microbial growth and activity compared to treatments without biochar (Song et al. 2021).

6.7.3.2 Co-Application of Biochar with Phytoremediation

Phytoremediation involves the use of plants to extract or stabilize contaminants from the soil (Ren et al. 2020). The addition of biochar to soils undergoing phytoremediation increases the efficacy of the process by improving the physical properties of the soil, enhancing nutrient availability, and decreasing contaminant toxicity to plants (Zhang et al. 2019). Furthermore, biochar provides microsites where root-associated microorganisms can proliferate, enhancing microbial activity in the rhizosphere (Dike et al. 2022). One successful example of biochar co-application with phytoremediation is the use of *Medicago sativa* (alfalfa) in soils contaminated with heavy metals, where biochar not only helped the plant grow in a toxic environment but also increased the stability of the metals in the soil, reducing their bioavailability (Zhang et al. 2019; Jan et al. 2023). This type of co-application is particularly useful in the remediation of heavy metal-contaminated soils, which are more challenging to degrade or remove (Dike et al. 2022; Haider et al. 2022b).

6.7.3.3 Co-Application of Biochar with Bio-stimulation

Bio-stimulation involves the addition of nutrients, such as P and N, to the soil to stimulate the growth of indigenous microorganisms (Dike et al. 2022). Biochar acts as a slow-release nutrient source, improving water retention and soil structure, which in turn favors the development of contaminant-degrading microbial communities (Lü et al. 2016). Studies have shown that the co-application of biochar and nutrients can increase hydrocarbon degradation rates by over 50%, compared to treatments using only biochar or only bio-stimulation (Ren et al. 2020; Dike et al. 2022).

In soils contaminated with polycyclic aromatic hydrocarbons, bio-stimulation combined with biochar has shown promising results, where biochar facilitates microbial activity by adsorbing and concentrating contaminants, and nutrients promote microbial multiplication, accelerating the biodegradation process (Lü et al. 2016; Haider et al. 2022b).

6.7.4 Importance and Benefits of Biochar Co-Application

The co-application of biochar with microorganisms and remediation techniques offers several benefits both in terms of contaminated soil remediation efficiency and the restoration of soil ecological functions (Dike et al. 2022; Zheng et al. 2022).

Increased Remediation Efficiency: Combining biochar with bioaugmentation, bio-stimulation, or phytoremediation maximizes contaminant degradation by increasing microbial activity and providing a favorable environment for microorganism action (Guo et al. 2021).

Improved Soil Health: Biochar helps improve the physical and chemical properties of the soil, such as water retention, porosity, and nutrient availability, enabling the recovery of degraded soils.

6.8 CONCLUSIONS

The application of biochar in soil management represents a powerful intersection of material science, environmental sustainability, and agricultural innovation. Biochars, with their diverse properties derived from various feedstocks and pyrolysis conditions, offer a unique set of physicochemical characteristics that can be finely tuned to address specific soil challenges. The structural properties of biochar, including its high surface area and porosity, combined with its chemical properties, such as cation exchange capacity and pH modulation, position biochar as a versatile tool for enhancing soil health.

Biochar amendment has demonstrated its ability to significantly improve problematic soil properties across a range of soil types. Whether mitigating issues in sandy soils by enhancing water retention and reducing nutrient leaching, or improving the structure and aeration of clayey soils, biochar's role in optimizing soil conditions is clear. Additionally, its capacity to immobilize heavy metals and organic contaminants offers a practical solution for soil remediation, protecting plant health and productivity in contaminated environments.

The physicochemical improvements provided by biochar further reinforce its value in sustainable agriculture. By enhancing soil structure, increasing nutrient retention, and promoting a healthy microbial ecosystem, biochar contributes to the long-term soil fertility and resilience. Its ability to sequester carbon also adds a critical dimension to climate change mitigation efforts, positioning biochar not only as a tool for enhancing crop production but also as a key component in broader environmental stewardship strategies.

In summary, biochar emerges as a multifaceted soil amendment with the potential to transform agricultural practices. By leveraging its unique properties and the wide array of benefits it offers, biochar can play a crucial role in addressing the complex challenges of soil management, from improving soil health and productivity to mitigating environmental impacts and contributing to climate resilience. As research and application techniques continue to evolve, the integration of biochar into agricultural and land management practices is poised to become an increasingly important strategy for sustainable development.

Biochar offers significant potential to improve soil health and enhance plant growth, particularly in degraded or problematic soils. By improving nutrient retention, water-holding capacity, and microbial activity, biochar contributes to sustainable agricultural practices, reducing the need for chemical inputs while increasing crop productivity. However, its application must be tailored to specific soil and environmental conditions to maximize benefits.

Ongoing research and field trials are essential to fully understand the role of biochar in promoting sustainable crop production, particularly in the face of climate change and increasing pressures on global food systems.

The co-application of biochar with soil microorganisms and other remediation techniques such as bioaugmentation, phytoremediation, and biostimulation is a highly efficient strategy for tackling soil contamination and promoting ecological restoration. Biochar's ability to adsorb contaminants, improve soil structure, and serve as a shelter for microorganisms makes its combined use with bioremediation techniques essential for optimizing remediation processes.

NOMENCLATURE AND ACRONYMS

Al	:	Aluminum
As	:	Arsenic
B	:	Boron
Ca	:	Calcium
Cd	:	Cadmium
CEC	:	Cation exchange capacity
Cl	:	Chloride
CPs	:	Chlorophenols
Cr	:	Chromium
Cu	:	Copper
DCP	:	Dichlorophenols
DO	:	Oxygen availability
DOC	:	Dissolved organic compounds
EC	:	Electrical conductivity
ESP	:	Exchangeable sodium percentage
Fe	:	Iron
H	:	Hydrogen

Hg	:	Mercury
HV	:	Heavy metals
K	:	Potassium
MCP	:	Monochlorophenols
MeHg	:	Methylmercury
Mg	:	Magnesium
Mn	:	Manganese
Mo	:	Molybdenum
N	:	Nitrogen
Na	:	Sodium
Ni	:	Nickel
O_2	:	Oxygen
P	:	Phosphorus
PAH	:	Polycyclic aromatic hydrocarbons
Pb	:	Lead
TCP	:	Trichlorophenols
Zn	:	Zinc

REFERENCES

Adhikari, S., W. Timms and M.P. Mahmud. 2022. Optimising water holding capacity and hydrophobicity of biochar for soil amendment–A review. Science of the Total Environment. 851: 158043.

Agegnehu, G., A.K. Srivastava and M.I. Bird. 2017. The role of biochar and biochar-compost in improving soil quality and crop performance: A review. Applied Soil Ecology. 119: 156–170.

Alkharabsheh, H.M., M.F. Seleiman, M.L. Battaglia, A. Shami, R.S. Jalal, B.A. Alhammad, et al. 2021. Biochar and its broad impacts in soil quality and fertility, nutrient leaching and crop productivity: A review. Agronomy. 11(5): 993.

Angon, P.B., M.S. Islam, Shreejana, KC, A. Das, N. Anjum, A. Poudel, et al. 2024. Sources, effects and present perspectives of heavy metals contamination: Soil, plants and human food chain. Heliyon. 10 (7): e28357.

Baiamonte, G., G. Crescimanno, F. Parrino and C. De Pasquale. 2019. Effect of biochar on the physical and structural properties of a desert sandy soil. Catena. 175: 294–303.

Banik, C., M. Lawrinenko, S. Bakshi and D.A. Laird. 2018. Impact of pyrolysis temperature and feedstock on surface charge and functional group chemistry of biochars. Journal of Environmental Quality. 47(3): 452–461.

Brewer, C.E., V.J. Chuang, C.A. Masiello, H. Gonnermann, X. Gao, B. Dugan, et al. 2014. New Approaches to Measuring Biochar Density and Porosity. Biomass and Bioenergy. 66: 176–185.

Brtnicky, M., R. Datta, J. Holatko, L. Bielska, Z.M. Gusiatin, J. Kucerik, et al. 2021. A critical review of the possible adverse effects of biochar in the soil environment. Science of the Total Environment. 796: 148756.

Carvalho, M.L., M.T. de Moraes, C.E.P. Cerri and M.R. Cherubin. 2020. Biochar amendment enhances water retention in a tropical sandy soil. Agriculture. 10(3): 62.

Chen, Y. and P. Barak. 1982. Iron nutrition of plants in calcareous soils. Advances in Agronomy. 34: 217–240.

Collin, S., A. Baskar, D.M. Geevarghese, M.N.V.S. Ali, P. Bahubali, R. Choudhary, et al. 2022. Bioaccumulation of Lead (Pb) and its effects in plants: A review. Journal of Hazardous Materials Letters. 3: 100064.

Couto, M.N.P., E. Monteiro and M.T.S. Vasconcelos. 2010. Mesocosm trials of bioremediation of contaminated soil of a petroleum refinery: comparison of natural attenuation, biostimulation and bioaugmentation. Environmental Science and Pollution Research. 17: 1339–1346.

Cui, Z., Y. Wang, L. Du and Y. Yu. 2022. Contamination level, sources, and health risk of polycyclic aromatic hydrocarbons in suburban vegetable field soils of Changchun, Northeast China. Scientific Reports. 12(1): 11301.

Dai, Y., H. Zheng, Z. Jiang and B. Xing. 2020. Combined effects of biochar properties and soil conditions on plant growth: A meta-analysis. Science of the Total Environment. 713: 136635.

Dike, C.C., I.G. Hakeem, A. Rani, A. Surapaneni, L. Khudur, K. Shah, et al. 2022. The co-application of biochar with bioremediation for the removal of petroleum hydrocarbons from contaminated soil. Science of the Total Environment. 849: 157753.

El-Naggar, A., S.S. Lee, J. Rinklebe, M. Farooq, H. Song, A.K. Sarmah, et al. 2019. Biochar application to low fertility soils: A review of current status, and future prospects. Geoderma. 337: 536–554.

Frankel, M.L., T.I. Bhuiyan, A. Veksha, M.A. Demeter, D.B. Layzell, R.J Helleur, et al. 2016. Removal and biodegradation of naphthenic acids by biochar and attached environmental biofilms in the presence of co-contaminating metals. Bioresource Technology. 216: 352–361.

Gale, N.V. and S.C. Thomas. 2019. Dose-dependence of growth and ecophysiological responses of plants to biochar. Science of the Total Environment. 658: 1344–1354.

García-Delgado, C., I. Alfaro-Barta and E. Eymar. 2015. Combination of biochar amendment and mycoremediation for polycyclic aromatic hydrocarbons immobilization and biodegradation in creosote-contaminated soil. Journal of Hazardous Materials. 285: 259–266.

Godwin, R.J., D.R. White, E.T. Dickin, M. Kaczorowska-Dolowy, W.A.J. Millington, E.K. Pope, et al. 2022. The effects of traffic management systems on the yield and economics of crops grown in deep, shallow and zero tilled sandy loam soil over eight years. Soil and Tillage Research. 223: 105465.

Golia, E.E., P.S.C. Aslanidis, S.G. Papadimou, O.D. Kantzou, M.A. Chartodiplomenou, K. Lakiotis, et al. 2022. Assessment of remediation of soils, moderately contaminated by potentially toxic metals, using different forms of carbon (charcoal, biochar, activated carbon). Impacts on contamination, metals availability and soil indices. Sustainable Chemistry and Pharmacy. 28: 100724.

Graves, A.R., J. Morris, L.K. Deeks, R.J. Rickson, M.G. Kibblewhite, J.A. Harris, et al. 2015. The total costs of soil degradation in England and Wales. Ecological Economics. 119: 399–413.

Guo, J., S. Yang, Q. He, Y. Chen, F. Zheng, H. Zhou, et al. 2021. Improving benzo (a) pyrene biodegradation in soil with wheat straw-derived biochar amendment: performance, microbial quantity, CO_2 emission, and soil properties. Journal of Analytical and Applied Pyrolysis. 156: 105132.

Guo, X., H. Liu and J. Zhang. 2020. The role of biochar in organic waste composting and soil improvement: A review. Waste Management. 102: 884–899.

Gworek, B., W. Dmuchowski and A.H. Baczewska-Dąbrowska. 2020. Mercury in the terrestrial environment: A review. Environmental Sciences Europe. 32(1): 128.

Haider, F.U., J.A. Coulter, C.A.I. Liqun, S. Hussain, S.A. Cheema, W.U. Jun, et al. 2022a. An overview on biochar production, its implications, and mechanisms of biochar-induced amelioration of soil and plant characteristics. Pedosphere. 32(1): 107–130.

Haider, F.U., M. Farooq, M. Naveed, S.A. Cheema, N.U. Ain, M.A. Salim, et al. 2022b. Influence of biochar and microorganism co-application on stabilization of cadmium (Cd) and improved maize growth in Cd-contaminated soil. Frontiers in Plant Science. 13: 983830.

Imam, A., P.K. Kanaujia, A. Ray and S.K. Suman. 2021. Removal of petroleum contaminants through bioremediation with integrated concepts of resource recovery: A review. Indian Journal of Microbiology. 61: 250–261.

Ismayilov, A.I., A.I. Mamedov, H. Fujimaki, A. Tsunekawa and G.J. Levy. 2021. Soil salinity type effects on the relationship between the electrical conductivity and salt content for 1:5 soil-to-water extract. Sustainability. 13(6): 3395.

Jan, T., M. Arif, S. Anwar and D. Muhammad. 2023. Co-application of effective microbes with biochar and farmyard manure enhances maize growth and yield. Journal of Xi'an Shiyou University, Natural Science Edition. 19: 1.

Johnson, R., K. Vishwakarma, M.S. Hossen, V. Kumar, A.M. Shackira, J.T. Puthur, et al. 2022. Potassium in plants: Growth regulation, signaling, and environmental stress tolerance. Plant Physiology and Biochemistry. 172: 56–69.

Jones, D.L., J. Rousk, G. Edwards-Jones, T.H. DeLuca and D.V. Murphy. 2012. Biochar-mediated changes in soil quality and plant growth in a three year field trial. Soil Biology and Biochemistry. 45: 113–124.

Kannan, P., M. Paramasivan, S. Marimuthu, C. Swaminathan and J. Bose. 2021. Applying both biochar and phosphobacteria enhances *Vigna mungo* L. growth and yield in acid soils by increasing soil pH, moisture content, microbial growth and P availability. Agriculture, Ecosystems & Environment. 308: 107258.

Kavitha, B., P.V.L. Reddy, B. Kim, S.S. Lee, S.K. Pandey and K.H. Kim. 2018. Benefits and limitations of biochar amendment in agricultural soils: A review. Journal of Environmental Management. 227: 146–154.

Keyte, I.J., R.M. Harrison and G. Lammel. 2013. Chemical reactivity and long-range transport potential of polycyclic aromatic hydrocarbons – A review. Chemical Society Reviews. 42(24): 9333.

Kim, P., A. Johnson, C.W. Edmunds, M. Radosevich, F. Vogt, T.G. Rials, et al. 2011. Surface functionality and carbon structures in Lignocellulosic-Derived Biochars Produced by Fast Pyrolysis. Energy and Fuels. 25(10): 4693–4703.

Kroulík, M., F. Kumhála, J. Hůla and I. Honzík. 2009. The evaluation of agricultural machines field trafficking intensity for different soil tillage technologies. Soil and Tillage Research. 105(1): 171–175.

Kuppusamy, S., T. Palanisami, M. Megharaj, K. Venkateswarlu and R. Naidu. 2016. In-situ remediation approaches for the management of contaminated sites: A comprehensive overview. Reviews of Environmental Contamination and Toxicology. 236: 1–115.

Liu, P., C.J. Ptacek and D.W. Blowes. 2019. Release of nutrients and trace elements from Wood-, Agricultural Residue- and Manure-Based Biochars. International Journal of Environmental Research. 13(4): 747–758.

Lü, F., C. Luo, L. Shao and P. He. 2016. Biochar alleviates combined stress of ammonium and acids by firstly enriching Methanosaeta and then Methanosarcina. Water Research. 90: 34–43.

Lu, K., X. Yang, G. Gielen, N. Bolan, Y.S. Ok, N.K. Niazi, et al. 2017. Effect of bamboo and rice straw biochars on the mobility and redistribution of heavy metals (Cd, Cu, Pb and Zn) in contaminated soil. Journal of Environmental Management. 186: 285–292.

Lü, H., C.H. Mo, H.M. Zhao, L. Xiang, A. Katsoyiannis, L.W. Li, et al. 2018. Soil contamination and sources of phthalates and its health risk in China: A review. Environmental Research. 164: 417–429.

McDowell, R.W. and A.N. Sharpley. 2001. Approximating Phosphorus Release from Soils to Surface Runoff and Subsurface Drainage. Journal of Environmental Quality. 30(2): 508–520.

Naz, M., Z. Dai, S. Hussain, M. Tariq, S. Danish, I.U. Khan, et al. 2022. The soil pH and heavy metals revealed their impact on soil microbial community. Journal of Environmental Management. 321: 115770.

Ndede, E.O., S. Kurebito, O. Idowu, T. Tokunari and K. Jindo. 2022. The potential of biochar to enhance the water retention properties of sandy agricultural soils. Agronomy. 12(2): 1–12.

Nguyen, K.T., M.B. Ahmed, A. Mojiri, Y. Huang, J.L. Zhou and D. Li. 2021. Advances in as contamination and adsorption in soil for effective management. Journal of Environmental Management. 296: 113274.

Nguyen, B.T., G.D. Dinh, T.X Nguyen, D.T.P. Nguyen, T.N. Vu, H.T.T Tran, et al. 2022. The potential of biochar to ameliorate the major constraints of acidic and salt-affected soils. Journal of Soil Science and Plant Nutrition. 22(2): 1340–1350.

Oladoye, P.O., T.O. Ajiboye, E.O. Omotola and O.J. Oyewola. 2022. Methylene blue dye: Toxicity and potential elimination technology from wastewater. Results in Engineering. 16: 100678.

Pagalan Jr., E., M. Sebron, S. Gomez, S.J. Salva, R. Ampusta, A.J. Macarayo, et al. 2020. Activated carbon from spent coffee grounds as an adsorbent for treatment of water contaminated by aniline yellow dye. Industrial Crops and Products. 145: 111953.

Palansooriya, K.N., Y.S. Ok, Y.M. Awad, S.S. Lee, J.K. Sung, A. Koutsospyros, et al. 2019. Impacts of biochar application on upland agriculture: A review. Journal of Environmental Management. 234: 52–64.

Qian, S., X. Zhou, Y. Fu, B. Song, H. Yan, Z. Chen, et al. 2023. Biochar-compost as a new option for soil improvement: Application in various problem soils. Science of the Total Environment. 870: 162024.

Rahim, H.U., W.A. Akbar and J.M. Alatalo. 2022. A comprehensive literature review on cadmium (Cd) status in the soil environment and its immobilization by biochar-based materials. Agronomy. 12(4): 877.

Ren, H., B. Huang, V. Fernández-García, J. Miesel, L. Yan and C. Lv. 2020. Biochar and rhizobacteria amendments improve several soil properties and bacterial diversity. Microorganisms. 8(4): 502.

Richardson, A.E., J.M. Barea, A.M. McNeill and C. Prigent-Combaret. 2009. Acquisition of phosphorus and nitrogen in the rhizosphere and plant growth promotion by microorganisms. Plant and Soil. 321(1–2): 305–339.

Rodriguez Ortiz, L., E. Torres, D. Zalazar, H. Zhang, R. Rodriguez and G. Mazza. 2020. Influence of pyrolysis temperature and bio-waste composition on biochar characteristics. Renewable Energy. 155: 837–847.

Saghir, S., C. Pu, E. Fu, Y. Wang and Z. Xiao. 2022. Synthesis of high surface area porous biochar obtained from pistachio shells for the efficient adsorption of organic dyes from polluted water. Surfaces and Interfaces. 34: 102357.

Sánchez, E., R. Zabaleta, M.P. Fabani, R. Rodriguez and G. Mazza. 2022. Effects of the amendment with almond shell, bio-waste and almond shell-based biochar on the quality of saline-alkali soils. Journal of Environmental Management. 318: 115604.

Saravanan, A., P.S. Kumar, D.V.N. Vo, S. Jeevanantham, S. Karishma and P.R. Yaashikaa. 2021. A review on catalytic-enzyme degradation of toxic environmental pollutants: Microbial enzymes. Journal of Hazardous Materials. 419: 126451.

Schmidt, H.P., C. Kammann, C. Niggli, M.W. Evangelou, K.A. Mackie and S. Abiven. 2014. Biochar and biochar-compost as soil amendments to a vineyard soil: Influences on plant growth, nutrient uptake, plant health and grape quality. Agriculture, Ecosystems & Environment. 191: 117–123.

Shaheb, M.R., R. Venkatesh and S.A. Shearer. 2021. A review on the effect of soil compaction and its management for sustainable crop production. Journal of Biosystems Engineering. 46(4): 417–439.

Song, L., X. Niu, N. Zhang and T. Li. 2021. Effect of biochar-immobilized *Sphingomonas* sp. PJ2 on bioremediation of PAHs and bacterial community composition in saline soil. Chemosphere. 279: 130427.

Spokas, K.A. 2010. Review of the stability of biochar in soils: Predictability of O:C molar ratios. Carbon Managers. 1(2): 289–303.

Stavi, I., N. Thevs and S. Priori. 2021. Soil salinity and sodicity in drylands: A review of causes, effects, monitoring, and restoration measures. Frontiers in Environmental Science. 9: 712831.

Tang, J., X. Li, Y. Luo, G. Li and S. Khan. 2016. Spectroscopic characterization of dissolved organic matter derived from different biochars and their polycylic aromatic hydrocarbons (PAHs) binding affinity. Chemosphere. 152: 399–406.

Tang, X., G. Ran, J. Li, Z. Zhang and C. Xiang. 2021. Extremely efficient and rapidly adsorb methylene blue using porous adsorbent prepared from waste paper: Kinetics and equilibrium studies. Journal of Hazardous Materials. 402: 123579.

Thi, D.P., N.N.T. Hang, O.T. Viet, L.N. Van, A.N. Viet, P.D.T. Lan, et al. 2021. Sandy soil reclamation using biochar and clay-rich soil. Journal of Ecological Engineering. 22(6): 26–35.

Tomczyk, A., Z. Sokołowska and P. Boguta. 2020. Biochar physicochemical properties: pyrolysis temperature and feedstock kind effects. Reviews in Environmental Science and Bio/Technology. 19(1): 191–215.

Torres, E., L.A. Rodriguez-Ortiz, D. Zalazar, M. Echegaray, R. Rodriguez, H. Zhang, et al. 2020. 4-E (environmental, economic, energetic and exergetic) analysis of slow pyrolysis of lignocellulosic waste. Renewable Energy. 162: 296–307.

Uchimiya, M., D. Bannon, H. Nakanishi, M.B. McBride, M.A. Williams and T. Yoshihara. 2020. Chemical speciation, plant uptake, and toxicity of heavy metals in agricultural soils. Journal of Agricultural and Food Chemistry. 68(46): 12856–12869.

Wang, W., H. Wang, Y. Feng, L. Wang, X. Xiao, Y. Xi, et al. 2016. Consistent responses of the microbial community structure to organic farming along the middle and lower reaches of the Yangtze River. Scientific Reports. 6(1): 35046.

Wang, K., N. Peng, G. Lu and Z. Dang. 2020. Effects of pyrolysis temperature and holding time on physicochemical properties of swine-manure-derived biochar. Waste and Biomass Valorization. 11(2): 613–624.

Wang, L., J. Xin, H. Nai and X. Zheng. 2021. Effects of different fertilizer applications on nitrogen leaching losses and the response in soil microbial community structure. Environmental Technology & Innovation. 23: 101608.

Wang, H., Q. Nan, M. Waqas and W. Wu. 2022. Stability of biochar in mineral soils: Assessment methods, influencing factors and potential problems. Science of the Total Environment. 806: 150789.

Wei, X., P. Peng, Y. Meng, T. Li, Z. Fan, Q. Wang, et al. 2021. Degradation performance of petroleum-hydrocarbon-degrading bacteria and its application in remediation of oil contaminated soil. *In*: IOP Conference Series: Earth and Environmental Science. IOP Publishing, p. 012096.

Wołejko, E., A. Jabłońska-Trypuć, U. Wydro, A. Butarewicz and B. Łozowicka. 2020. Soil biological activity as an indicator of soil pollution with pesticides – A review. Applied Soil Ecology. 147: 103356.

Wu, Q., R. Lian, M. Bai, J. Bao, Y. Liu, S. Li, et al. 2021. Biochar co-application mitigated the stimulation of organic amendments on soil respiration by decreasing microbial activities in an infertile soil. Biology and Fertility of Soils. 57(6): 793–807.

Wu, C., D. Zhi, B. Yao, Y. Zhou, Y. Yang and Y. Zhou. 2022. Immobilization of microbes on biochar for water and soil remediation: A review. Environmental Research. 212(Part B): 113226.

Xiang, L., J.D. Harindintwali, F. Wang, M. Redmile-Gordon, S.X. Chang, Y. Fu, et al. 2022. Integrating biochar, bacteria, and plants for sustainable remediation of soils contaminated with organic pollutants. Environmental Science & Technology. 56(23): 16546–16566.

Xu, Y., B. Seshadri, B. Sarkar, H. Wang, C. Rumpel, D. Sparks, et al. 2018. Biochar modulates heavy metal toxicity and improves microbial carbon use efficiency in soil. Science of the Total Environment. 621: 148–159.

Yadav, S., S. Kumar and A.K. Haritash. 2023. A comprehensive review of chlorophenols: Fate, toxicology and its treatment. Journal of Environmental Management. 342: 118254.

Yang, C., J. Liu and S. Lu. 2021a. Pyrolysis temperature affects pore characteristics of rice straw and canola stalk biochars and biochar-amended soils. Geoderma. 397: 115097.

Yang, Q., Y. Wang and H. Zhong. 2021b. Remediation of mercury-contaminated soils and sediments using biochar: A critical review. Biochar. 3(1): 23–35.

Yang, T., N. Lupwayi, S.A. Marc, K.H.M. Siddique and L.D. Bainard. 2021c. Anthropogenic drivers of soil microbial communities and impacts on soil biological functions in agroecosystems. Global Ecology and Conservation. 27: e01521.

Yang, Y., K. Sun, L. Han, Y. Chen, J. Liu and B. Xing. 2022. Biochar stability and impact on soil organic carbon mineralization depend on biochar processing, aging and soil clay content. Soil Biology and Biochemistry. 169: 108657.

Yang, W., Y. Zhao, Y. Yang, M. Zhang, X. Mao, Y. Guo, et al. 2023. Co-application of biochar and microbial inoculants increases soil phosphorus and potassium fertility and improves soil health and tomato growth. Journal of Soils and Sediments. 23(2): 947–957.

Yu, H., W. Zou, J. Chen, H. Chen, Z. Yu, J. Huang, et al. 2019. Biochar amendment improves crop production in problem soils: A review. Journal of Environmental Management. 232: 8–21.

Zalazar-Garcia, D., A. Fernandez, L. Cavaliere, Y. Deng, J. Soria, R. Rodriguez, et al. 2022. Slow pyrolysis of pistachio-waste pellets: Combined phenomenological modeling with environmental, exergetic, and energetic analysis (3-E). Biomass Conversion and Biorefinery. 14: 9197–9215.

Zhang, B., L. Zhang and X. Zhang. 2019. Bioremediation of petroleum hydrocarbon-contaminated soil by petroleum-degrading bacteria immobilized on biochar. RSC Advances. 9(60): 35304–35311.

Zhang, Q., Y. Song, Z. Wu, X. Yan, A. Gunina, Y. Kuzyakov, et al. 2020a. Effects of six-year biochar amendment on soil aggregation, crop growth, and nitrogen and phosphorus use efficiencies in a rice-wheat rotation. Journal of Cleaner Production. 242: 118435.

Zhang, Y., J. Ding, H. Wang, L. Su and C. Zhao. 2020b. Biochar addition alleviate the negative effects of drought and salinity stress on soybean productivity and water use efficiency. BMC Plant Biology. 20(1): 1–11.

Zhang, J., J.E. Amonette and M. Flury. 2021. Effect of biochar and biochar particle size on plant-available water of sand, silt loam, and clay soil. Soil and Tillage Research. 212: 104992.

Zhang, G., J. Bai, Y. Zhai, J. Jia, Q. Zhao, W. Wang, et al. 2024. Microbial diversity and functions in saline soils: A review from a biogeochemical perspective. Journal of Advanced Research. 59: 129–140.

Zheng, S.J. 2010. Crop production on acidic soils: Overcoming aluminium toxicity and phosphorus deficiency. Annals of Botany. 106(1): 183–184.

Zheng, X., W. Xu, J. Dong, T. Yang, Z. Shangguan, J. Qu, et al. 2022. The effects of biochar and its applications in the microbial remediation of contaminated soil: A review. Journal of Hazardous Materials. 438: 129557.

Zhou, Y., S. Qin, S. Verma, T. Sar, S. Sarsaiya, B. Ravindran, et al. 2021. Production and beneficial impact of biochar for environmental application: A comprehensive review. Bioresource Technology. 337: 125451.

Chapter 7

Biochar as a Substrate in Soilless Crops

Daniela Zalazar-García*[a], Eliana Sanchez[a], Maria Paula Fabani[a, b], Ana Laura Navas[a], Angie Espinosa[c] and Diego Yepes Maya[d]

[a]Instituto de Ingeniería Química, Fac. de Ingeniería, Grupo Vinculado al PROBIEN (Universidad Nacional de San Juan, CONICET, Universidad Nacional del Comahue), San Juan, Argentina

[b]Instituto de Biotecnología, Facultad de Ingeniería, Universidad Nac. de San Juan, Argentina

[c]Mechanical Engineering Institute, Federal University of Itajuba, Brazil

[d]Excellence Group in Thermal Power and Distributed Generation (NEST), Institute of Mechanical Engineering, Federal University of Itajubá, MG, Brazil

7.1 INTRODUCTION

Soilless culture, defined as any method of growing plants without using soil as a rooting medium, is an intensive and efficient practice utilized in modern agriculture (Putra and Yuliando 2015). Although often considered a recent technique, it has historical roots dating back to the Egyptians who implemented similar methods several centuries B.C. (Raviv et al. 2019). Currently, this method is widely used for the production of vegetable and ornamental plants, especially in greenhouses and nurseries, where plant roots grow in growing media known as "substrates" (Barrett et al. 2016a).

Soilless cultivation has emerged as an innovative and sustainable alternative to traditional farming methods, offering a variety of benefits, including more

*Corresponding author: Daniela Zalazar-García (dzalazar@unsj.edu.ar)

DOI: 10.1201/9781032726694-7

efficient water use, reduced soil-borne diseases, and the ability to grow crops in areas unsuitable for conventional agriculture (Resh 2022a).

In soilless closed systems, minimal substrate is required, significantly mitigating the environmental impact of substrate disposal (Putra and Yuliando 2015). In this sense, the circular economy represents an economic model centered on the principles of sharing, leasing, reuse, repair, refurbishment, and recycling within a nearly closed-loop system. This approach aims to retain the highest possible utility and value of products, components, and materials at all times (Bourguignon 2016). The application of such a model to agribusinesses presents a potential paradigm shift, enabling a more balanced coexistence of ecological and economic systems. Specifically in agriculture, concerns about the sustainability of global food chains, especially in intensive production systems, have intensified (Kiss et al. 2019). Over the past two decades, this discussion has also extended to ornamental plant production, with growing interest from both the scientific community and policymakers in adopting circular economy principles.

Additionally, consumer preference for short supply chains and direct producer-to-consumer sales is rising, reflecting a broader environmental awareness, particularly among vegetable consumers (Verain et al. 2016). Moreover, the selection of growing media has been driven primarily by performance and cost considerations (Barrett et al. 2016a). Among the most commonly used organic materials in soilless systems are peat, coir, compost, biochar, and wood by-products (Raviv 2013). In this context, despite its widespread use, peat is facing increasing restrictions due to environmental concerns, leading to gradual replacement by recycled organic materials, especially in light of rising peat costs (Raviv 2013). Therefore, recent shifts in focus emphasize reducing the environmental footprint of plant production by selecting more eco-friendly substrates (Schmilewski 2014). Many of these substrates, chosen for their environmental benefits, are composed primarily of organic materials (Raviv et al. 2019), which are typically low-cost, widely available, renewable, and easier to dispose of sustainably (Raviv 2013). Coir, often used in combination with pumice or perlite, is one of the most popular substrates in professional vegetable production across the Mediterranean and for potted ornamental shrubs (Massa et al. 2020; Gruda 2019). This material exemplifies the recycling of organic waste; however, transportation distance may contribute to its environmental impact. Wood fibers offer a shorter recycling chain, producing substrates from local wood refuse. Biochar and composted materials, on the other hand, are particularly valuable for reducing chemical fertilizer use, as they contain significant levels of essential nutrients for plant growth (Lopez-Mondejar et al. 2010; Jones Jr. 2016). However, this approach also presents challenges and disadvantages, including the need for a significant initial investment, more complex technical management, and greater dependence on technology and energy (Savvas and Gruda 2018).

Within the broad spectrum of soilless cultivation, substrates play a crucial role. The selection of the appropriate substrate depends on several factors, including physicochemical and biological properties that directly affect plant growth and health. For example, the water-holding and aeration capacity of a substrate affects the availability of oxygen to roots, while its cation exchange capacity affects the availability of nutrients (Gruda 2009).

Substrate production requires careful consideration of these properties to optimize water retention, aeration, and nutrient availability (Barrett et al. 2016b). In addition, issues such as the sustainability and environmental impact of the materials used must be considered. For example, although rock wool is effective as a substrate, its production and disposal raise environmental concerns (Baudoin et al. 2013). In contrast, materials such as coconut are more sustainable but can vary in quality and availability (Carlile et al. 2019).

Soilless cultivation can also benefit from the use of hydroponic and aeroponic systems, where plants receive nutrients directly from aqueous solutions or sprays (Pardossi et al. 2011). These systems allow even greater control over the growing environment and can be implemented in a variety of configurations, from small domestic facilities to large commercial greenhouses (Kalossaka et al. 2021).

Biochar, a product of biomass pyrolysis, has gained interest as a potential substrate in soilless cropping systems due to its beneficial properties. Pyrolysis is a process of thermal decomposition of biomass in the absence of oxygen, resulting in the production of biochar, a carbon-rich material. This material improves the structure of the substrate, allowing for better aeration and drainage, while contributing to nutrient retention and pH balance (Lehmann and Joseph 2024). The porosity of biochar increases water and nutrient retention capacity, which can reduce the need for fertilizer and irrigation, making cultivation more efficient and sustainable (Atkinson et al. 2010).

In addition, the co-application of biochar with microorganisms can enhance the benefits of biochar by promoting a beneficial microbiota that improves the resilience and productivity of the cropping system. These microorganisms, including beneficial bacteria and fungi, can colonize the surface of biochar, creating a favorable environment for plant growth. The interaction between biochar and microorganisms can enhance nutrient availability, suppress soil pathogens, and increase plant tolerance to biotic and abiotic stresses (Elad et al. 2011).

The concept of microbially enriched biochar is presented as a promising innovation. This approach maximizes the beneficial effects of biochar and also integrates the role of microorganisms in promoting plant growth and suppressing pathogens. The ability of biochar to act as a habitat and support beneficial microorganisms enables the controlled release of nutrients and bioactive compounds, thereby improving overall plant health (Kammann et al. 2012). Studies have shown that the use of biochar enriched with microorganisms can increase crop yields and improve soil quality over the long term (Lehmann et al. 2011).

Thus, the combination of biochar and microorganisms can significantly transform the practice of soilless cultivation, providing a more efficient and sustainable growth medium. This approach can reduce reliance on chemical fertilizers and pesticides, thereby reducing the environmental impact of agriculture. In addition, the use of biochar as a substrate can contribute to climate change mitigation by sequestering carbon in the soil (Woolf et al. 2010).

This chapter will examine the various aspects of soilless cultivation, including its advantages and disadvantages, as well as the characteristics and production of substrates. The use of biochar and its interaction with microorganisms will be analyzed, highlighting its potential to revolutionize modern agriculture through case studies and practical examples of successful implementation in soilless systems. Figure 7.1 emphasizes the approach discussed in this chapter concerning functional relationships, highlighting the potential of various characteristics. It details improvements in habitat conditions for microorganisms, their applications, and the beneficial actions of microorganisms in soilless cultivation systems.

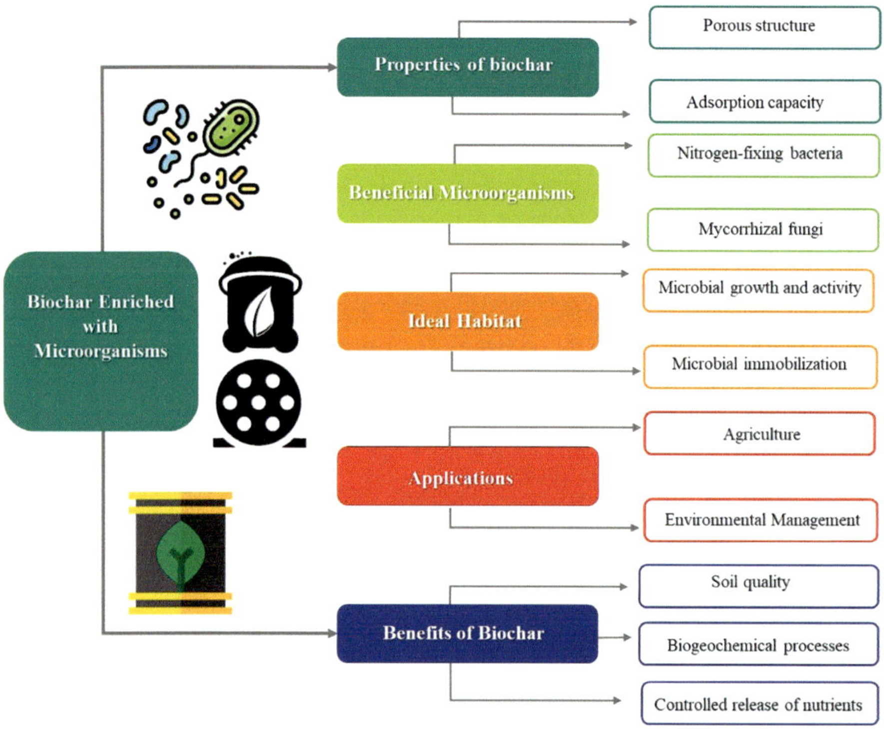

Figure 7.1 Functional relationships in soilless cultivation. Enhancing microbial habitat conditions and application.

7.2 SOILLESS CULTIVATION: ADVANTAGES AND DISADVANTAGES

Soil is an invaluable, non-renewable resource that supports countless ecosystems and provides essential food (Resh 2022b). In recent years, growing populations and increasing food demands have changed agricultural practices and accelerated land degradation, particularly in fertile regions, due to unsustainable agricultural practices (Jones Jr. 2016). Soil degradation can be chemical, physico-biological, and ecological. In semi-arid and arid areas, soils are already low in fertility, and intensive agriculture has further reduced their fertility. In response to these conditions, protected cropping systems have emerged, including soilless production, which has become competitive with open-field production (Savvas and Passam 2002).

Soilless cultivation systems (SCS) are increasingly being adopted in the modern greenhouse industry due to the high demand for water resources and food supply (Savvas and Gruda 2018). These systems include various innovative agricultural technologies that seek new materials to conserve resources, produce healthy plants, and contribute to sustainable practices. Soilless growing can be done using a variety of substrates such as peat or perlite ("soil growing") or simply water ("hydroponics").

Typical soilless growing systems include aeroponics, aquaponics, and hydroponics (Figure 7.2). In hydroponics, plants are grown in a nutrient-rich liquid solution or nutrient medium, considered the most robust and tolerant system (Savvas and Gruda 2018).

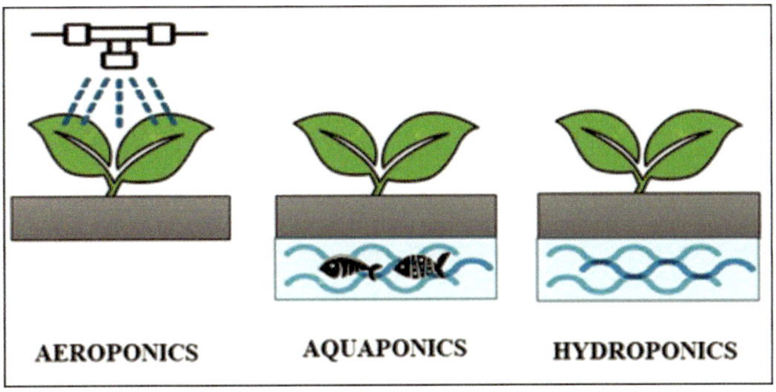

Figure 7.2 Soilless cultivation categories examples. Adapted from (Kalossaka et al. 2021).

Aquaponics integrates a hydroponic system with aquaculture, creating a closed-loop system in which aquatic animals provide nutrients for plants. Aeroponics, a newer method, exposes plant roots to the air and periodically mists them with a nutrient solution. Aquatic farming systems are primarily used to produce leafy vegetables such as lettuce (Savvas and Gruda 2018). Although they are currently being tested with various horticultural crops, floriculture, and even fruit production under greenhouse conditions (Di Lorenzo et al. 2013).

Despite its obvious advantages, soilless cultivation faces certain challenges and limitations. The following is an analysis of the main advantages and disadvantages of this innovative agricultural technique in comparison to traditional tillage methods.

One of the main advantages of soilless agriculture is its water-use efficiency, which uses approximately 5% of the water and a fraction of the land required to produce the same number of products as traditional agriculture. A study by Al Shrouf (2017) in the San Joaquin Valley, California, demonstrated that water use efficiency for tomato production is significantly higher with hydroponic irrigation, reaching 45 kg/m^3 in open hydroponic systems and up to 66 kg/m^3 in closed systems.

Table 7.1 presents consumption, yield, and productivity parameters for the various agricultural soilless systems described above. Comparisons between soilless and conventional agricultural systems indicate that the aeroponic system achieves the greatest efficiency in water productivity. Closed systems generally demonstrate superior water use efficiency compared to open systems, using between 5 and 20 times less water than traditional soilless agriculture. In terms of output, closed systems outperform open systems by almost 5%. In addition, open systems use between 15% and 17% more water than closed systems, which have lower evapotranspiration (Kalossaka et al. 2021).

Table 7.1 Consumption, yield, and productivity parameters for different soilless agricultural systems. Data adapted from (Al Shrouf 2017)

Parameters	**Hydroponics**				**Aeroponics**	**Aquaponics**
	Inert Substrate System		**Nutrient Solution System**			
	Open	**Closed**	**Open**	**Closed**		
% Irrigation water saving	80	85	85	90	95	80–85
% Fertilizer saving	55	80	68	85	85	85–99
% Productivity increase	100	150	200	250	300	100–150
% Water productivity	1,000	1,600	2,000	3,500	8,000	1,000–1,600

These systems allow precise control over growing conditions, such as temperature and root aeration, thereby optimizing the distribution of water and nutrients. Soilless cultivation also frees plants from soil-related problems such as salinization, soil-borne diseases, and the need for crop rotation (El-Kazzaz and El-Kazzaz 2017). However, these systems have disadvantages, such as the need for significant investment in specialized technology and equipment. In addition, management of these systems requires professional skills and advanced technical knowledge. Soilless systems also rely heavily on technology to control growing conditions, which can be a disadvantage in the event of technical failure (Al Shrouf 2017). Overall, soilless cultivation offers many advantages, particularly in terms of efficient use of water and land, high productivity, and better control of growing conditions. However, it also presents significant challenges, such as high initial investment and the need for advanced technical knowledge. As technologies develop and improve, these systems will continue to be a viable and sustainable alternative to traditional agriculture.

In the context of soilless agricultural systems, biochar stands out for its potential to enhance sustainability. Biochar's porous structure, high surface area, and ion exchange capacity make it suitable for use as a growing medium, although its chemical properties vary depending on the feedstock (Sohi et al. 2010; Evans et al. 2017; Atkinson et al. 2010). Rich in potassium and other essential nutrients, biochar can serve as both a substrate and a fertilizer (Grace et al. 2016; Massa et al. 2019). Given its thermo-sterilization during pyrolysis, biochar is pathogen-free, a significant advantage over other substrates like coir, which may carry microbial contaminants if not properly treated (Gruda 2019;

Zulfiqar et al. 2019). Thus, biochar exhibits a variety of characteristics that render it a valuable component in soilless systems. These properties play a crucial role in determining its effectiveness as a growth medium and its interaction with plants and microorganisms. The following sections will delve deeper into these characteristics and their implications for the use of biochar. About the advantages and disadvantages mentioned above, Figure 7.3 provides a visual representation of these aspects.

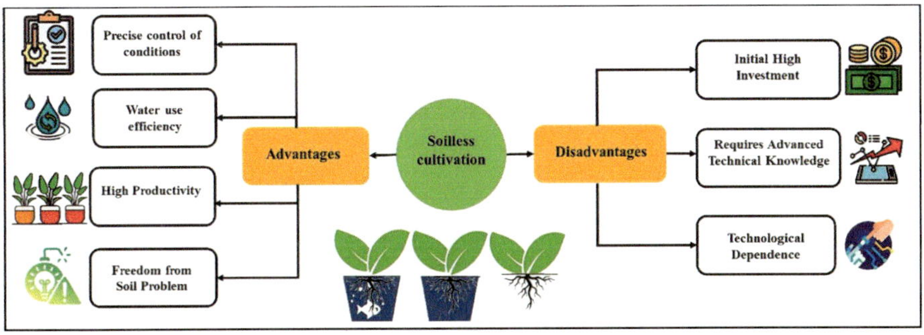

Figure 7.3 Overview of the advantages and disadvantages of soilless cultivation systems.

7.3 DIFFERENT SUBSTRATES USED IN SOILLESS CROPS

In soilless culture, substrate selection is crucial for plant growth and productivity. In general, soilless cultivation uses a mixture of different substrates (organic and inorganic) and additives, such as fertilizers, liming agents, biological control agents, or humectants. One way of classifying substrates is whether they are derived from inorganic or organic sources, each with specific characteristics and properties, and different applications (Bihari et al. 2023; Fussy and Papenbrock 2022).

7.3.1 Inorganic Growing Media

These are mainly derived from natural sources and generally do not involve industrial processes before use. Different substrates are listed below:

7.3.1.1 Rockwool

It is used as an insulation material in the construction sector and is currently used as a substrate for hydroponics, greenhouse fruit and vegetable production, due to its ability to retain water and air in a balanced manner. It is an inert inorganic material derived from volcanic rocks (such as basalt or dolomite) which, when melted at high temperatures (1,500–2,000°C), are transformed into wool-like fibers. These fibers are compressed into blocks, slabs, cubes, or blankets that are used as growing media. It is light (80–90 kg/m^3), easy to handle, inert, and does not contribute nutrients. It has excellent water and air-holding capacity, allowing the roots to receive sufficient oxygen while maintaining the necessary moisture. As it usually has an initial pH of 7–8, it needs to be rinsed or conditioned before use, after which it will not affect the pH of the medium. It also has a stable structure as it does not degrade or change over time. Its disadvantages include final disposal, which can be an environmental problem if not handled properly; it must be handled with care as it can be irritating to the skin and lungs (gloves and mask should be worn); and its high price compared to other substrates due to the cost of production.

Its use as a substrate is varied:

Hydroponics: This is one of the most commonly used substrates for growing vegetables such as tomatoes, peppers, and strawberries.

Propagation of Cuttings: Rock wool cubes are used for seed germination and rooting of cuttings.

Soilless Cultivation: Usually used in greenhouses to grow ornamental plants or vegetables, where it is necessary to control the nutrients supplied for their correct development.

In summary, rockwool is an ideal substrate for hydroponics and is very effective where precise control of water and nutrients is required. However, it is important to adjust the pH before use and dispose of it correctly to minimize its environmental impact (Fussy and Papenbrock 2022).

7.3.1.2 Perlite

It is a well-known inorganic substrate that is widely used in agriculture, gardening, and horticulture as a substrate or component of growing media. It is a siliceous volcanic mineral with the rare property of expanding greatly when heated to high temperatures (850–1,000°C). It is very light (125 kg/m^3), has a neutral pH (6.5–7.5), is sterile, and free of pests, diseases, and weeds. It is a growing medium that provides good aeration and drainage, as well as optimum moisture retention, which it releases gradually. It does not decompose, and its structure remains stable over time. Its disadvantages are that it does not provide nutrients and requires energy to produce, which makes it expensive in some countries.

Its use as a substrate is varied:

Hydroponics: It is used alone, as the main medium, or mixed with other substrates such as coconut fiber.

Soil: Combined with peat, vermiculite, or compost to improve aeration and drainage.

Rooting of Cuttings: Promotes rooting due to its ability to retain moisture without waterlogging.

Growing in Pots and Seed Beds: Enables healthy growth by preventing compaction of the substrate.

In short, perlite is an excellent substrate for crops that require an aerated and well-drained mix, especially in hydroponics, cuttings, and seedbeds. However, as it does not provide any nutrients, it is essential to combine it with other substrates or apply fertilizer to ensure proper plant development (Grillas et al. 2001).

7.3.1.3 Vermiculite

It is a naturally occurring mineral of the phyllosilicate family, found in sheet form and, when heated to high temperatures (> 1,000°C), expands considerably, forming small, light, spongy granules. It has a light volume weight (80–120 kg/m^3), nutrient retention capacity, due to its structure, which allows it to store and release nutrients, and good water holding ability, good pH buffering capacity, and good aeration. It is chemically and biologically inert, which makes it suitable as a substrate for use in plant cultivation. The disadvantage is that it compacts when it is too wet, and the cost of its production makes it expensive.

Its use as a substrate is varied:

Hydroponics: It acts as a substrate due to its ability to retain water and nutrients.

Soil: It is used as a soil amendment to improve aeration and moisture retention.

7.3.1.4 Sand Natural

It is a material composed mainly of mineral particles, the most common of which is quartz (silicon dioxide and SiO_2). However, this can vary depending on its origin and the minerals present in the area where it is mined. It has a particle size of 0.05–2.0 mm. Relatively cheap, also high-volume weight (1,400–1,600 kg/m^3). The sand has good drainage properties and low retention capacity, preventing waterlogging. Its structure is aerated, which facilitates air circulation and promotes oxygenation of the roots. Due to its granular texture, it provides a good surface for root growth. Its low density, neutral pH, and chemical inertness make it less susceptible to certain pests and soil diseases. It does not decompose over time. It is worth noting that it does not retain nutrients and should be mixed with other substrates that provide organic matter.

Organic Growing Media: These are used to grow plants sustainably, without the use of synthetic chemicals, and are therefore composed of natural raw materials that support plant growth and improve soil health (Fussy et. al. 2022). Different substrates are listed below:

A) Coconut Fibre: Derived from coconut husk, obtained by a natural anaerobic process. It is a lightweight and porous substrate. It has excellent water retention and aeration properties, a neutral pH, low microbial activity, and is biodegradable. The disadvantage of its use is that it is a finite resource, and the cost of production and its contraction can cause water repellence.

B) Rice Hull: A vegetable by-product of the rice milling industry. As a substrate, they are very useful for improving drainage, as they have a low weight. Slow breakdown is relatively resistant to decomposition

C) Compost: Decomposed organic matter that can be used in mixtures with other substrates. It provides nutrients and improves the health of the substrate.

***D) Vermicompost*:** This is a fertilizer produced by earthworms and is very rich in nutrients. It increases the fertility of the substrate and promotes a healthy root environment.

***E) Peat*:** This is an organic material from bogs and is the most widely used component of horticultural substrates, which can be used in substrate mixes. It has a high water and nutrient holding capacity, although its extraction from the site where it is produced can cause environmental problems (Durand et. al. 2021). Another disadvantage is that it becomes hydrophobic during drying, which degrades its water retention properties (Durand et al. 2021).

F) Bark: It is a waste by-product of the wood industry. It has a high air and water content, as well as a high total pore space, and the pH range is slightly acidic to neutral.

G) Biochar: Charcoal is used to improve nutrient and water-holding capacity. It has the property of increasing microbial activity and improving the structure of the substrate (Amery et al. 2021). Biochar contains nutrients from the raw material it comes from, which is why its chemical and physical composition is so varied.

Organic substrates are generally preferred for use in soilless cultivation because a) they promote sustainability by reducing the use of synthetic chemicals through the use of organic sources, thereby promoting more sustainable agricultural practices; b) they improve nutrient retention, which has a positive effect on plant growth; and c) they improve plant health by promoting a healthy environment for root development, resulting in stronger and more productive plants. It should be noted that some organic media may decompose over time and change the levels of nutrients available, requiring the addition of fresh substrate.

7.4 MICROORGANISM-ENRICHED BIOCHAR: APPLICATION IN COMBINATION

Biochar is outstanding for its physical properties and its ability to host beneficial microorganisms, including nitrogen-fixing bacteria and mycorrhizal fungi. In fact, due to its porous structure, an ideal habitat is provided that enhances the growth and activity of such organisms. Thus, it is a perfect matrix for microbial immobilization, finding applications in agriculture and environmental management.

Thanks to its high porosity and adsorption capacity, biochar provides a solid base for microorganisms, allowing controlled release. Moreover, biochar enhances soil quality by stimulating beneficial biogeochemical processes. It will also facilitate organic matter degradation of the remediation of contaminated soils and, at the same time, help improve plant growth (Bolan et al. 2023). Protection against pathogens strengthens health in agricultural ecosystems.

Integrating biochar with microorganisms improves crop productivity by reducing reliance on chemical fertilizers and pesticides, ensuring better food quality in agriculture.

The biochar impregnated with microorganisms acts as a way to help in the increased degradation of pollutants within the soil, acting as an effective bioremediation agent in the environment (Elhadidy 2019).

Different physical and chemical characteristics of biochar can support microbial immobilization and provide an optimal environment for the adhesion and growth of beneficial microorganisms (Wu et al. 2021). Essential properties include porosity, surface area, electrical charge, functional groups, and pH (Bolan et al. 2023).

7.4.1 Porosity

Porosity is one of the most crucial biochar characteristics, as it creates physical spaces for microbial attachments and proliferation (Hale et al. 2015). The more porous the biochar is, the greater the internal surface area is where microbes find shelter from adverse environmental conditions, such as drought or extreme temperature change.

7.4.1.1 Macropores, Mesopores, and Micropores

Biochar exhibits different pore types, which are classified based on their sizes. The pores above 50 nm are considered macropores; they are the pores that allow entry into nutrient and gas exchange, whereas mesopores in the range between 2–50 nm and micropores less than 2 nm serve as ideal refuges for microbial colonization. Micropores are crucial in protecting microorganisms from predators and extreme soil conditions (Mukherjee et al. 2022).

Figure 7.4 Schematic representation of a biochar showing the diversity of pore sizes.

In Figure 7.4, a schematic microscopic view displays the cross-section of a biochar cylinder, highlighting the various pore sizes. Different microorganisms, represented by colored, spherical, or textured shapes, inhabit the micropores and mesopores, illustrating how biochar provides refuge and an environment for microbial colonization.

7.4.1.2 Surface Area

The surface area is directly related to the biochar adsorption of microorganisms and nutrients. With increased surface area, there is more interaction with the microbial environment due to the greater number of contact points that microorganisms can adhere to (Leng et al. 2021). Generally, biochar produced at higher temperatures has a larger surface area due to the loss of volatile matter, forming a more carbonized and stable structure. That is the typical expectation for biochar derived from wood chips pyrolyzed at various temperatures (Tomczyk et al. 2020).

7.4.1.3 Advantages of Surface Area

Since biochar has a large surface area and is highly porous, it provides a better environment for microbial adhesion. It facilitates the development of symbiotic relationships between microorganisms and plants by improving nutrient and water retention capabilities. A large surface area is crucial for microbial adhesion, enhancing nutrients and water retention. This enhancement is vital for continued microbial activity, soil health, and plant growth (Bolan et al. 2023; Zhang et al. 2021).

Surface area aids in water retention in soil, maintaining moisture levels beneficial for microorganisms and plants (Xiang et al. 2022). This synergy is vital in sustainable agriculture, improving soil fertility and plant resilience.

7.4.1.4 Benefits of Surface Charge

The amount and nature of the electrical charge on the biochar surface are among the most critical factors in its interaction with microorganisms and their environmental applications. The surface charge of the biochar can be either positive or negative, depending on factors such as the surrounding pH, the nature of functional groups, and the pyrolysis temperature during production. Since most microorganisms have a negative charge on their cell walls, microbial cell immobilization is more effective when biochar has a positive surface charge. Electrostatic interactions between opposite charges promote adhesion, while electrostatic repulsion, which occurs when the biochar and microorganisms have the same charge, will impede immobilization. In this case, biochar has a negative surface charge, which might prevent the development of biofilms through its repelling effects on microbial attachment. This feature would be desirable in applications like microbial fuel cells (Zhao et al. 2023). Its surface allows it to serve as an effective vector for microbial inoculants, enhancing microbial life and activities in different environments (Bolan et al. 2023). This surface charge further improves nutrient retention capability in soil, a vital factor in encouraging soil fertility and plant development (Hossain et al. 2020).

In summary, biochar surface charge is crucial to its applications, influencing microbial interactions, nutrient dynamics, and environmental remediation efforts.

7.4.1.5 pH Biochar and Biochar Point of Zero Charge (pHPZC)

The pHPZC defines the polarity of the surface charge. It is the specific pH at which the biochar surface has no net charge, neither positive nor negative. When the surrounding pH is lower than the pHPZC, typically around 9.80, the surface charge of the biochar turns positive (Khandaker et al. 2018). This positive charge is highly effective because many microorganisms have negatively charged cell walls, which are attracted to biochar during microbial immobilization (Borba et al. 2019).

Therefore, interactions between biochar and microorganisms are crucial for applications such as microbial immobilization and the adsorption of contaminants or nutrients. Understanding surface charge dynamics determined by pH provides valuable insights into optimizing its use in agriculture and environmental contexts (Mukherjee et al. 2022).

If the condition's pH exceeds the pHPZC, the surface will have a negative charge, potentially reducing immobilization efficiency due to electrostatic repulsion between similarly charged microorganisms.

7.4.1.6 Pyrolysis Temperature

One of the most influential variables is the pyrolysis temperature during production, affecting crucial properties of biochar, such as its surface charge and the number of functional groups on its surface. At pyrolysis temperatures above 600°C, biochar loses many oxygenated functional groups, including carboxyl groups (Tomczyk et al. 2020), which reduces the sites available for microorganism interaction. Additionally, increased temperatures lead to biochar developing a more graphitic structure while its negative charge is reduced, potentially affecting microbial adhesion. This reduction in negative charge decreases the interaction sites available for microorganisms, thereby impacting microbial immobilization. That decreases the sites interacting with microorganisms.

Similarly, as temperatures increase, biochar becomes more graphitic, which reduces its negative charge and may affect microbial adhesion. Increasing temperatures during biochar production can enhance its graphitic structure while reducing its negative charge, which may decrease microbial adhesion to its surface. This reduction in negative charge decreases the interaction sites available for microorganisms, potentially affecting microbial immobilization (Khater et al. 2024). Higher pyrolysis temperatures also lead to biochar with greater surface area and microporosity, further influencing its interactions with microorganisms (Xiang et al. 2022). Developing a more graphitic structure at elevated temperatures modifies the surface properties of biochar, reducing the negative charge that typically inhibits microbial adhesion (Khater et al. 2024). Higher pyrolysis temperatures also result in biochar with a greater surface area and microporosity, further influencing its interactions with microorganisms (Xiang et al. 2022). Developing a more graphitic structure at elevated temperatures modifies the surface properties of biochar, reducing the negative charge that typically inhibits microbial adhesion (Khater et al. 2024).

7.4.1.7 Functional Groups

The hydroxyl (–OH), carboxyl (–COOH), and nitrogen (–NH_2) groups on the surface of biochar are crucial for microbial immobilization, as they influence both the surface charge of the biochar and its chemical bonding ability with microorganisms. These functional groups facilitate interactions with microbial cells, enhancing the biochar capacity for microbial adhesion and activity (Yaashikaa et al. 2020). In addition, they improve the chemical bonding between biochar and microorganisms, which is essential for effective microbial immobilization (Tan et al. 2021). Surface functional groups also contribute to its role as an efficient carrier for microbial inoculants, promoting microbial life and supporting its application in environmental remediation and agriculture (Bolan et al. 2023).

These groups regulate the adsorption capabilities of biochar and determine its suitability as a habitat for microorganisms, influencing its effectiveness in various applications.

7.4.1.8 Chemical Bond Formation

The carboxyl (–COOH) and hydroxyl (–OH) groups on the biochar's surface are crucial for forming chemical bonds with biopolymers in microbial cell walls, enhancing stability in microbial immobilization. These functional groups create hydrogen bonds and van der Waals forces, which are crucial for maintaining stability during the colonization of biochar by microorganisms (Mukherjee et al. 2022).

The bonds formed within this complex enhance structural integrity and durability, adding value to environmental remediation and soil management. This interaction significantly improves the performance of biochar in biotechnological and environmental applications by ensuring long-term stability and functionality.

7.4.1.9 Nutrient Content

Biochar produced at lower temperatures retains more nutrients like phosphorus, potassium, and nitrogen, which promote microbial growth and development. These nutrients also enhance biochar's properties in various agricultural applications and increase soil fertility (Kabir et al. 2023).

7.4.1.10 Modification Methods for Biochar

Biochar can be modified using a variety of techniques to optimize its properties for microbial immobilization.

Chemical Treatments

Acid or base treatments introduce new functional groups on the surface, enhancing the adsorption capacity for contaminants and microorganisms onto biochar. Such modifications are critical for altering surface charge and enhancing microbial interaction (Amalina et al. 2024).

— *Acid and Base Treatments:* These treatments introduce new functional groups on the surface of the biochar, significantly enhancing its ability to adsorb contaminants and microorganisms. Acid treatments, in particular, add acidic functional groups that increase adsorption capacity. These modifications play a crucial role in altering the surface charge of the biochar, improving its interactions with microorganisms. Base treatments have similar effects, fostering more efficient microbial colonization (Chen et al. 2024).

— *Oxidizing Agents:* The oxidizing agent's use, such as hydrogen peroxide, increases the presence of oxygenated functional groups like hydroxyls and carboxyls. These groups improve the chemical bonding between biochar, microorganisms, and contaminants, leading to stronger interactions and better immobilization (Huff and Lee 2016).

— *Metal Incorporation:* Adding metals such as iron or manganese to biochar promotes the formation of microbial biofilms and enhances the adsorption of specific contaminants, including heavy metals and essential nutrients. This modification is crucial for improving biochar's performance in applications that require microbial activity and contaminant removal (Hasan et al. 2024).

7.5. CONCLUSIONS

The use of biochar as a substrate in soilless cultivation systems provides a sustainable and efficient alternative for modern agriculture. Its physical properties, such as its high porosity, large surface area, and ion exchange capacity, make it suitable for promoting plant growth and hosting beneficial microorganisms. That not only improves water and nutrient retention but also reduces the reliance on chemical fertilizers. Furthermore, when combined with microorganisms, biochar enhances crop productivity, contributes to disease suppression, and increases resilience to environmental stresses. This synergy between biochar and microorganisms creates a more efficient, sustainable, and environmentally friendly agricultural system. Although biochar offers significant benefits, challenges remain, including variability in biochar properties depending on the feedstock and production methods. Further research is necessary to standardize its use and maximize its potential. However, the integration of biochar in soilless systems represents a promising innovation that could lead to more sustainable agricultural practices and contribute to mitigating environmental impacts.

REFERENCES

Al Shrouf, A. 2017. Hydroponics, aeroponic and aquaponic as compared with conventional farming. American Scientific Research Journal for Engineering, Technology, and Sciences (ASRJETS). 27: 247–255.

Amalina, F., S. Krishnan, A.W. Zularisam and M. Nasrullah. 2024. Pristine and modified biochar applications as multifunctional component towards sustainable future: Recent advances and new insights. Science of the Total Environment. 914: 169608.

Amery, F., J. Debode, S. Ommeslag, R. Visser, C. De Tender and B. Vandecasteele. 2021. Biochar for circular horticulture: Feedstock related effects in soilless cultivation. Agronomy. 11(4): 629.

Atkinson, C.J., J.D. Fitzgerald and N.A. Hipps. 2010. Potential mechanisms for achieving agricultural benefits from biochar application to temperate soils: A review. Plant Soil. 337: 1–18.

Barrett, G.E., P.D. Alexander, J.S. Robinson and N.C. Bragg. 2016a. Achieving environmentally sustainable growing media for soilless plant cultivation systems – A review. Scientia Horticulturae. 212: 220–234.

Baudoin, W., R. Nono-Womdim, N. Lutaladio, A. Hodder, N. Castilla, C. Leonardi, et al. 2013. Good Agricultural Practices for Greenhouse Vegetable Crops: Principles for Mediterranean Climate Areas. FAO.

Bihari, C., S. Ahamad, M. Kumar, A. Kumar, A.D. Kamboj, S. Singh, et al. 2023. Innovative soilless culture techniques for horticultural crops: A comprehensive review. International Journal of Environment and Climate Change. 13: 4071–4084.

Bolan, S., D. Hou, L. Wang, L. Hale, D. Egamberdieva, P. Tammeorg, et al. 2023. The potential of biochar as a microbial carrier for agricultural and environmental applications. Science of the Total Environment. 886: 163968.

Borba, L.L., R.M.F. Cuba, F.J.C. Terán, M.N. Castro and T.A. Mendes. 2019. Use of adsorbent biochar from pequi (Caryocar Brasiliense) husks for the removal of commercial formulation of glyphosate from aqueous media. Brazilian Archives of Biology and Technology. 62: e19180450.

Bourguignon, D. 2016. Closing the Loop: New Circular Economy Package. EPRS | European Parliamentary Research Service. https://www.europarl.europa.eu/%0ARegData/etudes/BRIE/2016/573899/EPRS_BRI(2016)573899_EN.pdf, accessed, 12 October 2024.

Carlile, W.R., M. Raviv and M. Prasad. 2019. Organic soilless media components. Soilless Culture. 303–378.

Chen, B., H. Zeng, F. Yang, Y. Yang, Z. Qiao, X. Zhao, et al. 2024. Functional biochar as sustainable precursors to boost the anaerobic digestion of waste activated sludge from a circular economy perspective: A review. Biochar. 6: 60.

Durand, S., B.E. Jackson, W.C. Fonteno and J.-C. Michel. 2021. The use of wood fiber for reducing risks of Hydrophobicity in Peat-Based Substrates. Agronomy. 11: 907.

Elad, Y., E. Cytryn, Y.M. Harel, B. Lew and E.R. Graber. 2011. The biochar effect: Plant resistance to biotic stresses. Phytopathologia Mediterranea. 50: 335–349.

Elhadidy, A. 2019. Performance of some new bioformulations against tomato fusarium wilt. Egyptian Journal of Desert Research. 69: 1–19.

El-Kazzaz, K.A. and A.A. El-Kazzaz. 2017. Soilless agriculture a new and advanced method for agriculture development: an introduction. Agricultural Research & Technology: Open Access Journal. 3: 63–72.

Evans, M.R., B.E. Jackson, M. Popp and S. Sadaka. 2017. Chemical properties of biochar materials manufactured from agricultural products common to the southeast United States. HortTechnology. 27: 16–23.

Fussy, A. and J. Papenbrock. 2022. An overview of soil and soilless cultivation techniques—Chances, challenges and the neglected question of sustainability. Plants. 11: 1153.

Grace, M., D. Esposito, M. Timmers, J. Xiong, G. Yousef, S. Komarnytsky, et al. 2016. Chemical composition, antioxidant and anti-inflammatory properties of pistachio hull extracts. Food Chemistry. 210: 85–95.

Grillas, S., M. Lucas, E. Bardopoulou, S. Sarafopoulos and M. Voulgari. 2001. Perlite based soilless culture systems: Current commercial applications and prospects. pp. 105–114. *In*: Acta Horticulturae. International Society for Horticultural Science (ISHS), Leuven, Belgium.

Gruda, N. 2009. Do soilless culture systems have an influence on product quality of vegetables? Journal of Applied Botany and Food Quality. 82: 141–147.

Gruda, N.S. 2019. Increasing sustainability of growing media constituents and stand-alone substrates in soilless culture systems. Agronomy. 9: 6.

Hale, L., M. Luth and D. Crowley. 2015. Biochar characteristics relate to its utility as an alternative soil inoculum carrier to peat and vermiculite. Soil Biology and Biochemistry. 81: 228–235.

Hasan, M., S. Chakma, X. Liang, S. Sutradhar, J. Kozinski and K. Kang. 2024. Engineered biochar for metal recycling and repurposed applications. Energies (Basel). 17: 4674.

Hossain, M.Z., M.M. Bahar, B. Sarkar, S.W. Donne, Y.S. Ok, K.N. Palansooriya, et al. 2020. Biochar and its importance on nutrient dynamics in soil and plant. Biochar. 2: 379–420.

Huff, M.D. and J.W. Lee. 2016. Biochar-surface oxygenation with hydrogen peroxide. Journal of Environmental Management. 165: 17–21.

Jones Jr., J.B. 2016. Hydroponics: A Practical Guide for the Soilless Grower. CRC Press.

Kabir, E., K.-H. Kim and E.E. Kwon. 2023. Biochar as a tool for the improvement of soil and environment. Frontiers in Environmental Science. 11: 1324533.

Kalossaka, L.M., G. Sena, L.M.C. Barter and C. Myant. 2021. Review: 3D printing hydrogels for the fabrication of soilless cultivation substrates. Applied Materials Today. 24: 101088.

Kammann, C., S. Ratering, C. Eckhard and C. Müller. 2012. Biochar and hydrochar effects on greenhouse gas (carbon dioxide, nitrous oxide, and methane) fluxes from soils. Journal of Environmental Quality. 41: 1052–1066.

Khandaker, S., Y. Toyohara, S. Kamida and T. Kuba. 2018. Adsorptive removal of cesium from aqueous solution using oxidized bamboo charcoal. Water Resources and Industry. 19: 35–46.

Khater, E.S., A. Bahnasawy, R. Hamouda, A. Sabahy, W. Abbas and O.M. Morsy. 2024. Biochar production under different pyrolysis temperatures with different types of agricultural wastes. Scientific Reports. 14: 2625.

Kiss, K., C. Ruszkai and K. Takács-György. 2019. Examination of short supply chains based on circular economy and sustainability aspects. Resources. 8: 161.

Lehmann, J., M.C. Rillig, J. Thies, C.A. Masiello, W.C. Hockaday and D. Crowley. 2011. Biochar effects on soil biota - A review. Soil Biology and Biochemistry. 43: 1812–1836.

Lehmann, J. and S. Joseph. 2024. Biochar for Environmental Management: Science, Technology and Implementation. Taylor & Francis.

Leng, L., Q. Xiong, L. Yang, H. Li, Y. Zhou, W. Zhang, et al. 2021. An overview on engineering the surface area and porosity of biochar. Science of the Total Environment. 763: 144204.

Lopez-Mondejar, R., A. Bernal-Vicente, M. Ros, F. Tittarelli, S. Canali, F. Intrigiolo, et al. 2010. Utilisation of citrus compost-based growing media amended with Trichoderma harzianum T-78 in Cucumis melo L. seedling production. Bioresource Technology. 101: 3718–3723.

Di Lorenzo, R., A. Pisciotta, P. Santamaria and V. Scariot. 2013. From soil to soil-less in horticulture: Quality and typicity. Italian Journal of Agronomy. 8: 255–260.

Massa, D., A. Bonetti, S. Cacini, C. Faraloni, D. Prisa, L. Tuccio, et al. 2019. Soilless tomato grown under nutritional stress increases green biomass but not yield or quality in presence of biochar as growing medium. Horticulture, Environment, and Biotechnology. 60: 871–881.

Massa, D., J.J. Magán, F.F. Montesano and N. Tzortzakis. 2020. Minimizing water and nutrient losses from soilless cropping in southern Europe. Agricultural Water Management. 241: 106395.

Mukherjee, S., B. Sarkar, V.K. Aralappanavar, R. Mukhopadhyay, B.B. Basak, P. Srivastava, et al. 2022. Biochar-microorganism interactions for organic pollutant remediation: Challenges and perspectives. Environmental Pollution. 308: 119609.

Pardossi, A., G. Carmassi, C. Diara, L. Incrocci, R. Maggini and D. Massa. 2011. Fertigation and Substrate Management in Closed Soilless Culture. Pisa: University of Pisa.

Putra, P.A. and H. Yuliando. 2015. Soilless culture system to support water use efficiency and product quality: A review. Agriculture and Agricultural Science Procedia. 3: 283–288.

Raviv, M. 2013. Composts in growing media: What's new and what's next? pp. 39–52. *In*: Acta Horticulturae. International Society for Horticultural Science (ISHS), Leuven, Belgium.

Raviv, M., J.H. Lieth and A. Bar-Tal. 2019. Soilless Culture: Theory and Practice: Theory and Practice. Elsevier Science.

Resh, H.M. 2022a. Hydroponic Food Production: A Definitive Guidebook for the Advanced Home Gardener and the Commercial Hydroponic Grower. CRC Press.

Savvas, D. and H. Passam. 2002. Hydroponic Production of Vegetables and Ornamentals. Embryo Publications Athens.

Savvas, D. and N. Gruda. 2018. Application of soilless culture technologies in the modern greenhouse industry - A review. European Journal of Horticultural Science. 83: 280–293.

Schmilewski, G. 2014. Producing growing media responsibly to help sustain horticulture. pp. 299–306. *In*: Acta Horticulturae. International Society for Horticultural Science (ISHS), Leuven, Belgium.

Sohi, S.P., E. Krull, E. Lopez-Capel and R. Bol. 2010. A review of biochar and its use and function in soil. Advances in Agronomy. 105: 47–82.

Tan, X.F., S.S. Zhu, R.P. Wang, Y. Di Chen, P.L. Show, F.F. Zhang, et al. 2021. Role of biochar surface characteristics in the adsorption of aromatic compounds: Pore structure and functional groups. Chinese Chemical Letters. 32: 2939–2946.

Tomczyk, A., Z. Sokołowska and P. Boguta. 2020. Biochar physicochemical properties: Pyrolysis temperature and feedstock kind effects. Reviews in Environmental Science and Bio/Technology. 19: 191–215.

Verain, M.C.D., S.J. Sijtsema and G. Antonides. 2016. Consumer segmentation based on food-category attribute importance: The relation with healthiness and sustainability perceptions. Food Quality and Preference. 48: 99–106.

Woolf, D., J.E. Amonette, F.A. Street-Perrott, J. Lehmann and S. Joseph. 2010. Sustainable biochar to mitigate global climate change. Nature Communications. 1: 56.

Wu, P., Z. Wang, A. Bhatnagar, P. Jeyakumar, H. Wang, Y. Wang, et al. 2021. Microorganisms-carbonaceous materials immobilized complexes: Synthesis, adaptability and environmental applications. Journal of Hazardous Materials. 416: 125915.

Xiang, L., J.D. Harindintwali, F. Wang, M. Redmile-Gordon, S.X. Chang, Y. Fu, et al. 2022. Integrating Biochar, Bacteria, and Plants for Sustainable Remediation of Soils Contaminated with Organic Pollutants. Environmental Science & Technology. 56: 16546–16566.

Yaashikaa, P.R., P.S. Kumar, S. Varjani and A. Saravanan. 2020. A critical review on the biochar production techniques, characterization, stability and applications for circular bioeconomy. Biotechnology Reports. 28.

Zhang, Y., Z. Zhang and Y. Chen. 2021. Biochar mitigates N_2O emission of microbial denitrification through modulating carbon metabolism and allocation of reducing power. Environmental Science & Technology. 55: 8068–8078.

Zhao, S., X. Wang, Q. Wang, T. Sumpradit, A. Khan, J. Zhou, et al. 2023. Application of biochar in microbial fuel cells: Characteristic performances, electron-transfer mechanism, and environmental and economic assessments. Ecotoxicology and Environmental Safety. 267: 115643.

Zulfiqar, F., S.E. Allaire, N.A. Akram, A. Méndez, A. Younis, A.M. Peerzada, et al. 2019. Challenges in organic component selection and biochar as an opportunity in potting substrates: a review. Journal of Plant Nutrition. 42: 1386–1401.

Chapter 8

Biochar as an Energy Vector

Anabel Fernandez*[a], Germán Mazza[b], Mathias Riveros-Gomez[a] and José Soria[b]

[a]Instituto de Ingeniería Química, Grupo Vinculado al PROBIEN (Universidad Nac. de San Juan, CONICET, Universidad Nac. Del Comahue), San Juan, Argentina

[b]Instituto de Investigación y Desarrollo en Ingeniería de Procesos, Biotecnología y Energías Alternativas, PROBIEN (CONICET-Universidad Nacional del Comahue), Neuquén, Argentina

8.1 INTRODUCTION

Biochar holds great potential for combustion and power generation, being an important renewable energy source with significant calorific value. It boasts advantages such as large reserves, low ash content, and carbon neutrality. Biomass combustion can significantly reduce the consumption of fossil fuels for electricity generation. Replacing fossil fuels like coal with biomass and developing coal-biomass co-firing technology is a low-cost and low-risk method for utilizing renewable energy. This approach can effectively address the shortage of fossil fuels and reduce emissions of traditional pollutants (such as SO_2 and NO_x) and greenhouse gases, thereby protecting the ecological environment (Sun et al. 2016).

The uses of biochar as fuel have certain benefits and limitations. The benefits highlight the reduction of greenhouse gas emissions and the use of waste to obtain biochar, contributing to a circular economy and reducing dependence on fossil resources. The main limitations of using biochar as fuel are its low calorific value compared to other solid fuels such as mineral coal and high production costs due to the pyrolysis process, which can require high quantities of energy, thereby

*Corresponding author: Anabel Fernandez (anafernandez@unsj.edu.ar)

DOI: 10.1201/9781032726694-8

increasing its production cost. Given the precedent, some of the challenges in using biochar as fuel are:

- Low Energy Density: Compared to fossil fuels, biochar has a lower energy density. This means more biochar is needed to produce the same amount of energy, potentially increasing transportation and storage costs.
- Ash Content: Biochar combustion produces ash, which needs to be managed properly.
- Competition With Other Uses: The demand for biochar as a soil amendment can compete with its use as a fuel.

While biochar holds promise as a renewable fuel, overcoming these challenges and conducting further research is essential to optimize its use and make it a commercially viable option.

8.2 BIOCHAR AS AN ENERGY CARRIER

8.2.1 Biochar Combustion

It is well known that biomass combustion has the potential to be CO_2-neutral. During the growth, biomass removes CO_2 from the atmosphere through photosynthesis, which is released again during combustion. On the other hand, biomass releases fewer gaseous pollutants, such as SO_2, NO_x, N_2O, and HCl, during the combustion process of biomass, due to its lower content of sulfur, nitrogen, and chlorine compared to fossil fuels. Once that biochar is produced, it can be burned directly for heat as shown in Figure 8.1 (Fernandez et al. 2024). It is important to determine the temperature at which the pyrolysis process would be carried out to obtain biochar. When pyrolysis is carried out in the range of 200–300°C (torrefaction), the combustion property of biomass biochar is not improved. However, higher temperatures favor the liquid and gaseous products, thereby decreasing the energy recovery in biochar (Chen et al. 2020c).

Biochar has domestic applications such as the use in stoves and ovens for cooking and heating homes. Its high carbon content means it is longer than traditional biomass like wood. Biochar also has industrial applications; it can be used in industrial furnaces and kilns, especially in industries like brickmaking or ceramics. Its high carbon content can contribute to higher temperatures.

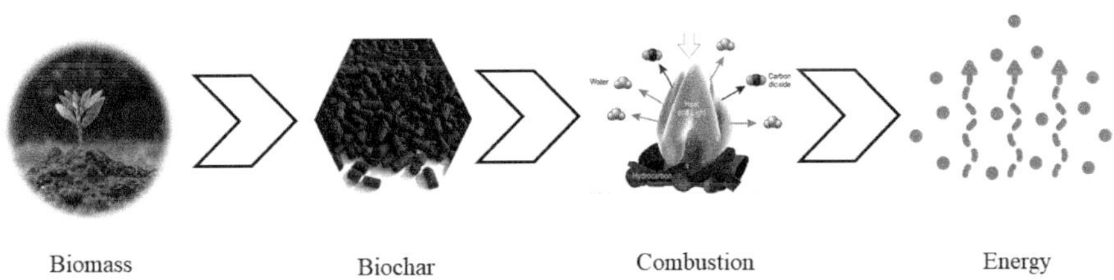

Figure 8.1 Biochar as an energy carrier.

8.2.2 Biochar Co-Firing

Co-firing involves burning two or more different fuels simultaneously in a combustion chamber. In the context of biochar, it typically involves mixing biochar with a traditional fossil fuel such as coal. This practice aims to reduce greenhouse gas emissions by utilizing biochar as a renewable energy source, reducing the overall carbon footprint of the plant while providing additional energy (Jeong et al. 2023).

In some cases, co-firing can create synergistic effects, where the combination of fuels results in better performance than when burning them separately. The high carbon content of biochar can accelerate the combustion process of other fuels, leading to higher temperatures and improved energy extraction. The addition of biochar can also help to reduce the amount of unburned carbon in the flue gas, improving combustion efficiency and reducing emissions. Co-firing with biochar can sometimes result in lower nitrogen oxide (NO_x) emissions due to changes in combustion conditions or the presence of nitrogen-containing compounds in the biochar. The physical properties of biochar can influence the formation and behavior of particulate matter, potentially leading to reductions in emissions. In some cases, biochar can act as a sulfur capture agent, reducing sulfur dioxide emissions (Jia et al. 2023; San Juan and Sy 2021).

The keys to a successful co-firing process are determining the optimal ratio of biochar to fossil fuel to balance environmental benefits and operational efficiency, as well as evaluating the overall environmental impact, including greenhouse gas emissions, air quality, and ash management.

8.2.3 Gasification

Gasification is a thermochemical conversion process that converts organic materials into a gaseous product called syngas. Syngas is a versatile fuel that can be used for various applications, including electricity generation, heat production, and chemical synthesis.

For syngas production, biochar acts as a solid fuel in gasifiers. When subjected to high temperatures and controlled amounts of oxygen or steam, biochar undergoes gasification to produce syngas, primarily composed of hydrogen, carbon monoxide, and carbon dioxide. The syngas composition depends on the gasification conditions and the properties of biochar. One of the most important challenges of gasification is the reduction of tar proportion. Tar is a complex mixture of organic compounds that can cause operational problems by clogging equipment, and the application of tar removal techniques is essential for successful biochar gasification. Some porous biochar can act as a catalyst for tar reduction, promoting the conversion of tar into gaseous products. The gasification of biochar can also contribute to carbon capture and storage by converting biomass-derived carbon into a gaseous form that can be further processed or sequestered (Pacioni et al. 2023; Yang et al. 2023).

The gasification process is significantly influenced by various operational parameters that directly impact the composition and properties of the resulting syngas, which must be considered for the gasifier design (Tian et al. 2021; Zaini et al. 2020):

- Temperature: Higher temperatures promote faster reaction rates, leading to increased production of H_2 and CO. However, excessive temperatures can lead to tar formation. Lower temperatures favor the formation of methane and higher hydrocarbons, thereby reducing syngas quality.
- Equivalence Ratio (ER): Higher ER (more oxygen) results in higher temperatures, increased CO_2 production, and lower H_2/CO ratio. Oppositely, lower ER (less oxygen) leads to lower temperatures, increased tar formation, and a higher H_2/CO ratio.
- Gasifying Agent: When steam is used in gasification, the H_2 production is increased due to water-gas shift production. O_2 provides heat for the process, but increases CO_2 production. Air is cheaper than O_2 but introduces nitrogen, diluting the syngas and reducing its calorific value.
- Pressure: Higher pressure shifts the equilibrium toward higher CO and H_2 concentrations, improving syngas quality. However, it also increases equipment and process costs. Lower pressure favors tar formation and lower syngas yield.
- Residence time: Longer residence time promotes complete conversion of feedstock and reduces tar formation. Nevertheless, shorter residence time increases tar formation and reduces syngas yield.
- Feedstock properties: The moisture content affects the energy balance and can influence gasification reaction; ash content can lead to catalyst poisoning and equipment fouling. It is important to make a proximate and ultimate analysis of feedstock to determine the potential syngas yield and composition.

8.2.4 Physicochemical Characteristics and Fuel Properties

The properties of biochar as a fuel are significantly influenced by the characteristics of the original biomass and the pyrolysis process used to produce it. These characteristics can impact the energy density of biochar, combustion efficiency, and emissions:

- Volatile Matter Content: Higher volatile matter results in higher ignition temperature and faster burning rates, leading to higher heat release rates. Lower volatile matter leads to slower combustion, lower heat release rates, and potentially high ash content.
- Fixed Carbon Content: A higher fixed carbon content indicates a higher energy content and slower burning rate, which can be beneficial for sustained heat release. A lower fixed carbon content results in a lower energy density and faster burning rates.
- Ash Content: Higher ash content can lead to equipment fouling, lower combustion efficiency, and potential slagging problems. Lower ash content improves combustion efficiency and reduces maintenance requirements.

- Particle Size and Density: Smaller particles increase the surface areas, promoting faster ignition and combustion.
- Porosity and Surface Area: Higher porosity and surface area can enhance oxygen diffusion, improving combustion efficiency.
- Oxygen and Hydrogen Composition: Higher oxygen content can increase the reactivity of biochar, leading to faster ignition and higher heat release. A higher hydrogen content can improve the quality of the gas produced if biochar is used in gasification processes.
- Mineral Composition: Certain minerals can act as catalysts, influencing combustion rates and emissions. For example, alkali and alkaline earth metals can promote ash-related problems.

Understanding these characteristics is essential for optimizing the use of biochar as a fuel. By carefully selecting the feedstock and optimizing the pyrolysis process, it is possible to produce biochar with specific properties tailored to the desired application.

8.3 BIOCHAR FOR ADVANCED ELECTROCHEMICAL/ ENERGY STORAGE APPLICATIONS

8.3.1 Heteroatom Doping in Biochar

Heat treatment of biowaste at high temperatures removes heteroatom functionalities derived from N, P, O, or S elements in the resulting biochar. These heteroatom functional groups provide the biochar surface with hydrophilicity, enhancing its wettability in aqueous solutions. To improve the biochar surface, heteroatom doping is performed, or the thermal treatment is adjusted to retain a certain number of heteroatoms. Heteroatom incorporation in biochar can be done indirectly or through self-doping. For indirect doping, external agents (inorganic or organic heteroatom sources) are introduced during activation or mixed with the feedstock before carbonization. In self-doping, the biochar retains its inherent heteroatoms.

Different types of biowaste are rich in N, P, or O heteroatoms, which are also present in their respective biochar. The oxygen content in biomass depends on the relative amounts of cellulose, hemicellulose, and lignin. Since carbohydrates have higher oxygen content than lignin, biomass with a higher holocellulose content will naturally have more oxygen. This oxygen can be released during carbonization as CO, CO_2, H_2O, or other volatile compounds containing oxygen functional groups. However, some oxygen can remain in the biochar as surface functional groups like phenolic moieties, carbonyl, carboxylic, ether, quinone, or lactone groups. Activating biochar in a steam and CO_2 atmosphere increases the O-containing functionalities on the biochar surface. Another method to increase oxygen content is through the ozonation of biochar. Kharel et al. (2019) reported that ozonation increased the oxygen content of biochar from 13.93 to 23.70%. This increase was mainly due to the development of carbonyl and hydroxy functional groups on the biochar surface.

Nitrogen self-doped biochar can be produced by optimizing process parameters using nitrogen-rich feedstocks. Gao et al. (2016) used human hair, rich in keratin, to prepare self-doped biochar. Controlled carbonization at 700°C resulted in biochar with 21.14% oxygen and 9.61% nitrogen. N doping with an external dopant was demonstrated using sawdust-derived biochar in a single-step synthesis, where urea phosphate acted as both a dopant and activator, yielding biochar with 4.24% nitrogen featuring pyridinic and pyrrolic N functionalities. Additionally, ammonium nitrate has been employed as an inorganic dopant to introduce nitrogen functionalities into biochar. It was found that increasing the carbonization temperature altered the distribution of N functionalities from pyrrolic to graphitic N in reed plant biochar. Organic precursors, such as melamine and urea, are commonly used for nitrogen doping in biochar. Mian et al. (2020) converted sewage sludge into N-doped biochar using melamine. By mixing nelamine with dried sewage sludge at a 2:1 ratio and carbonizing at 800°C, they achieved an N content of 9.5% on the biochar surface. Similarly, *in situ* pyrolysis of corn stalks with $NaHCO_3$ and urea at 700°C produced N-doped hierarchical porous carbon. The addition of urea developed micropores, resulting in a high nitrogen content of about 10%. Carbonizing corn straw in an ammonia atmosphere at 800°C resulted in the creation of microporosity and the incorporation of 8.81% nitrogen into the biochar surface.

P-doped biochar can be prepared using phosphorus precursors like phosphoric acid or phytic acid. S-doped biochar has been created from camellia flowers using a solvothermal pyrolysis technique in an H_2SO_4 solvent, resulting in the formation of S=O, $-SO_3H$, and SO_3^{2-} species on the biochar surface. Additionally, biochar can be doped with multiple heteroatoms. For example, N and P co-doped biochar was prepared by carbonizing phosphoric acid-doped biochar in an NH_3 atmosphere. In contrast, N and S co-doped biochar was produced by carbonizing biochar with thiourea.

8.3.2 Enhancing Graphitization in Biochar

The carbonization temperature has a significant influence on the degree of graphitization in biochar. High temperatures, exceeding 2,000°C, are required to produce a graphite-like carbon structure (Mian et al. 2020). As the carbonization temperature increases, the aromatic structural units evolve from amorphous to turbostratic, and ultimately, to graphitic carbon (Fan et al. 2012). Enhanced graphitization leads to the formation of ordered aromatic sp^2 domains, improving the electrical conductivity of the biochar (Tomczyk et al. 2020). Heat treatment of biochar at 950°C induces graphitic nanocrystal formation, substantially increasing its electrical conductivity (Gabhi et al. 2020). Additionally, the feedstock type impacts the electrical conductivity of the biochar. Figure 8.2 illustrates how increasing the carbonization temperature and using catalytic additives promote graphitization and enhance the electrical properties of biochar intended for use in Li-ion batteries.

Beyond thermal treatment, other methods can promote biochar graphitization. Transition metal catalysis lowers the required temperature by providing nucleation sites for the graphitization process. Coconut coir biochar carbonized at 500°C and subsequently heat-treated between 1,000°C and 1,300°C in the presence of nickel exhibited an ordered graphitic structure with interplanar spacing between 0.3369 nm and 0.3376 nm at 1,200°C. Similar Fe-catalyzed graphitization has been reported for biochar from sugarcane bagasse and chestnut shells. Controlled carbonization in the presence of ionic liquids, such as H_2PO_4, enhances graphitization and introduces heteroatom doping into the biochar.

However, it is important to note that many biochars contain randomly oriented graphitic domains with numerous structural defects and increased interlayer spacing (Gabhi et al. 2017). Even at temperatures exceeding 2,000°C, graphitization of such materials remains incomplete, with defective sites persisting alongside some ordered carbon domains. These are known as hard carbons and possess an amorphous structure. Despite their larger interlayer spacing compared to graphite and the presence of abundant structural defects, biowaste-derived hard carbons show significant promise for sodium-ion battery applications (Xie et al. 2020). Consequently, various techniques can be employed to modify the structural characteristics of the biochar and achieve the desired properties.

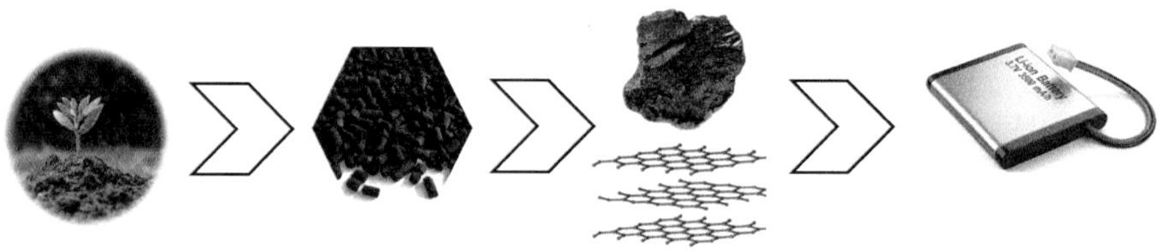

Figure 8.2 Graphitization in biochar for Li-ion battery.

8.3.3 Biochar-Based Material for Supercapacitor Application

Supercapacitors store energy through two main mechanisms: electric double-layer capacitance and pseudocapacitance. In electric double-layer capacitance, electrolyte ions electrostatically adsorb onto the electrode surface, forming an electric double layer when a potential is applied. Pseudocapacitance involves reversible Faradaic redox reactions that contribute to charge storage. While biochar with an enhanced porous structure generally exhibits better electric double-layer capacitance performance, a high specific surface area does not necessarily correlate with high specific capacitance. Instead, the electric double-layer capacitance behavior is influenced by the electrolyte-accessible surface area. Figure 8.3 illustrates the synergistic effects of pore structure and heteroatom doping on the electrochemical performance of biochar-based electrodes for supercapacitors. The type of pores in a material significantly impacts its charge storage capacity in supercapacitors. For instance, micropores with sizes matching those of electrolyte ions are essential for efficient ion adsorption onto active sites. Gadipelli et al. (2020) produced microporous biochar from pine cones with pore sizes ranging from less than 1 nm to a few nanometers. This material exhibited high specific capacitances of 320 and 230 F/g in aqueous and organic electrolytes, respectively (Thompson et al. 2021). Carbon materials with pores smaller than 1 nm can desolvate electrolyte ions upon adsorption, leading to exceptionally high specific capacitance.

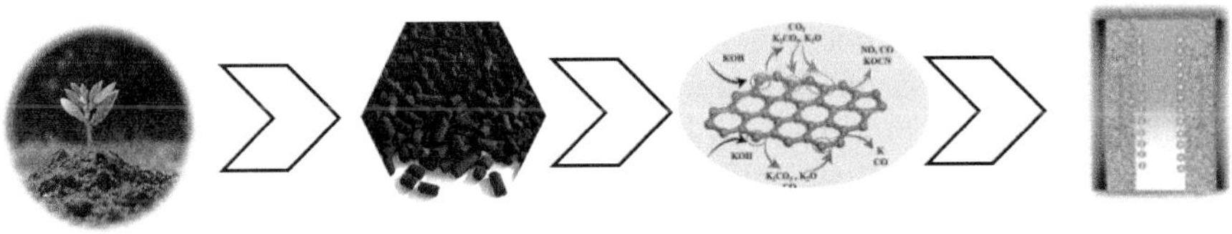

Figure 8.3 Biochar-based supercapacitor.

However, at higher current densities, ion diffusion through these constricted pores becomes hindered, reducing charge storage capacity. Mesoporous carbons overcome this micropore limitation. A multimodal porous structure is ideal for optimal supercapacitor performance. While micropores offer active sites for electrolyte adsorption, mesopores and macropores provide accessible diffusion pathways for electrolytes. This combination allows for high specific capacitance even at elevated current densities. Optimizing biochar porosity by KOH activation of peakon shells to achieve a micropore-to-mesopore volume ratio of 2.38 yielded a maximum specific capacitance of 447 F/g. In contrast, highly microporous carbons showed decreased specific capacitance retention at high current densities.

Incorporating heteroatoms into the porous biochar framework is another approach to enhance charge storage capacitance. The hydrophilic functional groups introduced by heteroatoms like nitrogen, phosphorus, oxygen, or sulfur improve the wettability of biochar-based electrodes. Additionally, these functional groups participate in reversible Faradaic redox reactions, contributing to charge storage through pseudocapacitance. The type of heteroatom in carbon influences charge storage behavior. Cai et al. (2022) found that N and O co-doped activated biochar exhibited higher specific capacitance than O-doped activated biocarbon. Similarly, N and S co-doped biochar prepared through controlled pyrolysis without additional activation steps achieved a specific capacitance of 201 F/g at 1 A/g, despite a modest specific surface area of 401 m^2/g due to its high nitrogen (8.2% wt) and sulfur (1.16% wt) content. Biochar derived from lignin and p-phenylenediamine exhibited a high nitrogen content of 14.3%, resulting in impressive gravimetric and volumetric capacitances of 480 F/g and 950 F/cm^3, respectively, at 1 A/g. Finally, a porous structure combined with a heteroatom-rich surface is crucial for successful biochar application as a supercapacitor electrode.

8.4 BIOCHAR-BASED MATERIAL FOR BATTERY ELECTRODES

Biochar has garnered considerable interest as a sustainable and economical material for battery electrodes. The unique characteristics of this material, such as its large surface area, porosity, and the existence of functional groups, make it a great option for use in lithium-ion and sodium-ion batteries. These properties improve the electrochemical efficiency of biochar and also respond to the increasing need for eco-friendly energy storage technologies (Baskar et al. 2023).

The synthesis and modification of biochar encompass a range of procedures that customize its characteristics to fulfill precise criteria for battery utilization. Pyrolysis and hydrothermal procedures are often used techniques to manufacture biochar with specific properties. In addition, the use of activation methods, such as physical and chemical activation, increases the surface area and porosity of biochar, hence improving its suitability as an electrode material (Lee et al. 2018). Moreover, recent progress in making composites from biochar and heteroatoms has shown encouraging enhancements in the electrochemical capabilities of biochar electrodes.

8.5 LI-ION BATTERY

Lithium-ion batteries are widely recognized for their high-energy charging capability and have emerged as one of the most successful electrochemical storage systems. A standard Lithium-ion battery consists of a carbon-based material serving as the anode and a lithium transition metal oxide working as the cathode. The theoretical charging capacity of graphitic carbon, which is frequently employed as the anode, is 372 mA/g (Lee et al. 2018). Nevertheless, it is necessary to develop new materials that can increase lithium-ion accumulation, as the total capacity of a lithium-ion cell is largely dependent on the lithium-ion storage capacity of the carbon structure. Although lithium metal anodes have a far larger capacity for storing energy, around 3,800 mAh/g, they have a limited lifespan because they react with both aprotic and protic solvents. This reaction causes electrochemical interactions with the electrolytes and can also create safety concerns. This highlights the necessity for more affordable and dynamic materials that can offer more capacity for ion accumulation while also ensuring stability.

Biochar-based materials are being recognized as potential candidates for anodes in lithium-ion batteries due to their structural features, such as a flexible pore structure, abundant surface functional groups, and diverse inorganic species. These characteristics are crucial for facilitating the spread of Li^+ ions and creating a large interface between the electrode and electrolyte, hence enhancing the electrochemical reactions within the battery (Liu et al. 2019). Moreover, biochar-based materials can be created using affordable and environmentally friendly techniques, which makes them a desirable substitute for traditional graphite anodes. In addition to having a greater specific capacity compared to conventional graphite, these materials also have a moderate surface area, which enhances safety and minimizes irreversible losses during the formation cycles.

The following sections provide a comprehensive analysis of the use of undoped biochar, heteroatom-doped biochar, biochar-based composites, and biochar-silicon composites as anode materials for lithium-ion batteries. The focus is placed on the specific properties, synthesis methods, and electrochemical performance of these biochar-based materials. This examination showcases the prospective contributions of these materials to the progress of next-generation lithium-ion batteries.

8.5.1 Undoped Biochar for Lithium-Ion Batteries

Undoped biochar has attracted considerable interest as a possible anode material for lithium-ion batteries because of its sustainable nature and intrinsic properties. For example, Niu et al. (2017) created porous carbons with a high number of defects and a prominent presence of mesopores by directly carbonizing cattle bones. The hydroxyapatites present in the cattle bones triggered self-activation, leading to the formation of carbons with a remarkably high specific surface area of 2,096 m^2/g, a substantial mesopore volume of 1.829 cm^3/g, and excellent electrical conductivity of 5,141 S/m. The improved carbon exhibited remarkable performance as an anode for lithium-ion batteries, with a substantial reversible capacity of 1,488 mA h/g after 250 cycles at 1 A/g and 661 mA h/g after 1,500 cycles at 10 A/g. Even when subjected to a current density of 30 A/g, it was able to maintain a capacity of 281 mA h/g, demonstrating its exceptional ability to store lithium.

Dos Reis et al. (2022) synthesized activated biochar from the bark of Norway spruce by employing $ZnCl_2$ and KOH for chemical activation. The $ZnCl_2$-activated biochar, referred to as Biochar-1, exhibited great characteristics such as 96.1% mesoporosity and a higher degree of graphitization, in contrast to the KOH-activated biochar, known as Biochar-2. Biochar-1 showed exceptional electrochemical performance as a lithium-ion battery anode. It achieved capacities of 370 mA h/g at a current density of 100 mA/g after 100 cycles, 332.4 mA h/g at 500 mA/g after 1,000 cycles, and 319 mA h/g at 1,000 mA/g after 5,000 cycles. Additionally, it maintained a Coulombic efficiency of 99.1%. The exceptional performance was ascribed to the high mesopore content, well-organized structure, and nitrogen functions of Biochar-1.

Dou et al. (2019) highlighted the effectiveness of porous carbons produced from jute fiber using $CuCl_2$ activation. The carbon materials exhibited a specific surface area of up to 2,043 m^2/g and achieved a specific capacity of 580.4 mA h/g at a rate of 0.2 C after undergoing 100 cycles. The high capacity was a result of the enlarged surface area of contact between the active electrode and electrolyte, higher transport of electrolyte, and improved diffusion of Li^+ ions facilitated by the macro-, meso-, and micropores. These studies jointly emphasize the promise of undoped biochar as a sustainable and high-performance material for anodes in lithium-ion batteries.

8.5.2 Heteroatom-Doped Materials for Lithium-Ion Batteries

Incorporating heteroatoms has proven to be a vital approach to improving the efficiency of biochar anodes in Lithium-ion batteries. The performance of batteries is considerably enhanced by including heteroatoms like nitrogen, oxygen, sulfur, and phosphorus into the carbon structure, which enhances the number of electrochemically active sites for lithium storage (Baskar et al. 2023). The selection of a dopant and the method of synthesis are crucial factors in influencing the electrochemical characteristics. Heteroatom doping is a versatile and efficient approach for enhancing the performance of biochar-based anodes in lithium-ion batteries.

Nitrogen doping stands out among the different heteroatoms due to its elevated electronegativity and its capacity to provide extra advantageous binding sites for lithium ions. Norouzi et al. (2019) highlighted the advantages of integrating nitrogen into carbon, as it promotes electronic conductivity, hence improving charge transfer and resulting in higher charge-discharge rates and durability in lithium-ion batteries. For instance, Zhang et al. (2016) utilized nitrogen-functionalized porous carbon produced from honey. This material facilitated extra lithium storage using chemisorption and enhanced the interlayer distance to allow for lithium intercalation. As a result, there was a significant reversible capacity of 1,359 mAh/g after 10 cycles at 100 mA/g, along with exceptional rate capability and cycling stability of 722 mAh/g after 200 cycles at 1 A/g. In their study, Wan and Hu (2019) devised a technique for synthesizing N-doped porous carbon (N/C) using bagasse biomass waste and melamine as a nitrogen precursor. Nitrogen doping increased the interaction between anchor sites and lithium ions, hence improving the adsorption and anchoring of lithium ions. After 100 cycles, the N/C anode demonstrated a remarkable reversible specific capacity of 530 mAh/g at 0.1 A/g.

Shrimp shells, egg yolk, and honeysuckle, which are high in nitrogen, have been utilized as sources of both carbon and nitrogen. When subjected to pyrolysis or hydrothermal carbonization, these materials transform into porous nitrogen-doped carbons (PNCs) that possess desirable characteristics, including expanded graphitic layers, graphitic flaws, edge sites, and a large surface area, all of which are advantageous for lithium-ion storage. For example, Chen et al. (2014) created hierarchically porous nitrogen-rich carbon (HPNC) using wheat straw. This material has many pores of different sizes, which increases the active surface area for storing Li-ion batteries. The presence of N atoms in the interconnecting micro-, meso-, and macropores enhances ion transit by establishing low-resistance ion channels. Additionally, this presence also increases the number of active sites available for Li storage.

Introducing oxygen into lithium-ion batteries can also improve their performance. Yu et al. (2016) synthesized oxygen-doped porous carbon using Enteromorpha, resulting in an initial reversible charge capacity of 1,709 mAh/g at a current density of 0.1 A/g. The combination of regulated hierarchical porosity and oxygen doping led to the formation of short ion diffusion routes and exceptional electrochemical performance.

In addition, the introduction of numerous heteroatoms, such as nitrogen, sulfur, and oxygen, through doping may result in synergistic effects, which further improve the electrochemical characteristics of biochar anodes. By including numerous dopants, a range of active sites is introduced, enhancing the rate at which particles diffuse and leads to a high capacity and consistent cycling performance. As an illustration, Gao et al. (2020) created a carbonized eggshell membrane

(CEM) anode for lithium-ion batteries that is both free-standing and binder-free. They utilized the naturally occurring nitrogen, oxygen, and sulfur in the eggshell membrane, which act as self-doped heteroatoms, to improve the performance of the anode. The carbon produced exhibited significant electrical conductivity, exceptional lithium-ion diffusivity, and considerable chemical stability.

8.5.3 Biochar-Based Composites for Lithium-Ion Battery

Transition metal oxides exhibit great theoretical capacities, but they experience limitations in terms of electronic conductivity and stability as a result of significant volume fluctuations during charge and discharge cycles (Baskar et al. 2023). The combination of biochar with transition metal oxides greatly improves the performance of lithium-ion battery anodes. This is achieved by enhancing electronic conductivity, ensuring structural stability, and increasing the availability of active sites for lithium storage. These composites exhibit high specific capacities, exceptional rate capabilities, and consistent cycle behavior, making them highly promising for advanced lithium-ion battery applications. For instance, electrodes based on MnO have demonstrated encouraging outcomes when paired with porous carbon. In their study, Chen et al. (2017) reported the successful development of a MnO/N-doped carbon composite featuring a three-dimensional porous structure. This composite demonstrated a remarkable reversible capacity of 513 mAh/g at a current density of 300 mA/g after undergoing 400 cycles. This configuration decreases the impedance of Li^+ migration, hence improving stability and efficiency. Similarly, Luo et al. (2020) created a core-shell MnO/C anode that successfully manages volume expansion, leading to a reliable electrode with exceptional electrochemical characteristics.

SnO_2 is a metal oxide that has a large theoretical capacity and operates at a low voltage, making it a good choice for lithium-ion batteries. The study conducted by Kwon et al. (2020) showed that SnO_2/porous carbon composites produced from pomelo peels exhibited a reversible specific capacity of 1,115 mAh/g at an applied current density of 0.2 A/g. The materials demonstrated excellent structural stability and maintained a capacity of 390 mAh/g even after undergoing 2,000 cycles at a current density of 5 A/g. Zhang et al. (2020c) employed wood flour to create nanoporous carbons that supported SnO_2 nanodots. This resulted in a remarkable reversible capacity of 1014.4 mAh/g at a current density of 156 mA/g. Integrating ZnO into biochar composites enhances the performance of the electrode. In their study, Li et al. (2018) found that the composites of ZnO/rice husk-based hollow carbonaceous nanospheres exhibited excellent cyclic stability and maintained a specific charge capacity of 920 mAh/g even after undergoing 100 cycles at a rate of 0.2 C.

Bimetallic oxides, such as $NiCo_2O_4$ and $ZnCo_2O_4$, have gained interest due to their improved electronic conductivity and decreased diffusion restrictions. Zhang et al. (2020a) demonstrated that the combination of $NiCo_2O_4$ with porous carbon derived from pomelo peels resulted in a specific capacity of 473.7 mAh/g after 210 cycles at a current density of 500 mA/g. Pan et al. (2020) fabricated

$ZnCo_2O_4$ nanocubes that included carbon produced from silk waste, resulting in a discharge capacity of 778 mAh/g after 100 cycles at a current density of 0.1 A/g.

8.5.4 Biochar-Silicon Composites for Lithium-Ion Battery

Adding silicon to biochar improves its effectiveness as an anode material for Li-ion batteries. Although silicon has a considerable theoretical potential for storing lithium, it experiences substantial variations in volume during the process of lithium absorption and release, which ultimately results in structural deterioration. The inclusion of silicon in a biochar matrix can alleviate these problems by offering a durable framework that is capable of handling fluctuations in volume (Baskar et al. 2023). Research has demonstrated that composites of biochar and silicon have enhanced stability throughout cycling and a greater ability to retain capacity when compared to anodes made solely of silicon. The porous structure of biochar offers abundant room for silicon expansion, while its conductivity enables efficient electron transport. The combination of biochar and silicon in composites shows great potential as materials for high-performance lithium-ion batteries (Dos Reis et al. 2023). In their study, Kwon et al. (2020) found that incorporating silicon nanoparticles into biochar microsized carbon balls resulted in a significant initial specific capacity of 1,800 mAh/g at a rate of 0.2 C. Furthermore, they observed that after 500 cycles, the capacity retention remained at 80%. This integration efficiently reduces fluctuations in silicon volume, hence preserving the structural integrity and improvingcycling stability.

Feng et al. (2020) synthesized SiO_2/N-doped carbon composites by utilizing rice husk and urea, resulting in enhanced electrical conductivity and long-term cycle stability. The BM-RH-SiO_2/NC anodes exhibit a reversible capacity of 780 mAh/g after 100 cycles at a current density of 0.1 A/g. The addition of a nitrogen-doped carbon coating minimized the irregularities in the structure throughout the process of alloying and dealloying reactions, resulting in improved electrochemical kinetics and overall performance. In addition, Li et al. (2021) devised a technique for isolating lignin-SiO_2 composites from Si-rich rice husks (RHs) by the process of alkali extraction and acid precipitation. Subsequently, the composite underwent carbonization, ball milling, magnesiothermic reduction, and the incorporation of additives to produce a Si/C composite with exceptional electrochemical characteristics. After 1,000 cycles, the Si/C anode material showed a stable cycling behavior with a high specific capacity retention of 572 mAh/g at 1 A/g.

8.6 NA-ION BATTERY

Sodium-ion batteries are being considered as a viable substitute for lithium-ion batteries, as sodium is more readily available and less expensive than lithium. Nevertheless, the substantial disparity in ionic radius between sodium (0.204 nm) and lithium (0.152 nm) poses considerable difficulties. Conventional electrode materials, which are designed for lithium ions, experience decreased capacity, slower ion diffusion, and poor cyclability due to the larger size of sodium ions. Graphite, commonly used as an anode material in LIBs, exhibits poor performance in sodium-ion batteries, with a capacity of less than 35 mAh/g (Baskar et al. 2023). This is because sodium ions are unable to efficiently insert themselves into the structure of graphite due to its narrower interlayer spacing. To address these difficulties, it is necessary to employ materials with interlayer distances larger than 0.37 nm.

One of the most promising candidates for sodium-ion battery anodes is biomass-derived porous carbon, particularly hard carbon. These materials have several benefits, including a high level of porosity and many active sites. These properties help reduce the increase in volume that occurs during charging and discharging cycles, allowing for high specific capacities. Extensive research and optimization have been conducted on hard carbon, which is obtained from biomass precursors, for sodium-ion storage. Hard carbons have a greater interlayer carbon spacing, which allows them to effectively accommodate larger sodium (Na) ions (Dos Reis et al. 2024). This leads to an improvement in the electrochemical performance of sodium-ion batteries.

The following sections provide a comprehensive analysis of the uses and progressions of materials based on biochar. Initially, undoped biochar will be analyzed, followed by an exploration of heteroatom-doped materials and biochar-based composites. This analysis will demonstrate how specific alterations to biochar can significantly improve the efficiency of sodium-ion batteries, therefore aiding the advancement of more sustainable and effective energy storage options.

8.6.1 Undoped Biochar for Sodium-Ion Battery

Undoped biochar exhibits considerable promise as an anode substance for sodium-ion batteries due to its elevated surface area, suitable pore structure, and cost-efficiency. Extensive research has been conducted on the capacity of hard carbon, a type of biochar, to incorporate sodium ions. Hard carbon is characterized by its non-graphitizable nature, indicating that it maintains a disordered structure even at elevated temperatures. This property results in greater spaces between layers, making it appropriate for sodium ion intercalation. This structure comprises a mixture of micro-, meso-, and macropores that enable the movement and storage of ions. Yu et al. (2020) produced hard carbon by carbonizing unburned charcoal at temperatures ranging from 1,100 to 1,400°C. They were able to achieve a specific capacity of 150 mAh/g at a current density of 0.1 A/g. Similarly, Zhu et al. (2018) documented the production of hard carbon from discarded apricot shells. This material displayed interconnected crosslinked structures and achieved a specific capacity of 400 mAh/g at a rate of 0.1 C after undergoing H_2 reduction to minimize defects and enhance the original Coulombic efficiency to 79%. In addition, Zhang et al. (2018) created a hard carbon using lotus stems. They enhanced its performance by subjecting it to high-temperature treatment and carefully controlling the arrangement of graphitic layers and pores. As a result, they achieved a reversible capacity of 350 mAh/g at a current density of 100 mA/g.

These examples illustrate how hard carbon derived from different biomass sources might enhance salt storage by optimizing its structural characteristics.

Recent progress in biochar production has been centered around improving the carbonization process and enhancing the structural characteristics of the resultant materials. In their study, Cong et al. (2020) found that subjecting corn stalks to hydrothermal treatment resulted in the production of hard carbon of exceptional quality. This hard carbon exhibited a reversible capacity of 270 mAh/g at a rate of 0.2 C. They attributed this high capacity to the large specific area of the materials and low concentration of inorganic impurities. In addition, the hierarchical porous carbon derived from rice husks demonstrated a reversible capacity of 274 mAh/g at a current density of 0.025 A/g. It maintained 92.7% of its capacity after undergoing 500 cycles, which can be attributed to its large specific surface area and low charge transfer resistance (Nie et al. 2020).

Molaiyan et al. (2023) analyzed the benefits of employing carbon materials obtained from biomass as an alternative to conventional graphite for anodes in sodium-ion batteries. The increased interlayer spacing in hard carbon derived from biomass enables more efficient sodium ion storage compared to graphite, which demonstrates subpar performance due to its inability to form stable intercalation compounds with sodium. Dos Reis et al. (2024) emphasized the significance of enhancing the porosity and surface area of hard carbons obtained from biomass for their use in sodium-ion battery applications. The study highlighted that the presence of hierarchical pore architectures, including micro-, meso-, and macropores, increases the accessibility of active sites for sodium ions. This leads to shorter distances for ion diffusion and improved kinetics for charge transfer. Optimizing the structure is essential for achieving a high storage capacity and maintaining steady cycling performance in sodium-ion batteries. These investigations highlight the adaptability and efficiency of undoped biochar as a sustainable and high-performing material for the anode in sodium-ion batteries. This opens up possibilities for future improvements in energy storage technology.

8.6.2 Heteroatom-Doped Materials for Sodium-Ion Batteries

Introducing heteroatoms into biochar anodes is a very successful method for improving the electrochemical performance of sodium-ion batteries. The addition of components such as nitrogen, oxygen, sulfur, and phosphorus to the carbon structure greatly increases the number of electrochemically active sites and enhances electronic conductivity, leading to improved sodium storage capacity.

The process of nitrogen doping has received extensive attention because of its capacity to generate favorable sites for the adsorption and storage of sodium ions. Romero-Cano et al. (2019) presented the production of nitrogen-doped porous carbon using grapefruit peels, yielding a reversible capacity of 180 mAh/g at a rate of 5 C. The addition of nitrogen functional groups improved the movement of sodium ions into the pores, increasing the overall storage capacity. Lu et al. (2019) created nitrogen-rich porous carbon aerogels with a large specific surface area. These aerogels were able to store and release energy with a capacity of 282 mAh/g when charged and discharged at a rate of 0.1 A/g. The activation process decreased the nitrogen concentration and produced surface defects, which improved the storage of sodium by adsorption.

Bartoli et al. (2024) examined the advantages of combining biomass with nitrogen-rich biopolymers like chitosan using co-pyrolysis to generate nitrogen-doped biochar. Dan et al. (2020) demonstrated this method by subjecting a combination of cellulose and chitosan to pyrolysis. This process yielded a hybrid material consisting of nitrogen-doped biochar and graphene oxide, which exhibited a capacity of 227 mAh/g at a current density of 5,000 mA/g. Oxygen doping is essential for enhancing the electrochemical performance of anodes in sodium-ion batteries. Chen et al. (2020b) synthesized wood-derived carbon material with a significant amount of C=O functional groups. This material exhibited a reversible capacity of 330 mAh/g at a current density of 40 mA/g. The controlled addition of oxygen functional sites not only expanded the space between layers but also created more locations for sodium storage, leading to rapid electrode kinetics and improved performance. Similarly, Zhang et al. (2020b) found that porous carbon with many micropores and oxygen functionalization exhibited a high specific capacity and excellent rate capability. This was attributed to the efficient electron transport aided by the graphitic carbon matrix.

Sulfur and phosphorus doping significantly improve sodium storage capacities. Zhao et al. (2017) successfully synthesized sulfur-doped activated porous carbon using a durian shell as the precursor material in their study. This carbon material exhibited a high reversible capacity of 345 mAh/g when tested at a current density of 0.1 A/g. In a study conducted by Zhu et al. (2019), it was found that adding phosphorus to carbon obtained from horse chestnut leaves increased the space between layers and created more active sites for sodium storage by adsorption. This led to a specific capacity of 310.4 mAh/g at a current density of 100 mA/g after 200 cycles.

8.6.3 Biochar-Based Composites for Sodium-Ion Battery

Biochar-based composites are being recognized as highly potential anode materials for sodium-ion batteries because they may overcome the drawbacks of pure transition metal oxides and sulfides, such as poor electronic conductivity and substantial volume fluctuations during cycling. MoS_2 and $Na_2Ti_3O_7$, which are transition metal oxides, have excellent theoretical capacities. However, they are hindered by inadequate conductivity and structural stability (Park et al. 2023). By incorporating carbon derived from biomass, these problems can be reduced, resulting in improved performance.

For example, the compounds MoS_2 and $MoSe_2$, which have high theoretical specific capacities, are frequently mixed with carbon obtained from biomass to enhance their electroconductivity and structural integrity. Recent research conducted by Xu et al. (2022) investigated the performance of a MoS_2/oxidized apple pomace carbon network (OAPC) in terms of charge capacity and cyclability.

The study found that the MoS_2/OAPC electrode exhibited a significantly greater charge capacity of 601.8 mAh/g at 200 mA/g after 50 cycles, compared to the original MoS_2 electrode. This indicates that the MoS_2/OAPC electrode has superior performance in terms of its ability to store and release electrical charge. Furthermore, a composite material consisting of $MoSe_2$, waste biomass-derived carbon, and carbon nanotubes showed remarkable long-term cycling retention of 81% after 250 cycles. This finding emphasizes the significant contribution of biomass carbon in improving electronic conductivity, as reported by Su et al. (2021). The addition of biochar also provides benefits to FeS_2. The FeS_2/carbon composites generated from Juncus demonstrated significant reversible capacities, achieving a specific capacity of 542.2 mAh/g at 0.1 A/g. This remarkable performance can be attributed to the reticulate structure of the carbon, which effectively mitigates volume expansion and enhances ion/electron kinetics, as reported by Chen et al. (2020a).

$Na_2Ti_3O_7$, a different anode material with potential, undergoes structural distortion and has low intrinsic electrical conductivity. Panda et al. (2021) enhanced the performance of the composite by using Tamarindus indica seed-derived biocarbon, leading to an increased initial reversible capacity of 232 mAh/g and improved cycling performance compared to the original $Na_2Ti_3O_7$ nanorods. In addition, the inclusion of Ni_3S_4 into porous carbon aerogels made from Carrageenan led to significant improvements in both reversible capacity (297 mAh/g after 100 cycles at 1 A/g) and structural stability. This enhancement can be attributed to the hierarchical structure of the aerogels, which enables efficient electron and ion transport (Guo et al. 2019). The combination of biomass carbon and metal sulfides or oxides improves the efficiency of sodium-ion batteries by increasing the number of active sites for Na^+ storage and reducing volume changes during cycling, resulting in enhanced electrochemical performance.

8.6.4 Biochar-Silicon Composites for Sodium-Ion Battery

Biochar-silicon composites have garnered significant attention as anode materials for sodium-ion batteries, as they have the potential to overcome the difficulties associated with pure silicon anodes, including the substantial variations in volume that occur during cycling. The significant volume expansion of silicon, which can reach up to 400%, limits its high theoretical capacity of 4,200 mAh/g. This expansion causes particle pulverization and results in rapid capacity fading. Compositing silicon with carbon materials provides a buffer to alleviate these volume changes and maintain structural integrity, thereby improving cycling stability and performance (Dos Reis et al. 2023).

The production of biochar-silicon composites encompasses a range of techniques, one of which is magnesiothermic reduction. For example, Sekar et al. (2020) utilized brown rice husk ashes to produce biomass carbon-silicon (C-Si) nanocomposites by this approach. The composites exhibited a discharge capacity of 378 mAh/g at a current density of 100 mA/g and a notable reversible capacity of 122 mAh/g at a current density of 0.2 A/g, accompanied by a Coulombic efficiency of 98% after 100 (Dos Reis et al. 2023). The performance of this system can be attributed to the successful containment of silicon nanoparticles within the highly conductive ultrathin C nanoflakes. This containment allows for the efficient movement of sodium ions in and out of the system.

In their study, Gong et al. (2022) showed that 2D silicon-embedded carbon sheets, produced by treating and carbonizing silicon, exhibited a high specific capacity, performed well at different rates, and maintained stable performance over multiple cycles of sodium-ion batteries. The anode demonstrated a specific capacity of 180 mAh/g at a current density of 100 mA/g, accompanied by a coulombic efficiency of 99.7% and a capacity retention rate of 94.6% after undergoing 1,000 cycles. The rapid reaction kinetics and high sodium ion diffusion coefficient of materials highlight their potential as a cost-effective and scalable electrode material for SIBs.

Using carbon sourced from biomass in Si/C composites will create opportunities for advancing sodium-ion battery applications and provide cost-effective, environmentally friendly ways of production. While advancement have been made in creating silicon/carbon anodes for sodium-ion batteries using biomass, further endeavors will be required in the future to exploit their capabilities and overcome current obstacles. Further investigation and refinement of these materials will be essential for the progress of SIB technology and to fulfill the increasing need for sustainable energy storage solutions (Dos Reis et al. 2023).

8.7 BIOCHAR FOR HYDROGEN STORAGE

Hydrogen is a prominent candidate for future energy solutions because of its exceptional energy density and positive impact on the environment. Hydrogen has a gravimetric energy output of 120 MJ/kg, which is almost three times greater than that of gasoline (44.5 MJ/kg). This makes hydrogen a clean energy source, as the sole byproduct of its burning is water (Liu et al. 2019). Nevertheless, the task of efficiently and securely storing hydrogen remains a major obstacle. Several storage strategies have been investigated, including chemical storage, physical adsorption, liquefaction, and gas compression. Carbon-based materials, specifically biochar, have become a focus of interest for physical adsorption due to their notable benefits, including high storage capacity, reversibility, and fast kinetics (Rawat et al. 2023).

Biochar shows great potential for hydrogen storage. In addition, activated biochar has a large surface area and a significant number of micropores, which are necessary for efficient hydrogen adsorption. The activated micropore structure of biochar improves hydrogen storage by utilizing capillary forces, which create more binding sites for hydrogen molecules. Several activation techniques, such as KOH, $ZnCl_2$, and steam activation, are used to increase the porosity and surface area of the material. The activation of KOH is particularly recognized for its efficacy in generating a large surface area and a well-developed micropore structure. Xiao et al. (2014) found that a biochar material made from melaleuca bark biomass and activated with KOH had a large surface area of 3,170 m^2/g and a micropore volume of 0.86 cm^3/g. As a result, it was able to store hydrogen at a capacity of 4.08 wt% at a temperature of 77 K and a pressure of 1 MPa. The effectiveness of this material can be related to its significant specific surface area and abundant micropore content, which enhance the efficiency of hydrogen adsorption. In a study conducted by Liang et al. (2023), the researchers examined the application of poplar sawdust biochar activated with KOH for storing hydrogen. They were able to achieve a specific surface area of 1,486.2 m^2/g and a significant micropore volume of 0.767 cm^3/g. The hydrogen storage capacity under normal pressure was measured to be 3.03 wt%, indicating excellent reversibility. In another study, Pedicini et al. (2020) examined the hydrogen storage potential of activated carbon obtained from Posidonia Oceanica and wood chips. The activation process using KOH yielded a surface area of 2,835 m^2/g and a hydrogen adsorption capability of 5 wt% at cryogenic temperatures.

In addition, the presence of hydroxyl and carboxyl groups on the surface of biochar enhances its hydrogen storage capacity through chemical adsorption mechanisms, hence improving its overall performance (Soni et al. 2024). Research has demonstrated that biochar treated with KOH and oxidized with ozone, resulting in the addition of more oxygen functions, displayed a 26% increase in hydrogen storage capacity compared to untreated biochar (Schaefer et al. 2020). Chen et al. (2019) produced activated biocarbon with heteroatom self-doping by synthesizing sword bean shell materials. This process resulted in a significant surface area of 2,838 m^2/g and a nitrogen concentration of 1.67 wt%. The study documented that at a temperature of 77 K, hydrogen storage capacities of 2.63 wt% and 5.74 wt% were observed at pressures of 1 bar and 40 bar, respectively. The hydrogen adsorption characteristics of biochar were greatly influenced by the presence of nitrogen and oxygen functional groups on its surface.

Biochar-based composites have demonstrated encouraging outcomes in the realm of hydrogen storage applications. Yeboah et al. (2020) showcased the utilization of palm kernel shell biochar as a milling aid in the production of Mg-biochar composites for hydrogen storage. By adding 20 wt% biochar to the ball milling process of magnesium, the conversion of Mg into MgH_2 increased to 93% and the decomposition temperature dropped by 40°C. This study emphasized the capacity of biochar to improve the hydrogen storage capabilities of metal hydrides when used as an addition.

Notwithstanding these progressions, there are still obstacles to be surmounted in the utilization of biochar for hydrogen storage. The ability of biochar to adsorb hydrogen at normal temperature is often restricted, generally being less than 1 wt%. To tackle this issue, scientists are investigating the use of metal-coated biochar materials, which can improve hydrogen storage by leveraging spillover effects. Metal nanoparticles, such as palladium and platinum, can enhance the separation of hydrogen molecules, causing them to transfer onto the biochar surface. This process enhances the ability of the biochar to adsorb hydrogen at normal temperatures. Although metal-coated biochar demonstrates encouraging outcomes, concerns such as increased sensitivity to air and moisture, as well as a diminished surface area resulting from metal loading, hinder its viability for practical use (Liu et al. 2019). Hence, it is crucial to conduct further investigation and refinement of techniques for activating biochar, modifying its surface, and integrating metal nanoparticles.

REFERENCES

Bartoli, M., A. Piovano, G.A. Elia, G. Meligrana, R. Pedraza, N. Pianta, et al. 2024. Pristine and engineered biochar as Na-ion batteries anode material: A comprehensive overview. Renewable and Sustainable Energy Reviews. 194: 114304.

Baskar, A.V., G. Singh, A.M. Ruban, J.M. Davidraj, R. Bahadur, P. Sooriyakumar, et al. 2023. Recent progress in synthesis and application of biomass-based hybrid electrodes for rechargeable batteries. Advanced Functional Materials. 33: 2208349.

Cai, X., Y. Xiao, W. Sun and F. Yang. 2022. Glucose-derived activated carbons for supercapacitors: comparison between single O doping and N/O co-doping. Electrochimica Acta. 406: 139861.

Chen, L., Y. Zhang, C. Lin, W. Yang, Y. Meng, Y. Guo, et al. 2014. Hierarchically porous nitrogen-rich carbon derived from wheat straw as an ultra-high-rate anode for lithium ion batteries. Journal of Materials Chemistry A. 2: 9684–9690.

Chen, L.F., S.X. Ma, S. Lu, Y. Feng, J. Zhang, S. Xin, et al. 2017. Biotemplated synthesis of three-dimensional porous MnO/C-N nanocomposites from renewable rapeseed pollen: An anode material for lithium-ion batteries. Nano Research. 10: 1–11.

Chen, T., Y. Zhou, L. Luo, X. Wu, Z. Li, M. Fan, et al. 2019. Preparation and characterization of heteroatom self-doped activated biocarbons as hydrogen storage and supercapacitor electrode materials. Electrochimica Acta. 325.

Chen, K., G. Li, Y. Wang, W. Chen and L. Mi. 2020a. High loading FeS_2 nanoparticles anchored on biomass-derived carbon tube as low cost and long cycle anode for sodium-ion batteries. Green Energy and Environment. 5: 50–58.

Chen, L., L. Bai, J. Yeo, T. Wei, W. Chen and Z. Fan. 2020b. Wood-derived carbon with selectively introduced C=O groups toward stable and high capacity anodes for sodium storage. ACS Applied Materials and Interfaces. 12: 27499–27507.

Chen, L., C. Wen, W. Wang, T. Liu, E. Liu, H. Liu et al. 2020c. Combustion behaviour of biochars thermally pretreated via torrefaction, slow pyrolysis, or hydrothermal carbonisation and co-fired with pulverised coal. Renewable Energy. 161: 867–877.

Cong, L., G. Tian, D. Luo, X. Ren and X. Xiang. 2020. Hydrothermally assisted transformation of corn stalk wastes into high-performance hard carbon anode for sodium-ion batteries. Journal of Electroanalytical Chemistry. 871: 114249.

Dan, R., W. Chen, Z. Xiao, P. Li, M. Liu, Z. Chen, et al. 2020. N-doped biomass carbon/reduced graphene oxide as a high-performance anode for sodium-ion batteries. Energy and Fuels. 34: 3923–3930.

Dos Reis, G.S., C. Mayandi Subramaniyam, A.D. Cárdenas, S.H. Larsson, M. Thyrel, U. Lassi, et al. 2022. Facile synthesis of sustainable activated biochars with different pore structures as efficient additive-carbon-free anodes for Lithium- and Sodium-Ion batteries. ACS Omega. 7: 42570–42581.

Dos Reis, G.S., P. Molaiyan, C.M. Subramaniyam, F. García-Alvarado, A. Paolella, H.P. de Oliveira, et al. 2023. Biomass-derived carbon–silicon composites (C@Si) as anodes for lithium-ion and sodium-ion batteries: A promising strategy towards long-term cycling stability: A mini review. Electrochemistry Communications. 153: 107536.

Dos Reis, G.S., S. Tuomikoski, D. Bergna, S. Larsson, M. Thyrel, H.P. de Oliveira, et al. 2024. Waste biomass conversion to energy storage material. *In*: Processing of Biomass Waste: Technological Upgradation and Advancement. 285–304.

Dou, Y., X. Liu, K. Yu, X. Wang, W. Liu, J. Liang, et al. 2019. Biomass porous carbon derived from jute fiber as anode materials for lithium-ion batteries. Diamond and Related Materials. 98: 107514.

Fan, C.L., H. He, K.H. Zhang and S.C. Han. 2012. Structural developments of artificial graphite scraps in further graphitization and its relationships with discharge capacity. Electrochimica Acta. 75: 311–315.

Feng, Y., L. Liu, X. Liu, Y. Teng, Y. Li, Y. Guo, et al. 2020. Enabling the ability of Li storage at high rate as anodes by utilizing natural rice husks-based hierarchically porous SiO_2/N-doped carbon composites. Electrochimica Acta. 359: 136933.

Fernandez, A., D. Zalazar-García, E. Torres, R. Torres-Sciancalepore, P. Parthasarathy, G. McKay, et al. 2024. Assessment of the use of biochar from the slow pyrolysis of walnut and almond shells as an energy carrier. Energy Sources, Part A: Recovery, Utilization, and Environmental Effects. 46: 15379–15394.

Gabhi, R., L. Basile, D.W. Kirk, M. Giorcelli, A. Tagliaferro and C.Q. Jia. 2020. Electrical conductivity of wood biochar monoliths and its dependence on pyrolysis temperature. Biochar. 2: 369–378.

Gabhi, R.S., D.W. Kirk and C.Q. Jia. 2017. Preliminary investigation of electrical conductivity of monolithic biochar. Carbon. 116: 435–442.

Gadipelli, S., C.A. Howard, J. Guo, N.T. Skipper, H. Zhang, P.R. Shearing, et al. 2020. Superior multifunctional activity of nanoporous carbons with widely tunable porosity: Enhanced storage capacities for carbon-dioxide, hydrogen, water, and electric charge. Advanced Energy Materials. 10: 1903649.

Gao, Y., S. Xu, Q. Yue, S. Ortaboy, B. Gao and Y. Sun. 2016. Synthesis and characterization of heteroatom-enriched biochar from keratin-based and algous-based wastes. Advanced Powder Technology. 27: 1280–1286.

Gao, S., Q. Jiang, Y. Shi, H. Kim, A. Busnaina, H.Y. Jung, et al. 2020. High-performance lithium battery driven by hybrid lithium storage mechanism in 3D architectured carbonized eggshell membrane anode. Carbon. 166: 26–35.

Gong, H., T. Du, L. Liu, L. Zhou, Y. Wang, H. Jia, et al. 2022. Self-source silicon embedded in 2D biomass-based carbon sheet as anode material for sodium ion battery. Applied Surface Science. 586: 152759.

Guo, R., D. Li, C. Lv, Y. Wang, H. Zhang, Y. Xia, et al. 2019. Porous Ni_3S_4/C aerogels derived from carrageenan-Ni hydrogels for high-performance sodium-ion batteries anode. Electrochimica Acta. 299: 72–79.

Jeong, Y., J.S. Kim, Y.E. Lee, D.C. Shin, K.H. Ahn, J. Jung, et al. 2023. Investigation and optimization of co-combustion efficiency of food waste biochar and coal. Sustainability. 15: 14596.

Jia, W., Y. Guo, F. Guo, H. Li, Y. Li, Y. Zhang, et al. 2023. Co-combustion of carbon-rich fraction from coal gasification fine slag and biochar: Gas emission, ash sintering, heavy metals evolutions and environmental risk evaluation. Chemical Engineering Journal. 471: 144312.

Kharel, G., O. Sacko, X. Feng, J.R. Morris, C.L. Phillips, K. Trippe, et al. 2019. Biochar surface oxygenation by ozonization for super high cation exchange capacity. ACS Sustainable Chemistry & Engineering. 7: 16410–16418.

Kwon, H.J., J.Y. Hwang, H.J. Shin, M.G. Jeong, K.Y. Chung, Y.K. Sun, et al. 2020. Nano/microstructured silicon-carbon hybrid composite particles fabricated with corn starch biowaste as anode materials for Li-ion batteries. Nano Letters. 20: 625–635.

Lee, H.W., Y.M. Kim, S. Kim, C. Ryu, S.H. Park and Y.K. Park. 2018. Review of the use of activated biochar for energy and environmental applications. Carbon Letters. 26: 1–10.

Li, Y., Y. Gao, H. Qi, K. Yu and C. Liang. 2018. A ZnO/rice husk-based hollow carbonaceous nanosphere composite as an anode for high-performance lithium-ion batteries. RSC Advances. 8: 33019–33024.

Li, Y., L. Liu, X. Liu, Y. Feng, B. Xue, L. Yu, et al. 2021. Extracting lignin-SiO_2 composites from Si-rich biomass to prepare Si/C anode materials for lithium ions batteries. Materials Chemistry and Physics. 262: 124331.

Liang, Y., Y. Wang, N. Ding, L. Liang, S. Zhao, D. Yin, et al. 2023. Preparation and hydrogen storage performance of poplar sawdust biochar with high specific surface area. Industrial Crops and Products. 200: 116788.

Liu, W.J., H. Jiang and H.Q. Yu. 2019. Emerging applications of biochar-based materials for energy storage and conversion. Energy and Environmental Science. 12: 1751–1779.

Lu, Y., D. Li, C. Lyu, H. Liu, B. Liu, S. Lyu, et al. 2019. High nitrogen doped carbon nanofiber aerogels for sodium ion batteries: Synergy of vacancy defects to boost sodium ion storage. Applied Surface Science. 496: 143717.

Luo, J.D., H. Zhang, X.T. Qi, J. Yu, Z. Zhang, J.C. Wei, et al. 2020. Agaric-assisted synthesis of core-shell MnO@C microcubes as super-high-volumetric-capacity anode for lithium-ion batteries. Carbon. 162: 36–45.

Mian, M.M., G. Liu and H. Zhou. 2020. Preparation of N-doped biochar from sewage sludge and melamine for peroxymonosulfate activation: N-functionality and catalytic mechanisms. Science of the Total Environment. 744: 140862.

Molaiyan, P., G.S. Dos Reis, D. Karuppiah, C.M. Subramaniyam, F. García-Alvarado and U. Lassi. 2023. Recent progress in biomass-derived carbon materials for Li-ion and Na-ion batteries—A review. Batteries. 9: 116.

Nie, W., X. Liu, Q. Xiao, L. Li, G. Chen, D. Li, et al. 2020. Hierarchical porous carbon anode materials derived from rice husks with high capacity and long cycling stability for Sodium-Ion batteries. ChemElectroChem. 7: 631–641.

Niu, J., R. Shao, J. Liang, M. Dou, Z. Li, Y. Huang, et al. 2017. Biomass-derived mesopore-dominant porous carbons with large specific surface area and high defect density as high performance electrode materials for Li-ion batteries and supercapacitors. Nano Energy. 36: 322–330.

Norouzi, O., P. Salimi, F. Di Maria, S.E.M. Pourhosseini and F. Safari. 2019. Synthesis and design of engineered biochars as electrode materials in energy storage systems. *In*: Fang, Z., Smith, Jr, R., Tian, XF. (eds). Production of Materials from Sustainable Biomass Resources. Biofuels and Biorefineries, vol 9. Springer, Singapore.

Pacioni, T.R., D. Soares, M. Di Domenico, J.L.F. Alves, E. Virmond, R. de F.P.M. Moreira, et al. 2023. Kinetic modeling of CO_2 gasification of biochars prepared from Brazilian agro-industrial residues: effect of biomass indigenous mineral content. Biomass Conversion and Biorefinery. 13: 6675–6688.

Pan, P., Y. Hu, K. Wu, Z. Cheng, Z. Shen, L. Jiang, et al. 2020. Growth of $ZnCo_2O_4$ nanocubes on flexible biochar substrate derived from natural silk waste fabric for lithium-ion battery anode. Journal of Alloys and Compounds. 814: 152306.

Panda, M.R., A.R. Kathribail, B. Modak, S. Sau, D.P. Dutta and S. Mitra. 2021. Electrochemical properties of biomass-derived carbon and its composite along with

$Na_2Ti_3O_7$ as potential high-performance anodes for Na-ion and Li-ion batteries. Electrochimica Acta. 392: 139026.

Park, S., J. Song, W.C. Lee, S. Jang, J. Lee, J. Kim, et al. 2023. Advances in biomass-derived electrode materials for energy storage and circular carbon economy. Chemical Engineering Journal. 470: 144234.

Pedicini, R., S. Maisano, V. Chiodo, G. Conte, A. Policicchio and R.G. Agostino. 2020. Posidonia Oceanica and Wood chips activated carbon as interesting materials for hydrogen storage. International Journal of Hydrogen Energy. 45: 14038–14047.

Rawat, S., C.T. Wang, C.H. Lay, S. Hotha and T. Bhaskar. 2023. Sustainable biochar for advanced electrochemical/energy storage applications. Journal of Energy Storage. 63: 107115.

Romero-Cano, L.A., H. García-Rosero, F. Carrasco-Marín, A.F. Pérez-Cadenas, L.V. González-Gutiérrez, A.I. Zárate-Guzmán, et al. 2019. Surface functionalization to abate the irreversible capacity of hard carbons derived from grapefruit peels for sodium-ion batteries. Electrochimica Acta. 326: 1–12.

San Juan, J. and C. Sy. 2021. Multi-objective target-oriented robust optimization of biomass co-firing networks under quality uncertainty. Journal of Sustainable Development of Energy, Water and Environment Systems. 9: 1–26.

Schaefer, S., A. Jeder, G. Sdanghi, P. Gadonneix, A. Abdedayem, M.T. Izquierdo, et al. 2020. Oxygen-promoted hydrogen adsorption on activated and hybrid carbon materials. International Journal of Hydrogen Energy. 45: 30767–30782.

Sekar, S., A.T.A. Ahmed, D.Y. Kim and S. Lee. 2020. One-pot synthesized biomass C-SI nanocomposites as an anodic material for high-performance sodium-ion battery. Nanomaterials. 10: 1–11.

Soni, K., N.L. Panwar and P.R. Lanjekar. 2024. Emergence of carbonaceous material for hydrogen storage: An overview. Clean Energy. 8: 147–168.

Su, C., Q. Ru, Y. Gao, Z. Shi, M. Zheng, F. Chen, et al. 2021. Biowaste-sustained $MoSe_2$ composite as an efficient anode for sodium/potassium storage applications. Journal of Alloys and Compounds. 850: 156770.

Sun, J., F. He, Y. Pan and Z. Zhang. 2016. Effects of pyrolysis temperature and residence time on physicochemical properties of different biochar types. Acta Agriculturae Scandinavica, Section B—Soil and Plant Science. 67: 12–22.

Thompson, M., Q. Xia, Z. Hu and X.S. Zhao. 2021. A review on biomass-derived hard carbon materials for sodium-ion batteries. Materials Advances. 2: 5881–5905.

Tian, H., Q. Hu, J. Wang, D. Chen, Y. Yang and A.V. Bridgwater. 2021. Kinetic study on the CO_2 gasification of biochar derived from Miscanthus at different processing conditions. Energy. 217: 119341.

Tomczyk, A., Z. Sokołowska and P. Boguta. 2020. Biochar physicochemical properties: Pyrolysis temperature and feedstock kind effects. Reviews in Environmental Science and Bio/Technology. 19: 191–215.

Wan, H. and X. Hu. 2019. Nitrogen doped biomass-derived porous carbon as anode materials of lithium ion batteries. Solid State Ionics. 341: 115030.

Xiao, Y., H. Dong, C. Long, M. Zheng, B. Lei, H. Zhang, et al. 2014. Melaleuca bark based porous carbons for hydrogen storage. International Journal of Hydrogen Energy. 39: 11661–11667.

Xie, F., Z. Xu, Z. Guo and M.M. Titirici. 2020. Hard carbons for sodium-ion batteries and beyond. Progress in Energy. 2: 042002.

Xu, Z., Y. Wang, F. Lu, X. Liu, J. Li and H. Fu. 2022. MoS_2 nanosheets fixed on network carbon derived from apple pomace for fast Na storage kinetics. Journal of Alloys and Compounds. 918: 165724.

Yang, G., Q. Hu, J. Hu, H. Yang, S. Yan, Y. Chen, et al. 2023. Hydrogen-rich syngas production from biomass gasification using biochar-based nanocatalysts. Bioresource Technology. 379: 129005.

Yeboah, M.L., X. Li and S. Zhou. 2020. Facile fabrication of biochar from palm kemel shell waste and its novel application to magnesium-based materials for hydrogen storage. Materials. 13: 625.

Yu, H.Y., H.J. Liang, Z.Y. Gu, Y.F. Meng, M. Yang, M.X. Yu, et al. 2020. Waste-to-wealth: low-cost hard carbon anode derived from unburned charcoal with high capacity and long cycle life for sodium-ion/lithium-ion batteries. Electrochimica Acta. 361: 137041.

Yu, W., H. Wang, S. Liu, N. Mao, X. Liu, J. Shi, et al. 2016. N, O-codoped hierarchical porous carbons derived from algae for high-capacity supercapacitors and battery anodes. Journal of Materials Chemistry A. 4: 5973–5983.

Zaini, I.N., Y. Gomez-Rueda, C. García López, D.K. Ratnasari, L. Helsen, T. Pretz, et al. 2020. Production of H2-rich syngas from excavated landfill waste through steam co-gasification with biochar. Energy. 207: 118208.

Zhang, Y., L. Chen, Y. Meng, J. Xie, Y. Guo and D. Xiao. 2016. Lithium and sodium storage in highly ordered mesoporous nitrogen-doped carbons derived from honey. Journal of Power Sources. 335: 20–30.

Zhang, N., Q. Liu, W. Chen, M. Wan, X. Li, L. Wang, et al. 2018. High capacity hard carbon derived from lotus stem as anode for sodium ion batteries. Journal of Power Sources. 378: 331–337.

Zhang, C., Z. Xie, W. Yang, Y. Liang, D. Meng, X. He, et al. 2020a. $NiCo_2O_4$/biomass-derived carbon composites as anode for high-performance lithium ion batteries. Journal of Power Sources. 451: 227761.

Zhang, J., Z. Chen, G. Wang, L. Hou and C. Yuan. 2020b. Eco-friendly and scalable synthesis of micro-/mesoporous carbon sub-microspheres as competitive electrodes for supercapacitors and sodium-ion batteries. Applied Surface Science. 533: 147511.

Zhang, W., Y. Xu, H. Li, C. Wang, B. Qin, Z. Li, et al. 2020c. Incorporating SnO_2 nanodots into wood flour-derived hierarchically porous carbon as low-cost anodes for superior lithium storage. Journal of Electroanalytical Chemistry. 856: 113654.

Zhao, G., G. Zou, H. Hou, P. Ge, X. Cao and X. Ji. 2017. Sulfur-doped carbon employing biomass-activated carbon as a carrier with enhanced sodium storage behavior. Journal of Materials Chemistry A. 5: 24353–24360.

Zhu, Y., M. Chen, Q. Li, C. Yuan and C. Wang. 2018. A porous biomass-derived anode for high-performance sodium-ion batteries. Carbon. 129: 695–701.

Zhu, Y., Y. Huang, C. Chen, M. Wang and P. Liu. 2019. Phosphorus-doped porous biomass carbon with ultra-stable performance in sodium storage and lithium storage. Electrochimica Acta. 321: 134698.

Chapter 9

Use of Biochar for Wastewater Treatment

Viviana Fernández-Maldonado*[a], Daniela Zalazar-García[a], María Celia Román[a], Ana Laura Navas[a] and Mario Herrera[b]

[a]Instituto de Ingeniería Química, Fac. de Ingeniería, Grupo Vinculado al PROBIEN (Universidad Nacional de San Juan, CONICET, Universidad Nacional del Comahue), San Juan, Argentina

[b]Grupo de Geobotánica y Fitogeografía, Instituto Argentino de Investigación de las Zonas Áridas (IADIZA), CCT-Mendoza, Av. Ruiz Leal s/n, Parque General San Martín, Mendoza CP 5500, Argentina

9.1 IMPACT OF PERSISTENT ORGANIC POLLUTANTS

POPs are synthetic chemical compounds highly resistant to environmental degradation (Fitzgerald and Wikoff 2014). This resistance allows them to persist in the environment for extended periods, accumulating in soils, sediments, water, and the tissues of living organisms. POPs are typically hydrophobic and lipophilic, facilitating their bioaccumulation in the food chain. Examples include organochlorine pesticides, polychlorinated biphenyls, and dioxins (Alharbi et al. 2018). The study of POPs is crucial due to their severe implications for the environment and human health. Due to their persistence, POPs can be transported long distances from their source, contaminating remote ecosystems (Fitzgerald and Wikoff 2014). Their toxic nature and ability to bioaccumulate means they can cause significant harm to wildlife and humans. POPs are linked to a variety of health problems, including cancer, neurological disorders, immunotoxicity, and endocrine disruption (Rokni et al. 2023).

*Corresponding author: Viviana Fernández-Maldonado (vivifernandezm@unsj-cuim.edu.ar)

DOI: 10.1201/9781032726694-9

POPs cause both direct and indirect effects on the environment due to interactions between biotic and abiotic elements. POP contamination disrupts ecological systems and threatens environmental integrity and the health of living organisms. During the 1980s and 1990s, several studies indicated the severe impacts of POPs on aquatic ecosystems due to these toxic contaminants in various species of the Great Lakes, the Arctic Ocean, the Baltic Sea, and the North Sea (Rokni et al. 2023). Thus, POP contamination in ecosystems results in the poisoning of birds, animals, and insects (Fitzgerald and Wikoff 2014), leading to imbalances in fertility, sex ratio changes, and physical abnormalities. Additionally, studies have documented the harmful impacts of POPs on wildlife, including eggshell thinning, the extinction of certain bird species, skeletal deformities, reproductive failures, and hormonal alterations in various wildlife species (Miljeteig et al. 2012).

9.1.1 Origin and Properties of POPs

9.1.1.1 Sources of POP Emissions

In agriculture, organochlorine pesticides and herbicides, such as DDT, have been extensively used for pest and disease control in crops. Although many of these products have been banned, their residues persist in soil and water (López-Benítez et al. 2024). Domestic products, such as flame retardants in furniture and textiles, personal care products, and cleaning products, may contain POPs like polybrominated diphenyl ethers (PBDEs) and perfluorinated compounds (PFCs). These POPs can be released into the environment through the everyday use and disposal of these products (Keswani et al. 2022).

9.1.1.2 Chemical Characteristics Contributing to Persistence and Bioaccumulation

POPs possess specific chemical characteristics that enable them to persist in the environment and bioaccumulate in living organisms. Carbon-halogen bonds, such as chlorine-carbon and bromine-carbon linkages, render POPs stable and resistant to decomposition. These bonds do not readily react with water or other common decomposition agents. In addition, POPs are typically hydrophobic, meaning they repel water, and lipophilic, meaning they dissolve in fats. This dual nature allows them to accumulate in the fatty tissues of living organisms, to bioaccumulate and to biomagnify through the food chain. Due to their chemical stability and resistance to degradation, POPs have a prolonged half-life in the environment. They can persist for years or even decades in soils, sediments, and water bodies (Keswani et al. 2022, López-Benítez et al. 2024).

9.1.1.3 Examples of Common POPs and Their Uses

DDT (Dichloro-Diphenyl-Trichloroethane) is used as a pesticide in agriculture for insect control. It has been banned in many countries due to its negative effects on wildlife and human health. PCBs (Polychlorinated Biphenyls) are employed in electrical transformers, capacitors, and as cooling fluids. They are known for their toxicity and persistence in the environment. PBDEs (Polybrominated Diphenyl Ethers) are used as flame retardants in furniture, textiles, and electronic products. These compounds can be released into the environment through the use of the product and its disposal. Dioxins and furans are by-products of industrial processes such as waste incineration and the production of certain pesticides. They are extremely toxic and persistent, and are associated with various serious health problems (Keswani et al. 2022; López-Benítez et al. 2024).

9.1.2 Health Impacts

POPs impact wildlife, fauna, and flora. Studies have documented that polar bears in the Arctic are particularly vulnerable to POPs due to their position at the top of the food chain. POPs, such as polychlorinated biphenyls (PCBs) and organochlorine pesticides, bioaccumulate in their fatty tissues, causing problems in the immune system, hormonal alterations and reproductive issues, and reduce the survival rate of cubs. Additionally, changes in the health of polar bears can be an early indicator of the overall health of the Arctic ecosystem. Seabirds are also severely affected by POPs. For instance, the thinning of eggshells in birds like the bald eagle and the brown pelican has been linked to DDT exposure, leading to a significant decline in the populations of these birds due to the low hatching rate of their eggs.

Humans are primarily exposed to POPs through diet, as these compounds bioaccumulate in the food chain. Exposure to POPs is associated with a range of health risks. Several POPs, including dioxins and some PCBs, are recognized as human carcinogens. Prolonged exposure can increase the risk of developing cancer, particularly in organs such as the liver and kidneys (Castillo Reinoso et al. 2023). POPs can affect neurological development, causing cognitive and behavioral problems, especially in children. Prenatal exposure to POPs, such as polychlorinated biphenyls, has been associated with lower IQ and attention deficits in childhood (Toledano et al. 2024). Additionally, POPs act as endocrine disruptors, interfering with the production and regulation of hormones. This can lead to reproductive problems, such as infertility and developmental abnormalities, in both humans and wildlife (Castillo Reinoso et al. 2023).

9.1.3 Environmental Impact: Relationship Between POPs and Climate Change

Climate change and POP pollution are interrelated and can exacerbate each other. Rising temperatures due to climate change can increase the volatilization of POPs, facilitating their long-distance transport. This can lead to the deposition of POPs in remote regions (far from their use or origin), such as the Arctic and Antarctica, affecting sensitive ecosystems in these areas. Changes in climate patterns can alter the carbon cycle, affecting the mobility and degradation of POPs in the environment. For example, increased precipitation can enhance the leaching of

POPs from soil into water bodies. These combined effects can have synergistic consequences that exacerbate environmental and health impacts. For instance, ice layer reduction can release POPs previously trapped in the ice, increasing the exposure of organisms to these contaminants.

The transport of POPs across different environmental media is a complex process influenced by various physical, chemical, and biological factors. Figure 9.1 illustrates key elements of this transport process, including atmospheric, terrestrial, and aquatic pathways, as well as biogeochemical interactions.

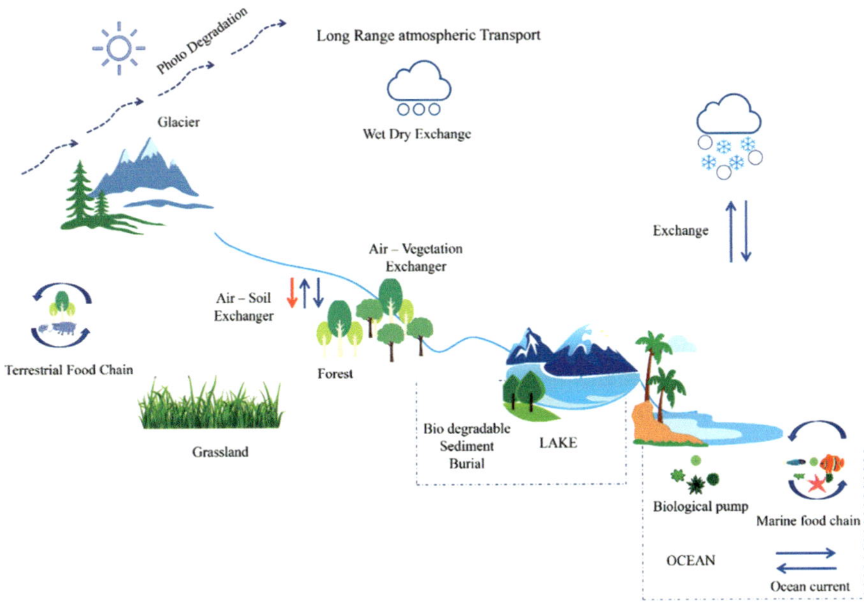

Figure 9.1 Transport mechanisms of POPs.

Photodegradation on glaciers occurs when sunlight induces the breakdown of POPs into less harmful substances through photochemical reactions facilitated by ultraviolet (UV) radiation. This process acts as a natural detoxification mechanism, reducing the environmental persistence of POPs (Klán et al. 2003). Long-Range Atmospheric Transport (LRAT) occurs when POPs volatilize into the atmosphere and are transported over long distances by prevailing winds (Luarte et al. 2023), leading to their deposition in remote regions far from their sources, such as the Arctic and Antarctic. LRAT serves as the main pathway for the global dissemination of POPs, contributing to their presence in polar regions far from their use or origin (Luarte et al. 2023).

POPs in the atmosphere can be deposited onto terrestrial and aquatic surfaces through wet deposition (precipitation) or dry deposition (settling of particles) (Teran et al. 2012). This process is crucial for transferring atmospheric POPs to land and water bodies, affecting their distribution and bioavailability in the environment. In forest ecosystems, POPs can be absorbed by plant foliage from the air, influenced by factors such as leaf surface area, cuticle composition, and atmospheric concentration of POPs. Plants can act as sinks or sources of POPs, depending on environmental conditions (Franzaring and Van Der Eerden 2000).

The atmosphere and soil can also exchange POPs. Soil properties, temperature, and the concentration gradient of POPs between air and soil drive this exchange. Soils can act as reservoirs for POPs, reintroducing them into the atmosphere under certain conditions. In terrestrial food chains, POPs can enter the system through soil ingestion, plant uptake, and the consumption of contaminated organisms. Bioaccumulation occurs when POPs concentrate in an organism's fatty tissues, and biomagnification leads to higher concentrations of POPs at higher trophic levels in the food chain (Weber et al. 2019).

In aquatic environments, sediments can bury POPs, which may then undergo biodegradation by microorganisms. This process reduces the availability of POPs to marine organisms and helps mitigate their toxic effects (Zacharia 2019). However, buried POPs can be remobilized under certain conditions, such as sediment disturbance (Josefsson et al. 2010). The biological pump in oceans refers to the process by which POPs are transported from the ocean surface to deeper waters as organic matter sinks (Zacharia 2019). Marine organisms incorporate POPs into their bodies, and when they die, the organic matter sinks, effectively sequestering POPs in deep ocean sediments. Bioaccumulation and biomagnification occur in aquatic species, posing risks to top predators and human consumers of seafood. Ocean currents play a significant role in the horizontal and vertical distribution of POPs in oceanic environments. Currents can transport POPs over long distances, influencing their global distribution and potential to cause environmental impact (De La Rocha and Passow 2014).

Biochar has a porous structure that provides a high specific surface area, which is crucial for the adsorption of contaminants. This porosity allows biochar to act like a sponge, capturing and retaining POPs in its pores (Ahmad et al. 2014). The high specific surface area offers numerous active sites for contaminant adsorption, thereby enhancing its capacity to trap and retain POPs (Mane et al. 2024). Many POPs are hydrophobic, meaning they repel water and tend to adhere to similarly hydrophobic surfaces. Due to its pyrogenic nature, biochar develops hydrophobic characteristics that favor the adsorption of POPs. This mechanism is fundamental because it facilitates the adhesion of contaminants to the surface of the biochar rather than allowing them to remain dissolved in water (Murtaza et al. 2022). Van der Waals forces are weak attractions between molecules that allow POPs to adhere to the surface of the biochar. Although these forces are relatively weak, their collective effect on the porous surface of biochar is significant. Additionally, POPs with aromatic rings can form pi-pi bonds with the aromatic structures present in biochar, increasing adsorption efficiency (Mukherjee et al. 2022). Biochar contains various functional groups on its surface, such as carboxyls (–COOH), hydroxyls (–OH), and carbonyls (C=O), which play a crucial role in the adsorption of POPs. These functional groups can chemically interact with POPs through several mechanisms. Carboxyl and hydroxyl groups can form hydrogen bonds with POPs, enhancing the adsorption capacity of the biochar. Carbonyl groups can interact with POPs through dipole-dipole interactions (Ahmad et al. 2014).

9.1.4 Effects of Biochar on POPs Bioavailability

Biochar can interact with the rhizosphere and the soil zone surrounding plant roots, and this interaction significantly influences the bioavailability of POPs. The rhizosphere is a crucial region where POPs can enter crops and the food chain. Root exudates, released during plant growth, play a fundamental role in modifying the bioavailability of POPs. These exudates can regulate the diffusion, migration, transformation, and metabolism of POPs at the root-soil interface (Beesley et al. 2011). Furthermore, the rhizosphere is the most active zone for soil-root-microbe interactions, which directly influence the microbial degradation of POPs in the soil. Therefore, the presence of biochar in the rhizosphere can enhance the immobilization of POPs, thereby reducing their bioavailability and the risk of entering the food chain (Kolton et al. 2017).

Root exudates are compounds released by plant roots that play a crucial role in maintaining the activity of the rhizosphere's microecosystems. Multiple mechanisms between these exudates and biochar significantly affect the bioavailability and degradation of POPs. First, root exudates can increase the bioavailability of POPs by acting as surfactants that promote the desorption of contaminants from biochar (Gul et al. 2015). For example, low molecular weight organic acids (LMWOAs) released by roots can increase the desorption of PAHs from biochar, thereby increasing their bioavailability (Zhang et al. 2021). Second, root exudates participate directly in the degradation of POPs by secreting enzymes that exhibit degradative activity, such as laccase and dehalogenase. Third, the combination of root exudates and biochar can alter the efficiency of POP degradation by regulating the structure and function of soil microorganisms, providing an easily utilizable source of organic carbon that promotes microbial growth and activity (Lehmann et al. 2011).

Several factors can influence the adsorption capacity of biochar for POPs, including the pyrolysis temperature, the type of biochar, and the specific characteristics of the POPs. The pyrolysis temperature during biochar production significantly impacts its adsorption capacity. Biochar produced at higher temperatures typically has a larger specific surface area and a more developed pore structure, which enhances its adsorption capacity (Zalazar-Garcia et al. 2024). For example, biochar produced at 850°C has a considerably higher adsorption capacity for specific compounds than biochar produced at 450°C (Kookana et al. 2011). However, excessively high temperatures can reduce the nutrient content in biochar, affecting its effectiveness in supporting the microbial degradation of POPs (Mukherjee and Zimmerman 2011). Additionally, the biomass used to produce biochar also influences its adsorption properties. Biochar derived from different materials, such as wheat straw, eucalyptus wood, or rice husks, can have distinct characteristics that affect their efficacy in adsorbing POPs. Chemical modifications of biochar, such as oxidation, can improve its capacity to adsorb specific contaminants (Chen and Chen 2009).

Moreover, the physicochemical properties of POPs, such as water solubility and molecular structure, also play a crucial role in their interaction with biochar. POPs with flat and aromatic structures can form stronger pi-pi interactions with biochar, which improves adsorption efficiency (Ni et al. 2020). These complex interactions and influencing factors highlight the importance of optimizing the conditions of production and use of biochar to maximize its effectiveness in the remediation of POP-contaminated sites. Biochar can be chemically modified to introduce or increase the amount of these functional groups, thereby optimizing its adsorption capacity. For example, biochar oxidation can increase the number of carboxyl groups, improving its efficiency in adsorbing specific POPs (Zhang et al. 2022).

9.1.5 Summary of Research on the Efficacy of Biochar in Reducing POPs in Contaminated Soils

Numerous studies have documented the efficacy of biochar in reducing POPs in contaminated soils (Ni et al. 2020). Research has demonstrated that the application of biochar can significantly reduce the bioavailability and mobility of POPs, such as polycyclic aromatic hydrocarbons (PAHs), polychlorinated biphenyls (PCBs), and organochlorine pesticides (Ni et al. 2020). For example, a study conducted by Shang et al. (2024) showed that rice husk-derived biochar addition to soils contaminated with PAHs resulted in a 70% reduction in the bioavailability of these contaminants. Another study by Zhang et al. found that biochar produced at high temperatures (850°C) had a significantly greater adsorption capacity for diuron, a common herbicide, than biochar produced at lower temperatures (450°C) (Zhang et al. 2022). These results highlight the potential of biochar as an effective tool in the remediation of POP-contaminated soils.

9.1.6 Comparison of Different Types of Biochar and Their Adsorption Capacity

The adsorption capacity of biochar can vary considerably depending on its origin and production process. Biochar from different materials, such as wheat straw, eucalyptus wood, and rice husks, has shown varying efficacy in adsorbing POPs (Guo et al. 2020). Furthermore, the pyrolysis temperature plays a crucial role in determining the structure and properties of biochar.

A comparative study by Kong et al. (2021) revealed that wheat straw biochar prepared at 500°C was more effective in removing PAHs than biochar from the same residue prepared at 700°C. This was due to the larger pore volume and positive influence on soil microbes. Similarly, Chatterjee et al. (2020) found that biochar pyrolyzed at 800°C had a higher adsorption capacity for diuron than biochar pyrolyzed at 400°C due to the presence of micropores and a larger specific surface area.

9.1.7 Analysis of the Interaction Between Biochar, POPs, and Soil Microorganisms

The interaction between biochar, POPs, and soil microorganisms is complex and multifaceted. Biochar not only adsorbs POPs but can also influence the soil microbial community, promoting the biological degradation of these contaminants. Ni et al. (2020) demonstrated that biochar application to soils contaminated with POPs can increase microbial activity, enhancing biodegradation. Root exudates and other compounds released by plants can interact with biochar, modulating its adsorption capacity and the bioavailability of POPs. For instance, Song et al. (2016) found that root exudates of *Lolium perenne* L. significantly increased the desorption of hexachlorobenzene immobilized in biochar. Additionally, biochar can serve as a habitat for POP-degrading microorganisms, providing anchorage sites and nutrients. This synergistic effect between biochar and microorganisms can improve the efficiency of soil remediation, reducing the toxicity and mobility of POPs.

In summary, case studies highlight the effectiveness of biochar in reducing POPs in contaminated soils, with its adsorption capacity varying based on the type and production process of the biochar. The interaction between biochar, POPs, and soil microorganisms plays a crucial role in mitigating these contaminants, presenting a promising solution for environmental remediation.

Despite its promise in mitigating POPs, researchers must address several limitations of biochar. One major limitation is the variability in the efficacy of biochar depending on its origin and production method. Different biomass types and pyrolysis conditions can produce biochars with distinct characteristics, affecting their adsorption capacity and interaction with POPs. Additionally, the large-scale application of biochar in contaminated soils can be costly and logistically challenging. Another limitation is the potential long-term release of POPs adsorbed in biochar under certain environmental conditions, which could reduce its effectiveness and pose secondary contamination risks.

To optimize biochar use for remediating POPs, researchers need to understand how different environmental conditions affect the interaction between biochar and POPs. This includes studies on the impact of variables such as soil pH, moisture, temperature, and other contaminants. Moreover, a deeper understanding of the specific mechanisms of POP adsorption and degradation in biochars produced from different materials and under various pyrolysis conditions is necessary. Long-term research is also essential to evaluate the stability and effectiveness of biochar in adsorbing POPs over time and under diverse environmental conditions.

The soil remediation field is constantly evolving, and there is potential for new technology development and approaches to enhance the efficacy of biochar and other remediation methods. Some promising areas include the development of chemically modified biochars to increase their adsorption capacity and affinity for specific POPs. This could involve impregnating biochar with chemical compounds that enhance its adsorption efficacy. Another approach is nanoparticle use in combination with biochar to improve the adsorption and degradation of POPs. Nanoparticles can provide additional adsorption sites and catalyze contaminant degradation (Egbosiuba 2022). Biochar-assisted bioremediation, which integrates biochar with POP-degrading microorganisms, can improve the biodegradation of contaminants in soil, increasing the efficiency of remediation and reducing the time required for cleaning contaminated sites. The development of biochar systems that not only adsorb POPs (Haider et al. 2022) but also release nutrients (Hou et al. 2022) and bioremediation agents (Singh et al. 2020) agents in a controlled manner to maintain microbial activity and soil health presents an innovative strategy for environmental remediation.

9.2 INORGANIC CONTAMINANTS: HEAVY METALS

Human activities and natural processes release heavy metals, elements with relatively high density and toxicity at low concentrations, into the environment. Common examples of heavy metals include lead, mercury, cadmium, and chromium. The study of heavy metals is crucial due to their persistence in the environment and their ability to cause significant adverse effects on ecosystems and human health. Because of their chemical stability, heavy metals do not decompose and can remain in soil, sediments, and water bodies for extended periods (Mitra et al. 2022). This persistence allows them to accumulate in living organisms, causing toxic effects throughout the food chain. Heavy metals can cause various types of harm to fauna and flora. Their toxicity can disrupt normal biological processes, causing cellular damage, organ dysfunction, and, in extreme cases, death. Heavy metals can affect the reproduction and development of aquatic organisms, reduce biodiversity, and alter food chains. In soils, heavy metals can inhibit plant growth and affect microbial activity, reducing soil fertility (Jaiswal et al. 2018).

9.2.1 Health Impact

Exposure to heavy metals can occur through contaminated food consumption, drinking water, and air. Chronic exposure to heavy metals can result in a range of severe health problems. Many heavy metals, such as cadmium and chromium, are recognized as human carcinogens and can increase the risk of developing cancer in various organs (Balali-Mood et al. 2021). Lead and mercury are neurotoxic and can cause significant damage to the nervous system, affecting cognitive development and behavior, especially in children. Exposure to heavy metals can cause damage to the kidneys and liver, affecting their ability to filter and detoxify the body. Some heavy metals can interfere with the endocrine system, causing hormonal imbalances and reproductive problems (Liu et al. 2023).

9.2.2 Environmental Impact

Climate change can influence the mobility and distribution of heavy metals in the environment. Variations in temperature, precipitation patterns, and rising sea levels can affect the release and transportation of heavy metals from soils and sediments into water bodies (Kibria et al. 2021). Additionally, extreme weather events, such as floods and droughts, can mobilize previously confined heavy metal deposits, increasing their availability and the risk of exposure (Xiao et al. 2024).

Various anthropogenic and natural sources release heavy metals, primarily categorizing them as industrial, agricultural, and domestic. Industrial activities significantly contribute to the presence of heavy metals. Processes such as mining, metal smelting, electronic manufacturing, and waste incineration can release heavy metals like lead, mercury, cadmium, and chromium into the environment (Xiao et al. 2024). These heavy metals can contaminate the air, water, and soil around industrial areas. In agriculture, pesticide and fertilizer use can introduce heavy metals into the soil. For example, some pesticides contain arsenic, and fertilizers may contain cadmium. These heavy metals can accumulate in the soil and be absorbed by crops, entering the food chain (Alengebawy et al. 2021). Domestic products can also be a source of heavy metals. Those include batteries, paints, cosmetics, and cleaning products. Improper disposal of these products releases heavy metals into the environment, contaminating soil and water.

Heavy metals possess specific chemical characteristics that allow them to persist in the environment and bioaccumulate in living organisms. Heavy metals do not degrade or decompose in the environment, allowing them to persist indefinitely. Due to their chemical stability, heavy metals can persist in the environment for years or even centuries once they enter (Jadaa and Mohammed 2023).

Examples and uses include lead (batteries, paints, pipes, formerly in leaded gasoline), mercury (thermometers, dental amalgams, industrial applications), cadmium (rechargeable batteries, pigments, coatings), and chromium (stainless steel production, coatings). Lead is toxic and can cause neurological and kidney damage. Mercury is highly hazardous and can cause severe neurological damage. Cadmium is harmful and can lead to kidney and bone damage. Hexavalent chromium is dangerous and can cause cancer and respiratory damage (Jaishankar et al. 2014).

9.2.3 Adsorption Mechanisms of Biochar for Heavy Metals

In the absence of oxygen, biochar formation, volatile gases, and bio-oils occur. Biochar has a specific porous structure that provides numerous active sites for the adsorption of contaminants, enhancing its ability to capture and retain heavy metals (Sokoto et al. 2020).

Heavy metals, often present as positively charged ions (cations), can be attracted and retained by the negatively charged surfaces of biochar. These electrostatic interactions are fundamental to the adsorption of heavy metals. Additionally, heavy metals can form complexes with functional groups on the surface of biochar, such as carboxyls (–COOH), hydroxyls (–OH), and carbonyls (C=O). Chemisorption involves the formation of chemical bonds between heavy metals and the active sites of biochar, resulting in more robust and stable adsorption (Kumkum and Kumar 2024).

Carboxyls (–COOH) can ionize in aqueous solutions, providing binding sites for metal cations through electrostatic interactions and complex formation. Hydroxyls (–OH) can form hydrogen bonds and participate in the adsorption of heavy metals through ion exchange and complexation mechanisms. Carbonyls (C=O) can interact with heavy metals through dipole-dipole interactions and complex formation. These properties make biochar a promising tool for the remediation of soils contaminated with heavy metals, reducing their mobility and toxicity in the environment.

Carboxyl and hydroxyl groups can form stable complexes with metal ions, increasing biochar's adsorption capacity. Carbonyl groups can interact with metal ions through coordination interactions, improving adsorption. These modifications and characteristics of biochar highlight its potential as an effective solution for mitigating contamination (Hama Aziz and Kareem 2023).

9.2.4 Thermochemical Pretreatments and Biochar Characterization

Different thermochemical pretreatments can produce biochar with distinct physical and chemical characteristics. Characterizing biochar is essential for assessing its suitability in various applications. For instance, biochar with low carbon content and high ash concentration is inappropriate for energy production, while biochar with low surface area and adsorption capacity is ineffective for treating wastewater contaminated with heavy metals (Jindo et al. 2014). Slow pyrolysis produces biochar with turbostratic graphite layers, whose spacing increases with reaction temperature. At 350°C, biochar is primarily aromatic carbon with small amounts of alkyl-O and alkyl-C groups, whereas, above 500°C, these groups convert to aromatic carbon, resulting in high-surface-area biochar (Kavan Kumar and Panwar 2024).

Chemical activation offers several advantages over physical activation, such as lower temperature requirements and higher surface area. However, it can generate secondary pollution, making it less preferred. For large-scale applications, physical methods are generally more suitable and do not require chemicals. Chemical modification can adjust the surface properties of biochar for specific applications, though it can be costly and requires contaminating chemicals. Biological methods can also modify biochar, but they typically need longer contact times (Gao et al. 2020).

There are various ways to modify biochar to enhance its capacity for heavy metal adsorption, classified into organically modified biochar, metal-modified biochar (Hama Aziz and Kareem 2023), non-metallic biochar (Amalina et al. 2024), and co-doped biochar (Murtaza et al. 2024). Co-pyrolysis of multiple biomasses is a promising technology to reduce waste and improve carbon conversion efficiency, increasing the production of volatiles and biochar with enhanced properties. The synergistic effect of co-pyrolysis depends on the type of biomass, operating conditions, and mixing ratio. For instance, coating corn cob biochar with MgO enhances its capacity to adsorb heavy metals such as Pb(II), As, and Cd(II) (Wantaneeyakul et al. 2021). Figure 9.2 highlights the significant impact of these modifications on biochar's capacity to adsorb heavy metals, exemplified by the enhanced adsorption efficiency of MgO-coated corn cob biochar for lead, arsenic, and cadmium.

Biochar activation increases its surface area and porosity, making it more effective as an adsorbent for heavy metal removal. In the physical activation process, biochar is heated with inert gases such as nitrogen or carbon dioxide, creating a porous structure. Chemical activation with agents such as potassium hydroxide (KOH) or phosphoric acid also develops porosity and introduces functional groups on the surface, increasing adsorption capacity (Hama Aziz and Kareem 2023).

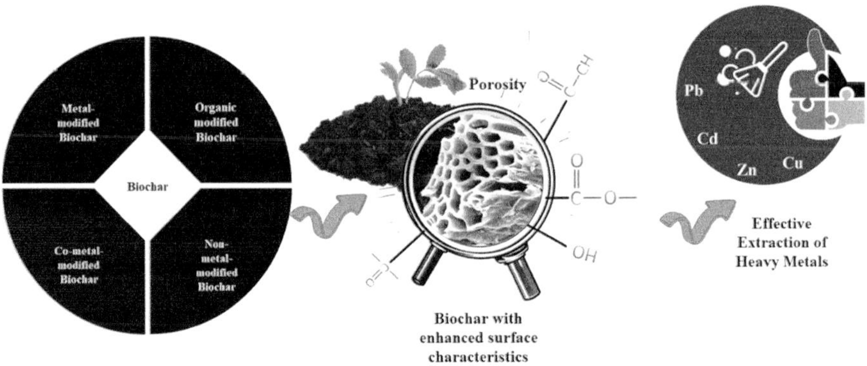

Figure 9.2 Types of biochar modification and their role in heavy metal adsorption.

Biochar can be chemically modified to introduce or increase the number of these functional groups, thus optimizing its adsorption capacity. For example, biochar oxidation can increase the number of carboxyl groups, enhancing its efficiency in adsorbing specific heavy metals (Jindo et al. 2014).

9.3 BIOCHAR ADSORPTION MECHANISM

Biochar, a carbonaceous material produced from the pyrolysis of biomass, possesses a large specific surface area and a porous structure, providing numerous adsorption sites that make it highly efficient for removing both organic and inorganic contaminants (Dong et al. 2024). The adsorption mechanisms of biochar can involve electrostatic interactions, ion exchange, pore filling, and precipitation, all of which depend on its physicochemical properties, such as dosage, pyrolysis temperature, and the pH of the medium or effluent (Ambaye et al. 2021a). These unique characteristics allow biochar to effectively remediate various pollutants in contaminated environments.

Several organic wastewater treatment technologies, including membrane technology, advanced oxidation processes, and biodegradation, have been developed to remove pollutants from water (Ma et al. 2021b). However, adsorption technology stands out as the most suitable due to its simple design, low cost, low energy consumption, and broad applicability. It has been widely used to control pollution from dyes, antibiotics, and phenols (Qiu et al. 2022).

In the case of inorganic pollutants, competition between toxic elements and specific contaminants can occur at active sites on biochar-based adsorbents. Thus, it is essential to evaluate the removal efficiency of biochar tailored for specific pollutants in multi-pollutant environments (Ambaye et al. 2021a). Adsorption, being a surface phenomenon, relies on the ability of certain solids to retain specific substrates (Barquilha and Braga 2021). The effectiveness of biochar in contaminant removal is determined by factors like pore size distribution, surface functional groups, and ion exchange capacity, which can be further enhanced through treatments such as acid, base, or metal impregnation.

There are two types of adsorption processes: physical and chemical. Physical adsorption, driven by weak forces like molecular polarity, depends on characteristics such as pore volume, pore size distribution, and specific surface area. Chemical adsorption, on the other hand, involves stronger forces, where molecules are bound to the adsorbent surface by valence forces similar to those between atoms in molecules (Barquilha and Braga 2021). These two adsorption mechanisms often interact or operate individually, making biochar a versatile and effective adsorbent for a wide range of contaminants in both aqueous and gaseous systems.

Given the widespread availability of raw materials and biochar's versatility, it has become a preferred choice among researchers for adsorption applications (Xu et al. 2019). The primary adsorption mechanisms in biochar are summarized in Table 9.1.

Table 9.1 Comparison between physical and chemical adsorption mechanisms

	Mechanisms of Physical Adsorption	**Mechanisms of Chemical Adsorption**
Nature of Interaction	Weak van der Waals forces (intermolecular forces) between the adsorbent and adsorbate are the primary mechanism driving physical adsorption, with no significant chemical bond formation.	Strong chemical bonds, either covalent or ionic, form between the adsorbent and adsorbate, involving the sharing or transfer of electrons.
Adsorption Heat	Low heat of adsorption, typically in the range of 20–40 kJ/mol. The energy required is small because interactions are weak.	High heat of adsorption, typically 40–400 kJ/mol. Much higher than physisorption due to strong interactions.
Temperature Dependence	Occurs at low temperatures and decreases as temperature increases due to desorption. Physisorption is generally reversible.	Occurs at higher temperatures where the activation energy barrier can be overcome. Chemisorption is usually irreversible.
Surface Coverage	Non-specific; it can occur on the entire surface of the adsorbent. Multilayer adsorption can occur since it is not limited to a single layer of adsorbate.	Specific; only occurs at active sites on the surface of the adsorbent. Generally, limited to a monolayer of adsorbate.
Activation Energy	Little or no activation energy is required. Fast process.	Requires activation energy because bonds need to be formed. Slower process compared to physisorption.
Effect of Pressure	Adsorption increases with pressure, following the Freundlich or Langmuir isotherm.	Less affected by pressure compared to physisorption. Usually follows a Langmuir isotherm, but is limited to a single layer.

In summary, physical adsorption is governed by weak interactions and is generally reversible, whereas chemical adsorption involves the formation of stronger chemical bonds, usually occurs at higher temperatures, and is typically irreversible.

9.4 CRITICAL FACTORS AFFECTING CONTAMINANT ADSORPTION IN BIOCHAR

Adsorption is an essential process in wastewater remediation, and the use of biochar as an adsorbent has gained popularity due to its low cost, availability, and efficiency in removing various contaminants. However, the adsorption capacity and efficiency depend on a range of factors that must be carefully considered to optimize the process (Sekulić et al. 2018). These factors include the physicochemical characteristics of biochar, the adsorbent dosage, contaminant concentration, contact time, agitation speed, solution pH, and temperature (Qambrani et al. 2017; Qiu et al. 2022). Additionally, interactions with other contaminants and dissolved substances can influence adsorption performance. The following sections provide an in-depth analysis of these factors.

9.4.1 Physicochemical Characteristics of Biochar

Biochar, a carbonaceous material produced by pyrolyzing biomass under limited or no oxygen conditions, has distinct physicochemical properties that greatly impact its adsorption capacity (Qiu et al. 2022). The structure and composition of biochar are determined by several factors, such as pyrolysis temperature, biomass type, activation methods, and post-production treatments. These factors influence its porosity, specific surface area, pore size distribution, surface chemistry, and the presence of functional groups (Qambrani et al. 2017).

9.4.1.1 Pyrolysis Temperature

The pyrolysis temperature is a critical factor that defines the textural and chemical characteristics of biochar. At temperatures between 300–400°C, biochar contains a higher amount of oxygenated surface functional groups, such as hydroxyl, carboxyl, and phenolic groups. These groups enhance the adsorption of polar contaminants, including heavy metals, due to their ability to form hydrogen bonds and electrostatic interactions (Chandra and Bhattacharya 2019). Biochar produced at these temperatures has been shown to adsorb metals effectively, such as cadmium (Cd) and lead (Pb), leveraging the affinity between the oxygen-containing groups and metal cations (Qambrani et al. 2017). At higher temperatures, typically above 500–700°C, biochar becomes more carbonized and aromatic. This results in an increase in hydrophobicity and a decrease in polar functional groups, favoring the adsorption of nonpolar organic contaminants such as polycyclic aromatic hydrocarbons (PAHs) and pesticides (Mohan et al. 2014). Studies have shown that biochar produced at 600°C exhibits high adsorption capacities for organic solvents and hydrophobic compounds due to its enhanced aromaticity and large surface area (Ahmad et al. 2014).

9.4.1.2 Biomass Origin

The type of biomass used for biochar production significantly affects its final properties. Biomass rich in cellulose and hemicellulose, such as agricultural residues (e.g., rice husks and corn stover), tends to produce biochar with higher porosity and surface area. These characteristics are favorable for the adsorption of large organic molecules and dyes (Inyang et al. 2016; Qambrani et al. 2017). Conversely, biomass with high lignin content, such as hardwood and certain forest residues, produces biochar with increased stability, mechanical strength, and resistance to degradation (Qiu et al. 2022). These properties are advantageous for long-term applications in soil remediation where sustained contaminant binding is required (Leng et al. 2021).

9.4.1.3 Biochar Activation

The activation process can further enhance the adsorption performance of biochar. Activation can be physical (using steam or CO_2 at high temperatures) or chemical (impregnation with activating agents like KOH, H_3PO_4, or $ZnCl_2$). Chemical activation often results in biochar with a greater specific surface area, microporosity, and the introduction of new active sites, which significantly increase its adsorption capacity (Xiang et al. 2020). For example, KOH-activated biochar has shown exceptional performance in removing toxic dyes such as methylene blue and heavy metals like arsenic (As) from water (Chen et al. 2020). The activation process not only increases the adsorption sites but also tailors the surface chemistry to interact more effectively with specific contaminants. Impregnation with nanoparticles, such as iron oxide (Fe_3O_4), has also been reported to enhance the selectivity of biochar for removing arsenic, chromium, and pharmaceuticals from complex water matrices (Ali et al. 2022).

9.4.1.4 Surface Functional Groups

The surface chemistry of biochar is crucial for determining the type of adsorption interactions. Functional groups like carboxylic, phenolic, and lactonic groups can participate in ion exchange and coordinate with metal ions, enhancing the adsorption of heavy metals and other polar contaminants (Sun et al. 2022a). Infrared spectroscopy (FTIR) and X-ray photoelectron spectroscopy (XPS) analyses are often used to characterize these functional groups and their changes post-adsorption. Biochars derived from nitrogen-rich biomasses, such as animal manure, have been found to possess nitrogen-containing groups that improve their capacity to adsorb anions like nitrate (NO_3^-) and phosphate (PO_4^{3-}) through electrostatic attraction and hydrogen bonding (Kasera et al. 2022).

9.4.2 Adsorbent Dosage

The amount of biochar added to a contaminated solution, known as the adsorbent dosage, is a key factor in determining the adsorption efficiency (Qiu et al. 2022). The relationship between the adsorbent dosage and adsorption capacity is complex and influenced by both the nature of the contaminant and the characteristics of the biochar itself (Qambrani et al. 2017).

9.4.2.1 Optimal Dosage

A higher dosage of biochar typically increases the number of available adsorption sites, thus improving the overall removal efficiency of the contaminant (Qambrani et al. 2017). However, this relationship is not linear; beyond a certain point, additional biochar may result in the aggregation of particles, which reduces the effective surface area and limits the accessibility of adsorption sites (Chen et al. 2011a). For example, studies on the adsorption of heavy metals like lead (Pb^{2+}) and copper (Cu^{2+}) have shown that an adsorbent dosage between 2–4 g/l achieves a high removal efficiency without significant particle aggregation (Sun et al. 2022a). For organic dyes such as methylene blue and rhodamine B, optimal dosages range from 1–3 g/l, depending on the initial concentration and solution characteristics (Huang et al. 2019).

9.4.2.2 Effect on Adsorption Kinetics

The dosage can also impact the kinetics of the adsorption process (Qiu et al. 2022). At higher dosages, the rate at which the contaminant is adsorbed often increases due to the greater availability of active sites. However, this can lead to a faster approach to equilibrium, which may mask the true kinetic behavior of the process. Kinetic models such as pseudo–first-order and pseudo–second-order models are often used to study these effects. The pseudo–second-order model, which assumes that chemisorption is the rate-limiting step, generally provides a better fit for systems involving higher adsorbent dosages (Yu et al. 2021).

9.4.2.3 Economic Considerations

While higher dosages may improve contaminant removal, the cost-effectiveness of the treatment must be taken into account. Using excessive amounts of biochar can lead to increased operational costs without proportional gains in adsorption capacity. Therefore, optimization studies often include economic analyses to balance adsorbent dosage with contaminant removal efficiency (Gai et al. 2014).

9.4.3 Contaminant Concentration

The initial concentration of the contaminant in the solution plays a significant role in the adsorption process. This factor determines the driving force required for mass transfer between the solution and the surface of the biochar (Qambrani et al. 2017).

9.4.3.1 Low vs. High Concentrations

At low contaminant concentrations, there is minimal competition for adsorption sites, allowing for rapid and efficient adsorption. However, as the initial concentration increases, the number of contaminant molecules exceeds the available active sites on the biochar surface, leading to a plateau in the adsorption capacity (Barquilha and Braga 2021). The Langmuir isotherm is often employed to describe such behavior, assuming monolayer adsorption on a homogeneous surface. The Freundlich isotherm, in contrast, is used for heterogeneous systems where multilayer adsorption can occur (Foo and Hameed 2010; Qiu et al. 2022).

9.4.3.2 Saturation and Equilibrium

The point at which no more contaminant can be adsorbed is known as saturation. At high concentrations, equilibrium is achieved more slowly due to the intense competition for available sites. This can be modeled using adsorption isotherms that account for both homogeneous and heterogeneous site distributions, such as the Langmuir-Freundlich hybrid model (Chen et al. 2011a).

9.4.3.3 Implications for Wastewater Treatment

In real-world scenarios, wastewater often contains high concentrations of multiple contaminants. The behavior of biochar under such conditions must be carefully studied to optimize its adsorption capacity. For example, biochar's performance in adsorbing heavy metals like arsenic (As) and mercury (Hg) is greatly influenced by their initial concentrations, with higher levels requiring more sophisticated treatment strategies or pre-treatment methods (Wang et al. 2018).

9.4.4 Contact Time

The contact time between the biochar and the contaminated solution determines how quickly equilibrium is reached in the adsorption process. This factor is crucial for the design of treatment systems that aim for optimal efficiency within a practical timeframe (Qiu et al. 2022).

9.4.4.1 Initial Rapid Phase

Adsorption typically occurs in two phases: an initial rapid phase followed by a slower approach to equilibrium. In the initial phase, the rate of adsorption is high due to the abundant availability of active sites and the high concentration gradient between the contaminant and the adsorbent. This phase can last anywhere from a few minutes to several hours, depending on the characteristics of the biochar and the contaminant (Wu et al. 2011).

9.4.4.2 Equilibrium Phase

The second phase involves the gradual saturation of the biochar surface, where the rate of adsorption slows down as fewer active sites remain available. The total contact time required to reach equilibrium can vary widely, from 30 minutes for certain dyes to 24 hours for complex heavy metals and multi-contaminant systems (Xiang et al. 2020). Kinetic studies often employ models such as pseudo–first-order, pseudo–second-order, and intraparticle diffusion models to describe these phases and predict the time needed for optimal adsorption (de Oliveira et al. 2024).

9.4.4.3 Impact on Kinetics and Mechanism

The rate of adsorption can provide insights into the underlying mechanism. For instance, a fit to the pseudo–second-order kinetic model indicates that chemisorption is the rate-limiting step, involving valence forces and electron sharing between the contaminant and the biochar. In contrast, a pseudo–first-order fit suggests a physisorption-dominated process (Tong et al. 2019).

9.4.5 Agitation Speed

The speed at which the solution containing the contaminant is agitated plays a role in ensuring the even distribution of biochar particles and contaminant molecules, enhancing contact and promoting mass transfer.

9.4.5.1 Effect on Adsorption Efficiency

Agitation improves the diffusion of contaminant molecules toward the surface of the biochar, preventing the formation of boundary layers that can impede mass transfer. Optimal agitation speeds have been shown to enhance the rate of adsorption by increasing the collision frequency between adsorbent and adsorbate (Ghosh et al. 2017a). For most batch adsorption systems, agitation speeds between 100–200 rpm are sufficient to maintain homogeneity without causing excessive fragmentation or desorption of particles.

9.4.5.2 Desorption Risks

While moderate agitation enhances adsorption, excessive stirring can have negative effects, such as the desorption of contaminants and the disintegration of biochar particles, leading to a reduction in adsorption capacity (Barquilha and Braga 2021). Careful control of agitation speed is essential for optimizing adsorption while maintaining the integrity of the biochar.

9.4.6 pH of the Solution

The pH of the solution affects the surface charge of the biochar and the speciation of the contaminants, influencing the adsorption mechanism.

9.4.6.1 Surface Charge and Ionization

The biochar surface carries different charges depending on the pH. At lower pH levels, the surface becomes protonated, favoring the adsorption of anionic contaminants through electrostatic attraction. Conversely, at higher pH levels, the surface deprotonates, increasing the attraction of cationic species (Mohammadi et al. 2014). The point of zero charge (pHpzc) of biochar is a critical parameter that indicates the pH at which the surface has a neutral charge. For most biochars, the pHpzc ranges from 6 to 9, impacting the adsorption of both metal cations and organic anions (Zhang et al. 2013).

9.4.6.2 Contaminant-Specific pH

The optimal pH for adsorption varies with the type of contaminant. For example, the adsorption of Cr(VI) is more effective at a low pH (around 2–3) due to the predominant species $HCrO_4^-$, which interacts electrostatically with the positively charged biochar surface. In contrast, the adsorption of metals like Pb(II) and Cu(II) is more favorable at pH levels above 5, where the surface is negatively charged, and hydrolyzed metal species can form complexes (Lu et al. 2012).

9.5 BIOCHAR AS A SUPPORT MEDIA DURING ANAEROBIC DIGESTION

Anaerobic digestion (AD) is a well-established biological method for processing organic waste. This process involves the decomposition of organic materials in the absence of oxygen, using microorganisms to transform the waste into carbon dioxide (CO_2), methane (CH_4), and other gases, which, after purification, are transformed into biogas. The microorganisms involved in AD are divided into two main groups: (1) acid producers, which include acidogens and acetogens, and (2) methane producers, known as methanogens. These groups differ in their physiology, growth rates, and sensitivity to operating conditions (Fagbohungbe et al. 2017). Therefore, an imbalance in the population of these microorganisms can lead to failure in the anaerobic digestion process (Fagbohungbe et al. 2017).

The generated biogas can be used to produce renewable energy sources for electricity or heat generation, or it can be used as vehicle fuel (Ošlaj and Muršec 2010). In addition, the residual material, known as digestate, can serve as a fertilizer as it has the potential to improve plant growth, soil health, and quality (Cavali et al. 2022). Therefore, anaerobic digestion is effective because it not only reduces the amount of waste ending up in landfills but also produces biogas, a renewable source of energy. However, AD faces challenges that limit its widespread application. These include low methane productivity, which can be insufficient due to substrate quality or suboptimal conditions in the digester (Rocha-Meneses et al. 2022). The pH imbalance can inhibit the activity of methanogenic bacteria, as changes in this parameter often result from the accumulation of volatile fatty acids (Alavi-Borazjani et al. 2020). Unstable performance is another significant problem, as variability in the composition of waste fed to the digester can cause fluctuations in biogas production, hindering predictability and system efficiency (Park et al. 2020). Furthermore, the degradation of recalcitrant substances, such as certain types of plastics or lignocellulosic waste, limits the overall process performance (Menzel et al. 2020). Finally, the presence of inhibitors, such as heavy metals or chemical residues, can negatively affect microbial activity, including methanogenic archaea (Kadam et al. 2022). Overcoming these challenges requires research and development to optimize operating conditions, improve substrate pretreatment, and explore technologies that increase the efficiency of anaerobic digestion.

Until now, various support materials have been explored to enhance methanogenesis, with carbon-based ones such as graphite, graphene, activated carbon, and biochar standing out (Kundu et al. 2023). However, in recent years, biochar has gained increased attention. Given its ability to enhance AD, including the promotion and regulation of methanogenesis, acidogenesis-acetogenesis, and hydrolysis, it improves digestate quality and biogas optimization in an economical and environmentally efficient way (Collins et al. 2023). Compared to other materials, biochar is cheaper, as it is produced from residual biomass. In addition, it presents physicochemical properties that benefit AD, such as a large surface area, high porosity, and excellent electron transfer capacity (Luz et al. 2018).

9.5.1 Impacts of Biochar on Anaerobic Digestion

9.5.1.1 Enhanced Hydrolysis

Hydrolysis, which refers to the conversion of macromolecular organic matter into soluble substances, is the first step in the complex AD process (Pan et al. 2019). The addition of biochar or hydrochar increases the presence of hydrolytic bacteria, promoting the production of hydrolytic enzymes and accelerating the first stage of AD (Ma et al. 2021a). However, Ma et al. (2021a) found that excessive addition of biochar inhibits hydrolysis, consequently hindering the efficient conversion of organic matter to methane.

9.5.1.2 Promotion of Acidogenesis and Acetogenesis

Acidogenic bacteria convert hydrolysates (amino acids, fatty acids, and sugars) into acetic acid, H_2, CO_2, and volatile fatty acids (VFAs). Acetogenesis from VFAs produces acetic acid (Jasińska et al. 2023). Both biochar and hydrochar support the growth of acidogenic and acetogenic bacteria, thus improving the buffering capacity of the system and promoting the syntrophic association of microorganisms involved in the direct electron transfer (DIET) process. Kobayashi and Kuramochi (2022) showed that the use of activated biochar, especially with treatments such as K_2CO_3, is effective in accelerating the production of fatty acids and acetic acid and improving the efficiency of anaerobic digestion. Masebinu et al. (2019) showed that acidogenesis conditions with a pH of 5, biochar enhanced the maximum H_2 production rate by 32.5% and improved H_2 yield by 31%.

9.5.1.3 Increased Methanogenesis

Methanogenesis, which runs in parallel with acetogenesis, converts CO_2 and H_2 into methane (Jasińska et al. 2023). The addition of BC or hydrochar enhances the community of archaea responsible for methanogenesis, resulting in higher amounts of methane (CH_4) and a reduced lag phase of the process. Tang et al. (2020) showed in their review that the incorporation of biochar in anaerobic digestion systems can increase biogas production by 20–30%. The following figure illustrates the processes of anaerobic digestion (Figure 9.3).

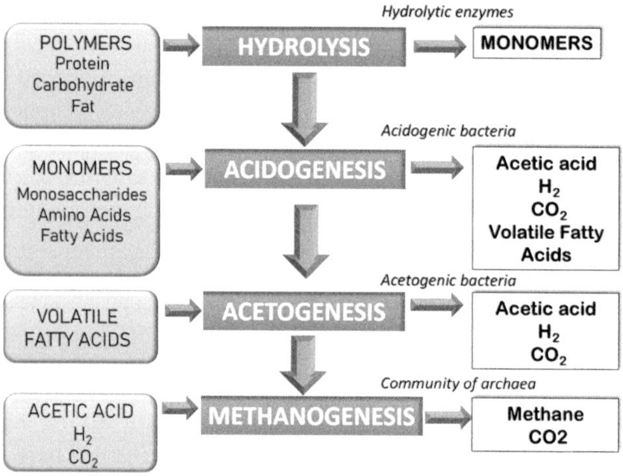

Figure 9.3 Stages of biomass decomposition through anaerobic digestion (AD), key compounds, and types of microorganisms.

9.5.2 Physicochemical Characteristics of Biochar and its Influence on AD

9.5.2.1 Surface Area and Porosity

These characteristics allow the immobilization of microorganisms and promote their interaction. Biochar generally has a greater surface area and porosity than hydrochar due to the higher temperatures used in its production, which result in a better porous structure and higher fixed carbon content (Quintana-Najera et al. 2021). These differences are due to the nature of the source biomass and the processing conditions used in its production.

9.5.2.2 Alkalinity

The accumulation of volatile fatty acids (VFAs) during anaerobic digestion (AD) can cause a significant drop in pH, affecting microbial activity and process efficiency. Although acetogens and methanogens help mitigate this problem by converting VFAs to methane and carbon dioxide, if VFA production exceeds their consumption, this can lead to system failure (Chiappero et al. 2020). Biochar and hydrochar with an appropriate pH can help buffer pH variations in systems handling high organic loads (Zhao et al. 2021). However, it is critical to find the optimal dosage of biochar, as excessive amounts can prove toxic.

9.5.2.3 Nutrient Retention and Conductivity

The presence of nutrients in biochar and hydrochar can favor microbial growth. Sun et al. (2022) showed that biochar enriched key microbial communities, such as Synergistia and Methanoculleus, which favored direct interspecies electron transfer (DIET) during anaerobic digestion. This transfer is crucial for improving the efficiency of methane production, as it allows microorganisms to collaborate more effectively in the decomposition of organic matter. Consequently, the addition of biochar not only increases biogas production but also optimizes the overall anaerobic digestion process (Sun et al. 2022b). Furthermore, the degree of carbonization, reflected by the H/C and O/C ratios, affects electrical conductivity, which is related to the DIET process. The addition of biochar in anaerobic digestion can influence the conductivity of the substrate, improving nutrient availability and microbial activity. Biochar, having conductive properties, can facilitate electron transfer between microorganisms, thereby optimizing biogas production. However, it is important to control the amount of biochar, as an excess can alter the conductivity and negatively affect the anaerobic digestion process (Song et al. 2021).

9.5.2.4 Increasing Buffering Capacity

Biochar contributes to pH stability in the anaerobic digester. By acting as a buffer, it helps maintain the acid-base balance of the system, which is critical for the optimal functioning of methanogenic microorganisms. Wang et al. (2018) showed that biochar has a remarkable pH buffering capacity, which helps counteract the decrease in pH caused by the accumulation of volatile fatty acids (VFAs) during the anaerobic digestion process. In groups treated with biochar, the pH drop was less pronounced, staying above 5.1, compared to the control group, which experienced a drop to 5.0. This buffering capacity is crucial, as low pH can negatively affect biogas production efficiency, and biochar contributes to stabilizing the environment, thus favoring microbial activity and methane production.

9.5.2.5 Adsorption of Inhibitors

Inhibition is one of the main causes of reduced biomethane production and instability in AD processes. Inhibitors, which can be direct (such as metals and organic compounds) or indirect (such as volatile fatty acids and ammonia), alter the microbial population and affect bacterial growth (Chiappero et al. 2020). In this context, biochar has been identified as a potential mitigator of ammonia inhibition through mechanisms such as cation exchange, chemical adsorption, and promotion of electron transfer, although further research is still required to fully understand these interactions. Furthermore, the particle size of biochar has been shown to influence its effectiveness; smaller particles (75 μm) are more efficient in producing and degrading VFAs. However, excessive doses can lead to propionate accumulation, inhibiting methanogenesis (Pan et al. 2019).

9.5.2.6 Electron Transfer Capacity

One of the most notable features of biochar is its ability to transfer electrons. This property is crucial in anaerobic digestion, as it enhances the electron transfer process between microorganisms, favoring methane production through the direct electron transfer (DIET) mechanism. Gao et al. (2021) showed that biochar modified with HNO_3 improved CH_4 production by 90%. This was because it facilitated interspecies electron transfer (DIET). In addition, it improves the stability of the AD system, but excess biochar can cause low pH inhibition.

9.5.2.7 Promotion of Syntrophic Catabolisms

It facilitates interactions between different types of microorganisms in the digester, promoting syntrophic catabolisms, which are essential for the complete decomposition of complex organic compounds. It acts as a support for the microbial community, promoting syntrophic interactions. However, an excess of biochar can cause inhibition of the process due to low pH (Gao et al. 2021).

9.5.2.8 Improving Digestate Quality

Digestate, the residual material resulting from the AD process, benefits from the addition of biochar. It can improve its properties as a fertilizer, providing additional nutrients and improving its structure. In addition, it mitigates the leaching of heavy metals and contaminants, as biochar acts as an adsorbent capable of retaining and reducing the toxicity of undesirable compounds in the system (Ambaye et al. 2021b).

9.5.2.9 Biogas Optimization

Raw biogas generated by anaerobic digestion (AD) contains mainly methane (CH_4) and carbon dioxide (CO_2), along with traces of other compounds. To meet the quality standards required for engines and pipelines, it is essential to purify the biogas, which can account for up to 55% of biomethane production costs (Chiappero et al. 2020). Conventional purification technologies include methods such as water scrubbing and cryogenic separation. The use of biochar (BC) has shown significant benefits in improving biogas quality, increasing methane concentration, and reducing pollutants such as CO_2 and H_2S, both in *in-situ* and *ex-situ* applications (Chiappero et al. 2020). *In situ* studies have reported increases in methane content and CO_2 removal efficiencies ranging from 54.9 to 86.3%, as well as H_2S levels below 5 ppb (Chiappero et al. 2020). The properties of BC, such as its high porosity and specific surface area, favor CO_2 capture mainly through physical adsorption, while H_2S removal involves chemical mechanisms due to functional groups on its surface. Furthermore, BC derived from potassium rich materials not only enhances biogas production, facilitating carbonate formation, but also promotes direct electron transfer between microorganisms, thereby increasing CH_4 production (Masebinu et al. 2019). Although positive results have been achieved in laboratory studies, the use of BC under mesophilic conditions, which represent the majority of AD installations, still requires further research. Most studies so far have been at the batch level, limiting the possibility of extrapolating these findings to large-scale AD plants.

9.6 BIOCHAR AS A FILTRATION SUPPORT MEDIA

Biochar is presented as a promising alternative for both water and air filtration. However, its most investigated use has been in water filtration. Several studies have evaluated water filters incorporating biochar for wastewater treatment, particularly in agricultural irrigation applications (Perez-Mercado et al. 2019).

One of the greatest challenges in these systems is reducing pathogen load. Several factors influence the effectiveness of biochar filters in removing pathogens. These include the hydraulic loading rate, the particle diameter of the biochar, the adsorption capacity of the filter material, and the characteristics of the biofilm that forms on the filter surfaces (Khiari et al. 2020). In addition, the porosity and charge heterogeneity on the biochar surface can significantly affect its ability to adsorb bacteria and viruses.

9.6.1 Water

9.6.1.1 Sand Filtration Column

Sand filtration columns are devices used in wastewater treatment, where sand acts as a filter medium to remove impurities and suspended particles such as microplastics. Water passes through a layer of sand, retaining contaminants by physical filtration and adsorption. This process can be gravity filtration, where water flows by the action of gravity, or pressure filtration, where pressure is applied to force the water through the medium. Biochar shows a high capacity to remove nitrates, pathogens, and microplastics, significantly outperforming conventional sand filters, with some studies reporting removal efficiencies of over 90% (Quispe et al. 2022). Its adsorptive capacity, low production cost, and utilization of agricultural residues make it a sustainable and economical option (Wang et al. 2020). Hsieh et al. (2022) added biochar in thin layers inside sand columns, which improved microplastic retention by reducing porosity and altering the electrostatic properties of the filter medium, thereby increasing the interaction between microplastics and the filter material. Dalahmeh et al. (2018) investigated the potential of biochar filters as a replacement or complement to sand filters for removing pharmaceutically active compounds from wastewater, and they found that biochar is a promising filter medium, especially for compounds such as carbamazepine and metoprolol.

9.6.1.2 Constructed Wetlands

Constructed wetlands are used for the remediation of contaminated waters, especially in decentralized communities and rural areas. One of the most relevant components in these wetlands is the substrate, which plays a fundamental role in the performance of wastewater treatment. In recent times, biochar has been incorporated as a material for biofiltration within wetlands, used in combination with other substrates and plants, such as LECA + biochar + *Typha latifolia* (Kasak et al. 2018). Various studies have shown that biochar significantly improves the removal of nutrients, organic pollutants, heavy metals, and pathogens in constructed wetlands, outperforming other traditional substrates (Deng et al. 2021). Furthermore, its incorporation not only optimizes wastewater treatment but also promotes the growth of macrophytes in the wetland (Deng et al. 2021). However, further research on the durability, recyclability, and potential adverse effects of biochar in these systems is needed to maximize its potential for practical applications.

9.6.1.3 Stormwater Biofilters

Bio-infiltration systems are stormwater management technologies that use permeable media, such as biofilters, to treat and manage water runoff. These systems allow water to infiltrate into the soil, where it filters through layers of material, helping to remove contaminants, including sediment and nutrients. However, biofilters with conventional media (such as sand, soil, mulch, or compost) do not reliably remove contaminants from stormwater. Biochar has been shown to significantly increase bacteria removal in bio-infiltration systems, depending on the type of biochar and the concentration of natural organic matter in the stormwater. In comparison, conventional biofilters achieve an average removal of approximately 55% of bacteria such as *Escherichia coli* (Boehm et al. 2020). Kranner et al. (2019) demonstrated that biochar improves the bacteria removal capacity of biofilters, especially during the early stages of conditioning. However, over time, a biofilm layer forms that progressively reduces the bacteria removal capacity of biochar, which represents a challenge for its long-term implementation.

9.6.1.4 Green Wall Systems

They are vertical structures incorporating vegetation, designed to improve aesthetics, thermal insulation, and air quality in urban environments. These can be used for greywater treatment, where plants and filter media work together to remove pollutants and reuse water in urban areas. Boano et al. (2021) showed that biochar, as an additive in greywater treatment, significantly improved the removal of pollutants such as total nitrogen and nitrate nitrogen (NO_3–N).

9.6.1.5 Ceramic and Biochar Filters

Ceramic filters are water purification devices that use a porous structure made of ceramic to remove contaminants and particulate matter. They work by allowing water to pass through the pores of the ceramic, which act as physical barriers that trap solids, bacteria, and other contaminants. These filters are especially useful in rural areas and developing countries, where access to clean water is limited. In this context, Chaukura et al. (2020) developed a ceramic filter combined with biochar, showing that this mixture is more effective in removing total hardness, total dissolved solids, and turbidity, reaching a 67% improvement compared to traditional ceramic filters. The results of their research concluded that the biochar-clay ceramic filter (BCF) not only improves the quality of the treated water but also optimizes the flow and adsorption capacity. In this way, the BCF is presented as a viable and sustainable solution for water treatment in communities with limited resources, highlighting its superior effectiveness compared to conventional ceramic filters.

9.6.2 Air

Air filtration systems are technologies designed to remove particulate matter, contaminants, and pathogens from the air, thereby improving indoor and outdoor air quality. These systems are essential in a wide variety of applications, from air treatment in industrial environments to air purification in enclosed spaces such as offices, hospitals, and homes. Depending on the type of contaminant to be removed, there are various types of filters, ranging from the simplest, capable of capturing large particles, to the most sophisticated that eliminate microscopic agents such as bacteria, viruses, or toxic gases (Gwenzi et al. 2021).

9.6.2.1 Electrostatic Precipitators

These devices remove solid and liquid particles from the air by applying an electric field, charging the particles, and causing them to adhere to oppositely charged collector plates. They are especially effective at removing particles such as ash, dust, and heavy metals, but less effective for gases such as elemental mercury (Hg_0) (Gwenzi et al. 2021).

9.6.2.2 Activated Carbon Injection

Activated carbon is used in air filtration to adsorb gaseous pollutants and volatile organic compounds (VOCs), such as odors and chemical vapors. It is used in industries where toxic gases or odors need to be removed from the air and is especially useful in air treatment in processing facilities and indoor air purification (Zhao et al. 2022).

9.6.2.3 Biochar Filters

Biochar, a porous material obtained by the pyrolysis of biomass, is used in filtration systems to remove particulate and gaseous pollutants from the air. Thanks to its high adsorption capacity, biochar is effective in capturing volatile organic compounds (VOCs), heavy metals, and toxic gases such as CO_2, SO_2, and H_2S (Gwenzi et al. 2021). Its porous structure and large specific surface area give it a high adsorption capacity, making it an efficient option for improving air quality. Biochar filters use this material as a filter medium, taking advantage of its properties to adsorb various gaseous pollutants. In addition, they can be chemically or physically modified to improve their performance, such as increasing porosity or the quantity of functional groups, which optimizes their capture capacity. Zhao et al. (2022) found that the results of several studies indicate that biochar filters can remove up to 77 mg/g of SO_2, in addition to other pollutants such as H_2S, CO_2, and VOCs, depending on the operating conditions and modifications applied to the biochar. Furthermore, being made from organic waste, biochar filters represent an ecological and sustainable option, contributing to the circular economy and reducing waste.

9.6.2.4 Fabric Filters

These consist of a permeable mesh or fabric that retains solid particles from the air, such as dust and suspended pollutants, as the air passes through it. They are commonly used in industrial ventilation systems, power plants, and manufacturing processes to control air pollution, as they are effective at capturing large and medium-sized particles such as dust and fibers, but may be less efficient at capturing toxic vapors and gases (Gwenzi et al. 2021). Their cost and maintenance can be limiting for large-scale implementation, which has led to the search for alternatives such as biochar filters.

9.7 ROLE OF BIOCHAR IN HETEROGENEOUS OXIDATION

Over recent decades, biochar has emerged as a versatile, sustainable material for environmental remediation applications, owing to its catalytic properties and ability to function in heterogeneous oxidation processes (Ahmad et al. 2014; Ghosh et al. 2017a). Produced via pyrolysis of biomass, biochar possesses high surface area, porosity, and various functional groups, enabling it to play a critical role in the degradation of persistent organic pollutants (POPs) and other toxic compounds (Chen et al. 2011b). Compared to traditional metal-based catalysts, biochar offers an economical and renewable option with comparable efficiency in environmental contaminant remediation (Mohan et al. 2014).

Biochar's role in heterogeneous oxidation centers on its function as a catalytic support, facilitating the generation of reactive oxygen species (ROS) that accelerate contaminant oxidation reactions (Ni et al. 2020; Mukherjee et al. 2022). Beyond its inherent physical and chemical properties, biochar can be chemically modified to optimize efficiency in these processes, broadening its applicability across diverse remediation conditions (Ahmad et al. 2014). This capacity to degrade and transform toxic compounds, such as POPs, is particularly relevant given these contaminants' high toxicity and environmental persistence (Castillo Reinoso et al. 2023).

Recent studies have highlighted that, compared with other catalysts, biochar not only demonstrates adequate efficiency but also represents a more accessible, sustainable alternative for industrial and environmental applications (Ghosh et al. 2017a; Mukherjee et al. 2022). Combining biochar with advanced oxidation processes may further enhance its performance, establishing it as a critical tool in green chemistry and the remediation of contaminated water and soil (Song et al. 2016).

This section will examine the catalytic properties of biochar, the factors influencing its performance in heterogeneous oxidation processes, and its applications in effluent remediation, aiming to provide a comprehensive perspective on biochar's potential as a sustainable and efficient catalyst for contaminant degradation.

9.7.1 Catalytic Properties of Biochar for Oxidation

The effectiveness of biochar as a catalyst in heterogeneous oxidation processes largely depends on its physicochemical properties, including porosity, surface area, and the presence of specific functional groups. These structural characteristics not only facilitate contaminant adsorption but also enhance the generation of reactive oxygen species (ROS), which are essential in oxidation processes (Chen et al. 2011b; Ahmad et al. 2014).

Porosity and surface area are critical for maximizing interaction between the catalyst and contaminants. Ghosh et al. (2017a) suggest that biochar with a higher surface area and appropriately sized pores allows for greater contaminant contact and active site availability, thus boosting catalytic efficiency. Furthermore, biochar's ability to withstand thermal and chemical modifications allows it to adapt to various reaction conditions, improving catalytic performance in targeted remediation applications (Mukherjee et al. 2022).

Chemical and thermal activation techniques are commonly employed to enhance biochar's catalytic properties. These modifications increase the number of active sites and strengthen biochar's ROS generation capabilities, essential for the degradation of persistent organic pollutants (POPs) (Song et al. 2016; Castillo Reinoso et al. 2023). Chemical modifications, specifically, can introduce additional functional groups, such as carboxyl and carbonyl groups, which elevate biochar's reactivity (Ahmad et al. 2014).

Analyzing biochar's structural characteristics and surface functional groups reveals their crucial role in oxidation mechanisms. According to Mohan et al. (2014), oxygenated functional groups on biochar's surface facilitate effective interaction with contaminants, enabling their decomposition through advanced oxidation mechanisms. These unique properties make biochar an adaptable and effective catalyst for oxidation in environmental applications, where contaminant degradation is a priority (Chen et al. 2011b).

9.7.2 Factors Affecting Catalytic Oxidation in Effluents

The efficacy of biochar as a catalyst in the oxidation of effluent contaminants is influenced by multiple factors impacting both its catalytic performance and overall process efficiency. Understanding these factors allows optimization of biochar's application in environmental remediation, particularly in the degradation of toxic compounds in wastewater (Ahmad et al. 2014).

One key factor in catalytic oxidation with biochar is the solution's pH. pH affects both biochar's surface charge and contaminant speciation, which together determine the oxidation reaction's efficiency (Mohammadi et al. 2014). Chen et al. (2011b) describe how, at low pH, biochar surfaces carry a positive charge, enhancing the adsorption of anionic species; at higher pH levels, the surface becomes negatively charged, promoting interaction with cations. These shifts directly influence ROS generation, which is crucial for the oxidation of effluent contaminants.

Biochar dosage also plays a significant role, as it determines the number of available active sites for the reaction. While optimal dosing maximizes process efficiency, excessive biochar can lead to particle aggregation, thereby reducing effective surface area and contaminant accessibility (Ghosh et al. 2017a). Identifying an appropriate dosage balances active site availability with contaminant removal efficiency (Ho and McKay 1999).

Agitation and mass transfer also significantly impact heterogeneous oxidation. Agitation enhances contact between biochar and contaminants, thereby preventing the formation of a boundary layer that can impede oxidation (Babel and Kurniawan 2003). Sun et al. (2020) report that moderate agitation improves contaminant diffusion to biochar's surface, increasing collision frequency and reaction rate. However, excessive agitation may cause contaminant desorption and biochar disintegration, reducing efficacy.

Finally, the contact time between biochar and contaminated solution dictates the time required for equilibrium in oxidation reactions. Initially, oxidation rates are rapid due to abundant active sites, though this rate decreases as biochar approaches saturation and equilibrium (Ho and McKay 1999). Studying these kinetics is essential for designing cost-effective and efficient effluent treatment systems (Wu et al. 2011).

9.7.3 Mechanism of Heterogeneous Oxidation

The heterogeneous oxidation mechanism with biochar involves a series of chemical reactions that generate reactive oxygen species (ROS), which are essential for breaking down contaminants in aqueous and soil systems. These ROS, such as hydroxyl radicals (^{-}OH) and hydrogen peroxide (H_2O_2), directly attack contaminant molecules, promoting their decomposition and reducing their toxicity (Wu et al. 2011; Mukherjee et al. 2022).

ROS generation in the presence of biochar is facilitated by its porous structure and oxygenated functional groups on its surface, which serve as active sites for oxidation reactions. Ahmad et al. (2014) explain that biochar's surface adsorbs contaminants, after which ROS are subsequently generated, enabling effective oxidation. These reactive species transform persistent organic pollutants (POPs) into lower-molecular-weight products, often with reduced toxicity, which is vital for mitigating environmental impact.

Modifying biochar through chemical activation or the addition of secondary catalysts, such as transition metals, can further increase ROS production efficiency. Mukherjee et al. (2022) highlight that these modifications enhance the number of active sites and improve biochar's capacity to function synergistically with other oxidizing agents. This synergy in advanced oxidation systems, known as "combined advanced oxidation," enables biochar to work alongside other catalysts to maximize contaminant degradation (Song et al. 2016).

Research into degradation pathways suggests that, under optimal conditions, biochar can initiate multiple oxidation routes for various contaminants. Chen et al. (2011b) describe how ROS catalyzed by biochar promote stepwise oxidation reactions that fragment complex molecules into less persistent compounds. This multi-pathway degradation process is crucial for improving heterogeneous oxidation efficiency in effluents contaminated with POPs (Castillo Reinoso et al. 2023).

9.7.4 Practical Applications and Case Studies in Effluent Treatment

The use of biochar in industrial and municipal effluent treatment has shown promising results due to its efficiency in the adsorption and oxidation of contaminants, as well as its sustainability and relatively low cost (Ahmad et al. 2014; Mukherjee et al. 2022). As industries and municipal water treatment facilities face rising demand for effective, eco-friendly solutions, biochar has established itself as a viable alternative for treating wastewater contaminated with heavy metals and persistent organic pollutants (POPs) (Chen et al. 2011b).

Case studies demonstrate biochar's significant contaminant reduction capacity in effluents. Ghosh et al. (2017b) report that biochar effectively adsorbs heavy metals, such as lead and cadmium, and facilitates organic compound oxidation in industrial wastewater systems, lowering effluent toxicity. These results underscore biochar's potential to act as a catalyst in real-world conditions, accommodating diverse wastewater matrices (Mohan et al. 2014).

In municipal effluent treatment, where organic contaminants and pathogenic microorganisms are common, biochar has also proven effective in contaminant removal. Mukherjee et al. (2022) show that chemically or thermally activated biochar not only enhances organic contaminant oxidation but also increases pathogen removal efficiency compared to conventional treatments. This versatility is especially valuable for municipal wastewater treatment plants seeking high-efficiency, low-cost solutions (Song et al. 2016).

Economic evaluations comparing biochar with traditional catalysts reveal that biochar is advantageous due to its renewable biomass source and lower production costs relative to metal-based catalysts. Chen et al. (2011b) conclude that, in industrial applications, biochar offers economic benefits while also supporting sustainable technology in effluent management. This dual benefit of cost-effectiveness and environmental sustainability makes biochar an appealing choice for industries aiming to minimize costs and integrate green practices.

9.7.5 Current Limitations and Future Perspectives

Despite its potential as a catalyst in heterogeneous oxidation, biochar faces several limitations that must be addressed to optimize its effectiveness and broaden its application in environmental remediation. A primary challenge is the regeneration and reusability of biochar after catalytic use. Accumulation of contaminants on its surface and degradation of catalytic properties over time reduce biochar's efficiency, limiting its sustainability and economic viability for industrial applications (Ahmad et al. 2014; Chen et al. 2011b).

Biochar regeneration, allowing for multiple oxidation cycles, is a critical area requiring technological advancement. Mohammadi et al. (2014) point out that current regeneration methods, such as thermal and chemical reactivation, can partially restore biochar's properties but are often costly and environmentally taxing. Research into more sustainable, efficient regeneration methods is essential for enhancing biochar's reusability and reducing operational costs (Mukherjee et al. 2022).

Modifying and improving biochar for future applications also remains a key research focus. Chemical modifications and the addition of metal nanoparticles have shown promise in increasing biochar's catalytic activity and ROS generation capacity, optimizing its performance in advanced contaminant oxidation (Song et al. 2016). However, Mukherjee et al. (2022) note that producing modified biochar remains expensive and presents challenges for large-scale applications, limiting its current feasibility in extensive water treatment systems.

In terms of environmental sustainability and regulatory considerations, biochar also presents challenges concerning its long-term environmental impact and safe disposal. As biochar applications expand in treatment systems, it is crucial to establish regulatory frameworks ensuring that biochar usage and disposal do not create additional environmental hazards (Ghosh et al. 2017b). Castillo Reinoso et al. (2023) underscore the lack of specific regulations governing biochar use in wastewater treatment, highlighting the need for standards that promote safe and responsible use.

Looking forward, biochar holds significant potential as a crucial tool in advancing sustainable environmental remediation practices. Advances in modification techniques and regeneration technology will enable this material to integrate more effectively into advanced oxidation systems, expanding its role in wastewater and effluent treatment. With continued investment in research and development, alongside regulatory frameworks that guarantee safe application, biochar may become an integral component in green chemistry and environmental management.

In conclusion, biochar presents itself as a versatile and sustainable catalyst, capable of transforming contaminated water treatment systems and contributing substantially to pollution mitigation efforts. The question remains: can biochar, considering technical and regulatory challenges, fulfill its promise as a large-scale, sustainable solution in environmental remediation?

NOMENCLATURE AND ACRONYMS

AD	:	Anaerobic digestion
As	:	Arsenic
BC	:	Biochar
BCF	:	Biochar-clay ceramic filter
C=O	:	Carbonyls
Cd	:	Cadmium
CH_4	:	Methane
CO_2	:	Carbon dioxide
–COOH	:	Carboxyls
Cu^{2+}	:	Copper
DDT	:	Dichloro-diphenyl-trichloroethane
DIET	:	Direct interspecies electron transfer
Fe_3O_4	:	Iron oxide
H/C	:	Hydrogen-to-carbon ratio
H_2	:	Dihydrogen

H_2O_2 : Hydrogen peroxide
H_2S : Hydrogen sulfide
H_3PO_4 : Phosphoric acid
$HCrO_4^-$: Hydrogen chromate ion
Hg : Mercury
IQ : Intelligence quotient
KOH : Potassium hydroxide
LRAT : Long-range atmospheric transport
MgO : Magnesium oxide
NO_3^- : Anions like nitrate
NO_3–N : Nitrate
O/C : Oxygen to Carbon Ratio
–OH : Hydroxyls
PAHs : Polycyclic aromatic hydrocarbons
Pb : Lead
Pb^{2+} : Lead(II) ion
PBDEs : Polybrominated diphenyl ethers
PCBs : Polychlorinated biphenyls
PFCs : Perfluorinated compounds
pHpzc : Point of Zero Charge
PO_4^{3-} : Phosphate
POPs : Persistent organic pollutants
ROS : Reactive oxygen species
SO_2 : Sulfur dioxide
VFAs : Volatile fatty acids
VOC : Volatile organic compounds
XPS : X-ray photoelectron spectroscopy
$ZnCl_2$: Zinc chloride

REFERENCES

Ahmad, M., A.U. Rajapaksha, J.E. Lim, M. Zhang, N. Bolan, D. Mohan, et al. 2014. Biochar as a sorbent for contaminant management in soil and water: A review. Chemosphere. 99: 19–33.

Alavi-Borazjani, S.A., I. Capela and L.A. Tarelho. 2020. Over-acidification control strategies for enhanced biogas production from anaerobic digestion: A review. Biomass Bioenergy. 143: 105833.

Alengebawy, A., S.T. Abdelkhalek, S.R. Qureshi and M.Q. Wang. 2021. Heavy metals and pesticides toxicity in agricultural soil and plants: ecological risks and human health implications. Toxics. 9(3).

Alharbi, O.M.L., A.A. Basheer, R.A. Khattab and I. Ali. 2018. Health and environmental effects of persistent organic pollutants. Journal of Molecular Liquids. 263: 442–453.

Ali, H., S. Ahmed, A. Hsini, S. Kizito, Y. Naciri, R. Djellabi, et al. 2022. Efficiency of a novel nitrogen-doped Fe_3O_4 impregnated biochar (N/Fe_3O_4@BC) for arsenic (III and V) removal from aqueous solution: insight into mechanistic understanding and reusability potential. Arabian Journal of Chemistry. 15(11): 104209. https://doi.org/10.1016/j.arabjc.2022.104209

Amalina, F., S. Krishnan, A.W. Zularisam and M. Nasrullah. 2024. Pristine and modified biochar applications as multifunctional component towards sustainable future: Recent advances and new insights. Science of the Total Environment. 914: 169608.

Ambaye, T.G., M. Vaccari, E.D. van Hullebusch, A. Amrane and S. Rtimi. 2021a. Mechanisms and adsorption capacities of biochar for the removal of organic and inorganic pollutants from industrial wastewater. International Journal of Environmental Science and Technology. 18: 3273–3294.

Ambaye, T.G., E.R. Rene, A.S. Nizami, C. Dupont, M. Vaccari and E.D. van Hullebusch. 2021b. Beneficial role of biochar addition on the anaerobic digestion of food waste: A systematic and critical review of the operational parameters and mechanisms. Journal of Environmental Management. 290: 112537.

Babel, S. and T.A. Kurniawan. 2003. Low-cost adsorbents for heavy metals uptake from contaminated water: A review. Journal of Hazardous Materials. 97: 219–243.

Balali-Mood, M., K. Naseri, Z. Tahergorabi, M.R. Khazdair and M. Sadeghi. 2021. Toxic mechanisms of five heavy metals: mercury, lead, chromium, cadmium, and arsenic. Frontiers in Pharmacology. 12: 643972.

Barquilha, C.E. and M.C. Braga. 2021. Adsorption of organic and inorganic pollutants onto biochars: Challenges, operating conditions, and mechanisms. Bioresource Technology Reports. 15: 100728. https://doi.org/10.1016/j.biteb.2021.100728

Beesley, L., E. Moreno-Jiménez, J.L. Gomez-Eyles, E. Harris, B. Robinson, T. Sizmur, et al. 2011. A review of Biochars' potential role in the remediation, revegetation and restoration of contaminated soils. Environmental Pollution. 159(12): 3269–3282.

Boano, F., E. Costamagna, A. Caruso, S. Fiore, M. Chiappero, A. Galvão, et al. 2021. Evaluation of the influence of filter medium composition on treatment performances in an open-air green wall fed with greywater. Journal of Environmental Management. 300: 113646.

Boehm, A.B., C.D. Bell, N.J. Fitzgerald, E. Gallo, C.P. Higgins, T.S. Hogue, et al. 2020. Biochar-augmented biofilters to improve pollutant removal from stormwater–can they improve receiving water quality? Environmental Science: Water Research & Technology. 6(6): 1520–1537.

Castillo Reinoso, A., A. Guzmán Guaraca, M.J. López Pino and E. Guanga Casco 2023. Persistent Organic Pollutants (POPs) in the environment: impacts, sources and mitigation strategies. Green World Journal 6(3): 092–092.

Cavali, M., N.L. Junior, R. de Almeida Mohedano, P. Belli Filho, R.H.R. da Costa and A.B. de Castilhos Junior. 2022. Biochar and hydrochar in the context of anaerobic digestion for a circular approach: An overview. Science of The Total Environment. 822: 153614. http://dx.doi.org/10.1016/j.scitotenv.2022.153614

Chandra, S. and J. Bhattacharya. 2019. Influence of temperature and duration of pyrolysis on the property heterogeneity of rice straw biochar and optimization of pyrolysis conditions for its application in soils. Journal of Cleaner Production. 215: 1123–1139.

Chatterjee, R., B. Sajjadi, W.Y. Chen, D.L. Mattern, N. Hammer, V. Raman, et al. 2020. Effect of pyrolysis temperature on physicochemical properties and acoustic-based amination of biochar for efficient CO_2 Adsorption. Frontiers in Energy Research. 8 (May): 1–18.

Chaukura, N., R. Chiworeso, W. Gwenzi, M.M. Motsa, W. Munzeiwa, W. Moyo, et al. 2020. A new generation low-cost biochar-clay composite 'Biscuit' ceramic filter for point-of-use water treatment. Applied Clay Science. 185: 105409.

Chen, B. and Z. Chen. 2009. Sorption of naphthalene and 1-Naphthol by biochars of orange peels with different pyrolytic temperatures. Chemosphere. 76 (1): 127–133.

Chen, X., G. Chen, L. Chen, Y. Chen, J. Lehmann, M.B. McBride, et al. 2011a. Adsorption of copper and zinc by biochars produced from pyrolysis of hardwood and corn straw in aqueous solution. Bioresource Technology. 102(19): 8877–8884. https://doi.org/10.1016/j.biortech.2011.06.078

Chen, B., Z. Chen, Z. and S. Lv. 2011b. A novel magnetic biochar efficiently sorbs organic pollutants and phosphate. Bioresource Technology, 102(2): 716 723. https://doi.org/10.1016/j.biortech.2010.08.067

Chen, T., Z. Zhou, H. Li, Z. Fan, J. Luo and S. Zhang. 2020. Enhancing the removal efficiency of heavy metals using chemically modified biochar. Environmental Science and Pollution Research. 27: 5894–5907.

Chiappero, M., O. Norouzi, M. Hu, F. Demichelis, F. Berruti, F. Di Maria, et al. 2020. Review of biochar role as additive in anaerobic digestion processes. Renewable and Sustainable Energy Reviews. 131: 110037.

Collins, B.A., C.H. Birzer, P.W. Harris, S.P. Kidd, B.K. McCabe and P.R. Medwell. 2023. Two-phase anaerobic digestion in leach bed reactors coupled to anaerobic filters: A review and the potential of biochar filters. Renewable and Sustainable Energy Reviews. 175: 113187.

Dalahmeh, S., L. Ahrens, M. Gros, K. Wiberg and M. Pell. 2018. Potential of biochar filters for onsite sewage treatment: Adsorption and biological degradation of pharmaceuticals in laboratory filters with active, inactive, and no biofilm. Science of the Total Environment. 612: 192–201.

De La Rocha, C.L. and U. Passow. 2014. The biological pump. pp. 93–122. *In*: D.H. Holland and K. Karl Turekian (eds). Treatise on Geochemistry: Second Edition. Elsevier. https://doi.org/10.1016/B978-0-08-095975-7.00604-5.

de Oliveira, J.T., D.C. Estumano and L.A. Féris. 2024. Resolution methods for adsorption models, batch and continuous mode, and its impact on process implementation: An experimental and statistical comparison. Journal of Water Process Engineering. 58: 104888.

Deng, S., J. Chen and J. Chang. 2021. Application of biochar as an innovative substrate in constructed wetlands/biofilters for wastewater treatment: Performance and ecological benefits. Journal of Cleaner Production. 293: 126156.

Dong, X., Y. Chu, Z. Tong, M. Sun, D. Meng and X. Yi. 2024. Mechanisms of adsorption and functionalization of biochar for pesticides: A review. Ecotoxicology and Environmental Safety. 272: 116019.

Egbosiuba, T.C. 2022. Biochar and bio-oil fuel properties from nickel nanoparticles assisted pyrolysis of cassava peel. Heliyon. 8(8): e10114.

Fagbohungbe, M.O., B.M. Herbert, L. Hurst, C.N. Ibeto, H. Li, S.Q. Usmani, et al. 2017. The challenges of anaerobic digestion and the role of biochar in optimizing anaerobic digestion. Waste management. 61: 236–249. http://dx.doi.org/10.1016/j.wasman.2016.11.028

Foo, K.Y. and B.H. Hameed. 2010. Insights into the modeling of adsorption isotherm systems. Chemical Engineering Journal. 156(1): 2–10. https://doi.org/10.1016/j.cej.2009.09.013

Fitzgerald, L. and D.S. Wikoff. 2014. Persistent organic pollutants. pp. 820–825. *In*: P. Wexler (ed.). Encyclopedia of Toxicology, Third Edition. Academic Press.

Franzaring, J. and L.J.M. Van Der Eerden. 2000. Accumulation of Airborne Persistent Organic Pollutants (POPs) in plants. Basic and Applied Ecology. 1(1): 25–30.

Gai, X., H. Wang, J. Liu, L. Zhai, S. Liu, T. Ren, et al. 2014. Effects of feedstock and pyrolysis temperature on biochar adsorption of ammonium and nitrate. PLoS ONE, 9(12): e113888. https://doi.org/10.1371/journal.pone.0113888

Gao, Y., Q. Yue, B. Gao and A. Li. 2020. Insight into activated carbon from different kinds of chemical activating agents: A review. Science of the Total Environment. 746: 141094.

Gao, B., Y. Wang, L. Huang and S. Liu. 2021. Study on the performance of HNO_3-Modified biochar for enhanced medium temperature anaerobic digestion of food waste. Waste Management. 135: 338–346.

Ghosh, P., I.S. Thakur and A. Kaushik. 2017a. Bioassays for toxicological risk assessment of landfill leachate: A review. Ecotoxicology and Environmental Safety. 141: 259–270. https://doi.org/10.1016/j.ecoenv.2017.03.023

Ghosh, D., A. Deb and S. Bhowmik. 2017b. Understanding the role of biochar in mitigating contamination in water and soil: A review. Environmental Technology Reviews. 6(1): 122–140. https://doi.org/10.1080/21622515.2017.1403445

Gul, S., J.K. Whalen, B.W. Thomas, V. Sachdeva and H. Deng. 2015. Physico-chemical properties and microbial responses in biochar-amended soils: Mechanisms and future directions. Agriculture, Ecosystems & Environment. 206: 46–59.

Guo, M., W. Song and J. Tian. 2020. Biochar-facilitated soil remediation: Mechanisms and efficacy variations. Frontiers in Environmental Science. 8: 521512.

Gwenzi, W., N. Chaukura, T. Wenga and M. Mtisi. 2021. Biochars as media for air pollution control systems: Contaminant removal, applications and future research directions. Science of the Total Environment. 753: 142249.

Haider, F.U., X. Wang, U. Zulfiqar, M. Farooq, S. Hussain, T. Mehmood, et al. 2022. Biochar application for remediation of organic toxic pollutants in contaminated soils: An update. Ecotoxicology and Environmental Safety. 248: 114322.

Hama Aziz, K.H. and R. Kareem. 2023. Recent advances in water remediation from toxic heavy metals using biochar as a green and efficient adsorbent: A review. Case Studies in Chemical and Environmental Engineering. 8: 100495.

Ho, Y.S. and G. McKay. 1999. Pseudo-second order model for sorption processes. Process Biochemistry. 34(5): 451–465.

Hou, J., A. Pugazhendhi, R. Sindhu, V. Vinayak, N.C. Thanh, K. Brindhadevi, et al. 2022. An assessment of biochar as a potential amendment to enhance plant nutrient uptake. Environmental Research. 214: 113909.

Hsieh, L., L. He, M. Zhang, W. Lv, K. Yang and M. Tong. 2022. Addition of biochar as thin preamble layer into sand filtration columns could improve the microplastics removal from water. Water Research. 221: 118783.

Huang, X., D. Wei, X. Zhang, D. Fan, X. Sun, B. Du, et al. 2019. Synthesis of amino-functionalized magnetic aerobic granular sludge-biochar for Pb (II) removal: Adsorption performance and mechanism studies. Science of the Total Environment, 685: 681–689.

Inyang, M., B. Gao, Y. Yao, Y. Xue, A. Zimmerman, A. Mosa, et al. 2016. A review of biochar as a low-cost adsorbent for aqueous heavy metal removal. Critical Reviews in Environmental Science and Technology. 46: 406–433.

Jadaa, W. and H.K. Mohammed. 2023. Heavy metals – definition, natural and anthropogenic sources of releasing into ecosystems, toxicity, and removal methods – An overview study. Journal of Ecological Engineering. 24(6): 249–271.

Jaishankar, M., T. Tseten, N. Anbalagan, B.B. Mathew and K.N. Beeregowda. 2014. Toxicity, mechanism and health effects of some heavy metals. Interdisciplinary Toxicology. 7(2): 60–72.

Jaiswal, A., A. Verma and P. Jaiswal. 2018. Detrimental effects of heavy metals in soil, plants, and aquatic ecosystems and in humans. Journal of Environmental Pathology, Toxicology and Oncology. 37(3): 183–197.

Jasińska, A., A. Grosser and E. Meers. 2023. Possibilities and limitations of anaerobic co-digestion of animal manure—A critical review. Energies. 16: 3885.

Jindo, K., H. Mizumoto, Y. Sawada, M.A. Sanchez-Monedero and T. Sonoki. 2014. Physical and chemical characterization of biochars derived from different agricultural residues. Biogeosciences. 11(23): 6613–6621.

Josefsson, S., K. Leonardsson, J.S. Gunnarsson and K. Wiberg 2010. Bioturbation-driven release of buried PCBs and PBDEs from different depths in contaminated sediments. Environmental Science & Technology. 44(19): 7456–7464.

Kadam, R., K. Khanthong, H. Jang, J. Lee and J. Park. 2022. Occurrence, fate, and implications of heavy metals during anaerobic digestion: A review. Energies. 15: 8618.

Kasak, K., J. Truu, I. Ostonen, J. Sarjas, K. Oopkaup, P. Paiste, et al. 2018. Biochar enhances plant growth and nutrient removal in horizontal subsurface flow constructed wetlands. Science of the Total Environment. 639: 67–74.

Kasera, N., P. Kolar and S.G. Hall. 2022. Nitrogen-doped biochars as adsorbents for mitigation of heavy metals and organics from water: a review. Biochar 4: 17. https://doi.org/10.1007/s42773-022-00145-2

Kavan Kumar, V. and N.L. Panwar. 2024. Pyrolysis technologies for biochar production in waste management: A review. Clean Energy. 8(4): 61–78.

Keswani, C., H. Dilnashin, H. Birla, P. Roy, R.K. Tyagi, D. Singh, et al. 2022. Global footprints of organochlorine pesticides: A pan-global survey. Environmental Geochemistry and Health. 44(1): 149–177.

Khiari, Z., K. Alka, S. Kelloway, B. Mason and N. Savidov. 2020. Integration of biochar filtration into aquaponics: Effects on particle size distribution and turbidity removal. Agricultural Water Management. 229: 105874.

Kibria, G., D. Nugegoda, G. Rose and A.K.Y. Haroon. 2021. Climate change impacts on pollutants mobilization and interactive effects of climate change and pollutants on toxicity and bioaccumulation of pollutants in estuarine and marine biota and linkage to seafood security. Marine Pollution Bulletin. 167: 112364.

Klán, P., J. Klánová, I. Holoubek and P. Čupr. 2003. Photochemical activity of organic compounds in ice induced by sunlight irradiation: The Svalbard project. Geophysical Research Letters. 30(6): 1313.

Kobayashi, T. and H. Kuramochi. 2022. Optimized production conditions and activation of biochar for effective promotion of long-chain fatty acid degradation in anaerobic digestion. Bioresource Technology. 358: 127393.

Kolton, M., E.R. Graber, L. Tsehansky, Y. Elad and E. Cytryn. 2017. Biochar-stimulated plant performance is strongly linked to microbial diversity and metabolic potential in the rhizosphere. New Phytologist. 213(3): 1393–1404.

Kong, L., B. Song, T. Zhang, K. Gao, J. Liu. 2021. Effects of soil organic matter on biochar application in developing the biodegradation potentials of Polycyclic Aromatic Hydrocarbons (PAHs). Applied Soil Ecology. 167: 104046.

Kookana, R.S., A.K. Sarmah, L. Van Zwieten, E. Krull and B. Singh. 2011. Biochar application to soil: Agronomic and environmental benefits and unintended consequences. Advances in Agronomy. 112: 103–143.

Kranner, B.P., A.N. Afrooz, N.J. Fitzgerald and A.B. Boehm. 2019. Fecal Indicator Bacteria and Virus Removal in Stormwater Biofilters: Effects of Biochar, Media Saturation, and Field Conditioning. PLoS One 14(9): e0222719.

Kumkum, P. and S. Kumar. 2024. A Review on biochar as an adsorbent for Pb(II) removal from water. Biomass. 243–272.

Kundu, R., B. Kunnoth, S. Pilli, V.R. Polisetty and R.D. Tyagi. 2023. Biochar symbiosis in anaerobic digestion to enhance biogas production: A comprehensive review. Journal of Environmental Management. 344: 118743.

Lehmann, J., M.C. Rillig, J. Thies, C.A. Masiello, W.C. Hockaday and D. Crowley. 2011. Biochar effects on soil biota – A review. Soil Biology and Biochemistry. 43(9): 1812–1836.

Leng, L., H. Huang, H. Li, J. Li, W. Zhou and Y. Dong. 2021. Biochar stability in the environment: Review of mechanisms and applications. Environmental Chemistry Letters. 19: 2277–2290.

Liu, D., Q. Shi, C. Liu, Q. Sun and X. Zeng. 2023. Effects of endocrine-disrupting heavy metals on human health. Toxics. 11(4): 322.

López-Benítez, A., A. Guevara-Lara, M.A. Domínguez-Crespo, J.A. Andraca-Adame and A.M. Torres-Huerta 2024. Concentrations of organochlorine, organophosphorus, and pyrethroid pesticides in rivers Worldwide (2014–2024): A review. Sustainability. 16(18): 8066.

Lu, H., W. Zhang, S. Wang, L. Zhuang, Y. Yang and R. Qiu. 2012. Characterization of sewage sludge-derived biochars from different feedstocks and pyrolysis temperatures. Journal of Hazardous Materials, 181(1-3), 37-43. https://doi.org/10.1016/j.jhazmat.2010.04.089

Luarte, T., V.A. Gómez-Aburto, I. Poblete-Castro, E. Castro-Nallar, N. Huneeus, M. Molina-Montenegro, et al. 2023. Levels of Persistent Organic Pollutants (POPs) in the Antarctic atmosphere over time (1980 to 2021) and estimation of their atmospheric half-lives. Atmospheric Chemistry and Physics. 23(14): 8103–8118.

Luz, F.C., S. Cordiner, A. Manni, V. Mulone and V. Rocco. 2018. Biochar characteristics and early applications in anaerobic digestion—A review. Journal of Environmental Chemical Engineering. 6: 2892–2909.

Ma, J., F. Chen, S. Xue, J. Pan, B. Khoshnevisan, Y. Yang, et al. 2021a. Improving anaerobic digestion of chicken manure under optimized biochar supplementation strategies. Bioresource Technology. 325: 124697.

Ma, Y., M. Li, P. Li, L. Yang, L. Wu and F. Zhang. 2021b. Hydrothermal synthesis of magnetic sludge biochar for tetracycline and ciprofloxacin adsorptive removal. Bioresource Technology. 319: 124199.

Mane, P.V., R.M. Rego, P.L. Yap, D. Losic and M.D. Kurkuri. 2024. Unveiling cutting-edge advances in high surface area porous materials for the efficient removal of toxic metal ions from water. Progress in Materials Science. 146: 101314.

Masebinu, S.O., E.T. Akinlabi, E. Muzenda and A.O. Aboyade. 2019. A review of biochar properties and their roles in mitigating challenges with anaerobic digestion. Renewable and Sustainable Energy Reviews. 103: 291–307.

Menzel, T., P. Neubauer and S. Junne. 2020. Role of microbial hydrolysis in anaerobic digestion. Energies. 13: 5555.

Miljeteig, C., G.W. Gabrielsen, H. Strøm, M.V. Gavrilo, E. Lie and B.M. Jenssen. 2012. Eggshell thinning and decreased concentrations of vitamin E are associated with contaminants in eggs of ivory gulls. Science of the Total Environment. 431: 92–99.

Mitra, S., A.J. Chakraborty, A.M. Tareq, T. Bin Emran, F. Nainu, A. Khusro, et al. 2022. Impact of heavy metals on the environment and human health: Novel therapeutic insights to counter the toxicity. Journal of King Saud University - Science. 34(3): 101865.

Mohammadi, A.S., S.H. Hosseini and S. Koushkbaghi. 2014. Adsorption of heavy metals by modified biochar: A review. Journal of Water Process Engineering. 3: 18–26. https://doi.org/10.1016/j.jwpe.2014.05.003

Mohan, D., H. Kumar and A. Sarswat. 2014. A review on the removal of pollutants from wastewater using various sorbents. Water Research. 50: 324–347.

Mukherjee, A., A.R. Zimmerman and W. Harris. 2011. Surface chemistry variations among a series of laboratory-produced biochars. Geoderma. 163(3–4): 247–255.

Mukherjee, S., B. Sarkar, V.K. Aralappanavar, R. Mukhopadhyay, B.B. Basak, P. Srivastava, et al. 2022. Biochar-microorganism interactions for organic pollutant remediation: Challenges and perspectives. Environmental Pollution. 308: 119609.

Murtaza, G., Z. Ahmed, D.Q. Dai, R. Iqbal, S. Bawazeer, M. Usman, et al. 2022. A review of mechanism and adsorption capacities of biochar-based engineered composites for removing aquatic pollutants from contaminated water. Frontiers in Environmental Science. 10: 1035865.

Murtaza, G., Z. Ahmed, M. Valipour, I. Ali, M. Usman, R. Iqbal, et al. 2024. Recent Trends and Economic Significance of Modified/Functionalized Biochars for Remediation of Environmental Pollutants. Scientific Reports. 14(1): 217.

Ni, N., D. Kong, W. Wu, J. He, Z. Shan, J. Li, et al. 2020. The role of biochar in reducing the bioavailability and migration of persistent organic pollutants in soil–plant systems: A review. Bulletin of Environmental Contamination and Toxicology. 104(2): 157–165.

Pan, J., J. Ma, L. Zhai, T. Luo, Z. Mei and H. Liu. 2019. Achievements of biochar application for enhanced anaerobic digestion: A review. Bioresource Technology. 292: 122058.

Park, J.G., D. Jiang, B. Lee and H.B. Jun. 2020. Towards the practical application of Bioelectrochemical Anaerobic Digestion (BEAD): Insights into electrode materials, reactor configurations, and process designs. Water Research. 184: 116214.

Perez-Mercado, L.F., C. Lalander, A. Joel, J. Ottoson, S. Dalahmeh and B. Vinnerås. 2019. Biochar filters as an on-farm treatment to reduce pathogens when irrigating with wastewater-polluted sources. Journal of Environmental Management. 248: 109295.

Qambrani, N.A., M.M. Rahman, S. Won, S. Shim and C. Ra. 2017. Biochar properties and eco-friendly applications for climate change mitigation, waste management, and wastewater treatment: A review. Renewable and Sustainable Energy Reviews. 79: 255–273.

Qiu, B., Q. Shao, J. Shi, C. Yang, H. Chu and L. Li. 2022. Application of biochar for the adsorption of organic pollutants from wastewater: Modification strategies, mechanisms and challenges. Separation and Purification Technology. 300: 121925.

Quintana-Najera, J., A.J. Blacker, L.A. Fletcher and A.B. Ross. 2021. The effect of augmentation of biochar and hydrochar in anaerobic digestion of a model substrate. Bioresource Technology. 321: 124494.

Quispe, J.B., L.C. Campos, O. Mašek and A. Bogush. 2022. Use of biochar-based column filtration systems for greywater treatment: A systematic literature review. Journal of Water Process Engineering. 48: 102908.

Rocha-Meneses, L., A. Hari, A. Inayat, A. Shanableh, M. Abdallah, C. Ghenai, et al. 2022. Application of nanomaterials in anaerobic digestion processes: A new strategy towards sustainable methane production. Biochemical Engineering Journal. 188: 108694.

Rokni, L., M. Rezaei, M. Rafieizonooz, E. Khankhajeh, A.A. Mohammadi and S. Rezania 2023. Effect of persistent organic pollutants on human health in South Korea: A review of the reported diseases. Sustainability. 15(14): 10851.

Sekulić, M.T., S. Pap, Z. Stojanović, N. Bošković, J. Radonić and T.Š. Knudsen. 2018. Efficient removal of priority, hazardous priority, and emerging pollutants with *Prunus armeniaca* functionalized biochar from aqueous wastes: Experimental optimization and modeling. Science of the Total Environment. 613: 736–750.

Shang, X., S. Wu, Y. Liu, K. Zhang, M. Guo, Y. Zhou, et al. 2024. Rice husk and its derived biochar assist phytoremediation of heavy metals and PAHs co-contaminated soils but differently affect bacterial community. Journal of Hazardous Materials. 466: 133684.

Singh, C., S. Tiwari and J.S. Singh. 2020. Biochar: a sustainable tool in soil pollutant bioremediation. pp. 475–494. *In*: Bioremediation of Industrial Waste for Environmental Safety: Volume II: Biological Agents and Methods for Industrial Waste Management, Singapore: Springer Singapore.

Sokoto, M.A., B. Biswas, J. Kumar and T. Bhaskar. 2020. Slow pyrolysis of defatted seeds cakes of African star apple and silk cotton for production of bio-oil. Renewable Energy. 146: 1710–1716.

Song, Y., Y. Li, W. Zhang, F. Wang, Y. Bian, L.A. Boughner, et al. 2016. Novel biochar-plant tandem approach for remediating hexachlorobenzene contaminated soils: Proof-of-concept and new insight into the rhizosphere. Journal of Agricultural and Food Chemistry. 64(27): 5464–5471.

Song, J., Y. Wang, S. Zhang, Y. Song, S. Xue, L. Liu, et al. 2021. Coupling biochar with anaerobic digestion in a circular economy perspective: A promising way to promote sustainable energy, environment, and agriculture development in China. Renewable and Sustainable Energy Reviews. 144: 110973.

Sun, Y., S. Zheng, D. Wang and X. Chen. 2020. Functional group influence on the adsorption properties of biochar: A review. Environmental Research. 187: 109631. https://doi.org/10.1016/j.envres.2020.109631

Sun, Y., B. Zeng, Y. Dai, X. Liang, L. Zhang, R. Ahmad, et al. 2022a. Modification of sludge-based biochar using air roasting-oxidation and its performance in adsorption of uranium (VI) from aqueous solutions. Journal of Colloid and Interface Science. 614: 547–555. https://doi.org/10.1016/j.jcis.2022.01.054

Sun, Z., L. Feng, Y. Li, Y. Han, H. Zhou and J. Pan. 2022b. The role of electrochemical properties of biochar to promote methane production in anaerobic digestion. Journal of Cleaner Production. 362: 132296.

Tang, S., Z. Wang, Z. Liu, Y. Zhang and B. Si. 2020. The role of biochar to enhance anaerobic digestion: A review. Journal of Renewable Materials. 8: 1033–1052.

Teran, T., L. Lamon and A. Marcomini. 2012. Climate change effects on POPs' environmental behaviour: A scientific perspective for future regulatory actions. Atmospheric Pollution Research. 3(4): 466–476.

Toledano, J.M., M. Puche-Juarez, J. Moreno-Fernandez, P. Gonzalez-Palacios, A. Rivas, J.J. Ochoa, et al. 2024. Implications of prenatal exposure to endocrine-disrupting chemicals in offspring development: A narrative review. Nutrients. 16(11): 1–23.

Tong, Y., P.J. McNamara and B.K. Mayer. 2019. Adsorption of organic micropollutants onto biochar: a review of relevant kinetics, mechanisms and equilibrium. Environmental Science: Water Research & Technology. 5(5): 821–838.

Wang, G., Q. Li, X. Gao and X.C. Wang. 2018. Synergetic promotion of syntrophic methane production from anaerobic digestion of complex organic wastes by biochar:

Performance and associated mechanisms. Bioresource Technology. 250: 812–820. https://doi.org/10.1016/J.BIORTECH.2017.12.004

Wang, Z., M. Sedighi and A. Lea-Langton. 2020. Filtration of microplastic spheres by biochar: Removal efficiency and immobilisation mechanisms. Water Research. 184: 116165.

Wantaneeyakul, N., K. Kositkanawuth, S.Q. Turn and J. Fu. 2021. Investigation of biochar production from copyrolysis of rice husk and plastic. ACS Omega. 6(43): 28890–28902.

Weber, R., L. Bell, A. Watson, J. Petrlik, M.C. Paun and J. Vijgen. 2019. Assessment of POPs contaminated sites and the need for stringent soil standards for food safety for the protection of human health. Environmental Pollution. 249: 703–715.

Wu, F., R.L. Tseng and R.S. Juang. 2011. Characteristics of pseudo-second-order kinetic model for adsorption systems. Chemical Engineering Journal. 151(1–3): 1–9. https://doi.org/10.1016/j.cej.2009.11.027

Xiang, W., X. Zhang, J. Chen, W. Zou, F. He, X. Hu, et al. 2020. Biochar technology in wastewater treatment: A critical review. Chemosphere. 252: 126539.

Xiao, W., Y. Zhang, X. Chen, A. Sha, Z. Xiong, Y. Luo, et al. 2024. The easily overlooked effect of global warming: Diffusion of heavy metals. Toxics. 12(6): 400.

Xu, J., Z. Cao, Y. Wang, Y. Zhang, X. Gao, M.B. Ahmed, et al. 2019. Distributing sulfidized nanoscale zerovalent iron onto phosphorus-functionalized biochar for enhanced removal of antibiotic florfenicol. Chemical Engineering Journal. 359: 713–722.

Yu, K.L., X.J. Lee, H.C. Ong, W.H. Chen, J.S. Chang, C.S. Lin, et al. 2021. Adsorptive removal of cationic methylene blue and anionic Congo red dyes using wet-torrefied microalgal biochar: Equilibrium, kinetic and mechanism modeling. Environmental Pollution. 272: 115986.

Zacharia, J.T. 2019. Degradation pathways of Persistent Organic Pollutants (POPs) in the environment. *In*: S.K. Donyinah (ed.). Persistent Organic Pollutants: Environmental Impacts and Management. IntechOpen, Rijeka.

Zalazar-Garcia, D., A. Fernandez, L. Cavaliere, Y. Deng, J. Soria, R. Rodriguez, et al. 2024. Slow pyrolysis of pistachio-waste pellets: Combined phenomenological modeling with environmental, exergetic, and energetic analysis (3-E). Biomass Conversion and Biorefinery. 14(8): 9197–9215.

Zhang, X., T. Zhong and H. Lu. 2013. Biochar-supported metal catalysts for environmental remediation. Environmental Chemistry. 10(6): 505–514. https://doi.org/10.1071/EN13118

Zhang, L., J. Qiao, H. Cui, M. Wang and X. Li. 2021. Using low molecularweight organic acids to enhance microbial degradation of polycyclic aromatic hydrocarbons: Current understanding and future perspectives. Water. 13(4): 446.

Zhang, P., W. Duan, H. Peng, B. Pan and B. Xing. 2022. Functional biochar and its balanced design. ACS Environmental Au. 2(2): 115–127.

Zhao, D., B. Yan, C. Liu, B. Yao, L. Luo, Y. Yang, et al. 2021. Mitigation of acidogenic product inhibition and elevated mass transfer by biochar during anaerobic digestion of food waste. Bioresource Technology. 338: 125531.

Zhao, Z., B. Wang, B.K. Theng, X. Lee, X. Zhang, M. Chen and P. Xu. 2022. Removal performance, mechanisms, and influencing factors of biochar for air pollutants: A critical review. Biochar. 4(1): 30.

Chapter 10

Soil Decontamination: Use of Biochar as Support for Biological Crusts

Ana Laura Navas*[a, b], Mario Herrera[b], Rosa Rodriguez[a] and Viviana Fernández-Maldonado[a,c]

[a]Instituto de Ingeniería Química, Fac. de Ingeniería, Grupo Vinculado al PROBIEN (Universidad Nacional de San Juan, CONICET, Universidad Nacional del Comahue), San Juan, Argentina

[b]Grupo de Geobotánica y Fitogeografía-IADIZA-CONICET-Mendoza

[c]Observatorio de Cambio Climático de San Juan, Secretaria de Ambiente y Desarrollo Sustentable, Gobierno de la Provincia de San Juan, Calle 5 y Pelegrini, Al Pie del Cerro Parkison, San Juan J5400ARL, Argentina

10.1 HEAVY METALS: ABUNDANCE AND TOXICITY IN THE ENVIRONMENT

Heavy metals (HMs) are naturally occurring elements, including metals and metalloids with high atomic weights or densities (greater than 5 g/cm^3), and are often toxic to the environment and humans (Rahman and Singh 2019). However, the definition of density is vague and not widely accepted because density is not always biologically or chemically relevant. Of the 90 naturally occurring elements, 53 are considered heavy metals, mostly transition elements, actinides, and lanthanides. Additionally, some p-block elements, such as arsenic, selenium, and aluminum, are also considered heavy metals due to their toxicity (Briffa et al. 2020). All HMs originate from the Earth's crust but can be released into the

*Corresponding author: Ana Laura Navas (anavas@unsj.edu.ar)

DOI: 10.1201/9781032726694-10

environment through human activities such as mining, agriculture, and industrial discharges (Sarker et al. 2023).

10.1.1 Main Properties That Affect the Environment

10.1.1.1 High Density and Toxicity

As mentioned above, one of the defining characteristics of heavy metals is their high density, usually greater than 5 g/cm^3. This gives them a considerably higher mass than other, lighter metals, which affects both their behavior in the environment and their ability to accumulate in organisms. Many heavy metals are highly toxic, even at low concentrations. This toxicity is related to their ability to interfere with various essential biological processes such as photosynthesis, plant growth, and reproduction, as well as cellular function in animals, affecting organs such as the brain, kidneys, and liver (Briffa et al. 2020; Priya and Shankar 2020). Recently, Khalid et al. (2021) demonstrated that the presence of microplastics in soil can increase the mobility and toxicity of heavy metals such as cadmium (Cd), which represents an additional risk for agricultural ecosystems. Furthermore, Naqash et al. (2020) showed how microplastic pollution in freshwater is an emerging environmental problem because the combination of microplastics and heavy metals can alter the behavior of organisms, including reduced predation and changes in the immune response, which affects the health and survival of species.

10.1.1.2 Persistence and Biodegradability in the Environment

Heavy metals are persistent environmental pollutants due to their non-biodegradable nature and long half-lives, which allow them to accumulate in organisms and ecosystems (Mitra et al. 2022). Their accumulation in the environment is a critical problem, as these metals cannot be easily removed through natural processes. This makes them persistent pollutants that can accumulate in soils, water, and organisms, generating long-term risks to ecosystems and human health (Edo et al. 2024).

10.1.1.3 Bioaccumulation

Heavy metals can accumulate in biological tissues over time, a process known as bioaccumulation, posing a significant risk to human and environmental health (Mitra et al. 2022). Fish, for example, can absorb metals through their gills or diet, and these metals can be transferred along the food chain, increasing in concentration among predators. This process, known as biomagnification, can have harmful effects on the health of both animals and humans (Ali et al. 2019). Furthermore, microplastics facilitate the bioaccumulation of heavy metals in the aquatic food chain, which may have repercussions for the health of end consumers, including humans (Khalid et al. 2021).

10.1.1.4 Chemical Reactivity and Formation of Toxic Compounds

Heavy metals are reactive, especially in the form of salts or compounds, and their reactivity depends on the pH and redox potential of the environment. They can form complexes with other substances, altering the chemistry of soil and water and increasing their toxicity, as well as interacting with biomolecules and causing toxicity in living organisms (Rahman and Singh 2019). For example, mercury can be converted into methylmercury (CH_3Hg) in aquatic environments, a much more toxic form that accumulates easily in aquatic organisms, especially fish and shellfish, which can pose a significant risk to human health when consumed (Sarker et al. 2023). The ability of heavy metals to form hazardous compounds increases their potential to cause ecological damage, affecting soil quality, inhibiting plant growth, contaminating water, causing bioaccumulation in the food chain, and leading to diseases and declines in wildlife population (Bharti and Sharma 2022).

10.1.1.5 Ionic and Metallic Forms

Heavy metals can exist in metallic or ionic forms, with the latter generally being more toxic and easily absorbed by organisms. Ionic nickel (Ni^{2+}) can cause cellular damage and cancer, ionic mercury (Hg^{2+}) affects neurological function and vital organs, and arsenic (As^{3+}) interferes with cellular processes and can induce cancer (Bharti and Sharma 2022).

10.1.1.6 Corrosion and Ease of Release

Corrosion of pipes and structures can release heavy metals such as lead and cadmium, which, under certain environmental conditions, transform into toxic ions, increasing their contaminating potential and posing a risk to public health. The release of these metals depends on their physicochemical properties and is exacerbated by industrial activities and river sedimentation, contributing to their toxicity and accumulation in ecosystems (Jadaa and Mohammed 2023).

10.1.2 Natural and Anthropogenic Sources

Their main natural sources include weathering of rocks containing these metals, volcanic eruptions, corrosion of metals in soil, evaporation of metals from water and soil, resuspension of sediments, and soil erosion (Ali et al. 2019). These processes contribute to the presence of heavy metals in the environment, although at concentrations generally lower than those caused by human activities (Jaishankar et al. 2014). However, their accumulation can generate negative impacts when they interact with organisms (Briffa et al. 2020).

Anthropogenic sources of heavy metals include human activities such as mining, industry, agriculture, and the use of chemicals. Specific examples are lead emissions from vehicles, industrial effluents containing heavy metals, and the use of pesticides that can introduce metals into the environment. These activities contribute significantly to soil, water, and air pollution, affecting human health and ecosystems (Jadaa and Mohammed 2023). In 2010, China was the largest producer of chemicals, followed by the United States, Japan, and Germany (Wu et al. 2016). Metals such as lead, thallium, cadmium, and antimony are common contaminants, with thallium being particularly toxic, although less abundant naturally (Rahman and Singh 2019). Exposure to heavy metals such as lead (Pb) and nickel (Ni) can cause serious health problems, including damage to the nervous system, developmental issues, and respiratory diseases. The toxicity of Pb is due to its ability to interfere with calcium and proteins, while Ni is carcinogenic. Additionally, the accumulation of these metals can induce oxidative stress and DNA damage (Rehman et al. 2018). To mitigate these harmful effects, it is crucial to reduce exposure by avoiding areas with high industrial emissions of heavy metals, which are common in soils, water, and air globally.

10.1.3 Effects of Heavy Meals on Soil, Air, and Water

10.1.3.1 Soil

Heavy metal contamination significantly impacts ecosystems by reducing the abundance and biodiversity of soil microorganisms and altering their community structure. These toxic metals can cause enzymatic damage, affect the integrity of cell membranes, and disrupt essential metabolic processes, which in turn inhibit the decomposition of organic matter and contribute to soil degradation (Lenart-Borón and Borón 2014). Additionally, the presence of heavy metals reduces plant growth, affecting seed germination, seedling height, and leaf area. They can also cause chlorosis, necrosis, loss of turgor, and damage to the photosynthetic apparatus, potentially leading to plant death and altering important metabolic processes such as water and nutrient absorption (Rubiya et al. 2018).

10.1.3.2 Air

Some heavy metals, such as mercury, can be released into the air as particles or vapors through industrial processes or the burning of fuels. These metals can be transported by wind over long distances, resulting in the contamination of regions far from the source and impacting ecosystems in remote areas (Cimboláková et al. 2019). Common air pollutants include lead (Pb), iron (Fe), arsenic (As), and mercury (Hg), which are primarily released from anthropogenic sources such as industrial and vehicular emissions, and contribute significant amounts of these metals to the environment (Bautista-Olivas et al. 2017).

10.1.3.3 Water

Heavy metal contamination in water sources is a growing global environmental problem, with sources including mining waste, industrial wastewater, and natural phenomena (Aziz et al. 2023). These metals can leach from the soil or be released through industrial activities, mining, or the burning of fossil fuels, and can seep into drinking water sources, affecting water quality and the well-being of aquatic species. Heavy metal contamination causes toxicity in various marine species, leading to population declines and disruptions in ecological communities. The accumulation of metals in marine organisms can reduce genetic diversity and the ability of species to adapt to environmental changes, while biomagnification throughout the food chain can impact top predators, including humans, and affect the health of marine ecosystems as a whole (Sharifuzzaman et al. 2016).

10.2 AMENDMENTS-ASSISTED PHYTOREMEDIATION

Amendments-assisted phytoremediation is a synergistic approach that combines the natural capabilities of plants to remediate contaminated soils with the application of external agents, such as organic, inorganic, or chemical amendments. This technique addresses the limitations of conventional phytoremediation, including slow remediation rates and limited bioavailability of contaminants, by enhancing soil quality and improving contaminant uptake. Amendments such as biochar, compost, and chelating agents play a crucial role in facilitating metal stabilization, increasing plant biomass, and optimizing the overall efficiency of soil remediation (Sharifuzzaman et al. 2016; Hasan et al. 2019; Manori et al. 2021).

The need for effective soil decontamination has grown due to industrial activities, agricultural practices, and urbanization, which have led to the widespread contamination of soils with heavy metals like cadmium (Cd), lead (Pb), and arsenic (As). These contaminants pose severe risks to ecosystems and public health due to their persistence and toxicity. Amendment-assisted phytoremediation offers a sustainable, low-cost, and environmentally friendly alternative to conventional methods such as excavation and chemical treatments (Ogundiran et al. 2018; Serrano et al. 2024).

10.2.1 Role of Amendments in Phytoremediation

10.2.1.1 Biochar as a Soil Amendment

Biochar, a carbon-rich material derived from the pyrolysis of organic biomass under low-oxygen conditions, has gained significant attention for its ability to improve soil properties and immobilize heavy metals. Key benefits of biochar application include:

- *Soil Property Enhancement:* Biochar improves soil structure, pH, and cation exchange capacity, which in turn enhances microbial activity and nutrient availability. For instance, studies show that biochar combined with hyperaccumulator plants like *Brassica napus* and *Bidens pilosa* significantly boosts biomass production and metal uptake (Gasco et al. 2019; Simieli et al. 2020; Manori et al. 2021).
- *Metal Immobilization:* The high porosity and active functional groups in biochar enable it to adsorb and stabilize heavy metals like Cd, Pb, and As, reducing their mobility and toxicity. In a study with *Salix purpurea*,

biochar effectively decreased metal bioavailability while supporting robust root development (Lebrun et al. 2018; Simiele et al. 2020).

- *Synergistic Effects on Plants:* Biochar assists plants in tolerating contaminated soils by mitigating oxidative stress and providing essential nutrients. For example, the addition of biochar enhanced the phytoextraction capacity of *Lolium perenne* by improving root and shoot growth (Zhang et al. 2019; Serrano et al. 2024).

10.2.1.2 Organic Amendments

Organic materials, including compost and manure-derived amendments, are widely used to improve soil fertility and support plant growth. Compost, in particular, has shown remarkable potential in phytoremediation applications:

- *Nutrient Enrichment:* Compost adds organic matter and nutrients to degraded soils, enhancing their fertility and water retention capacity. Studies involving *Moringa oleifera* indicate that compost improves root and shoot growth, enabling the plant to tolerate higher levels of Pb contamination (Ogundiran et al. 2018).
- *Phytostabilization Support*: Compost reduces the bioavailability of heavy metals by forming stable complexes, limiting their mobility in the soil. For instance, compost-amended soils treated with *Moringa oleifera* showed reduced Pb leaching, supporting long-term stabilization (Ogundiran et al. 2018; Serrano et al. 2024).

10.2.1.3 Chemical Amendments

Chemical agents such as ethylenediaminetetraacetic acid (EDTA), phosphorus-loaded biochar, and iron sulfates are frequently used to enhance phytoremediation efficiency by increasing the solubility and bioavailability of contaminants:

- *Chelating Agents:* EDTA and similar compounds enhance the solubility of heavy metals, allowing their uptake by plants. However, excessive use can lead to secondary pollution, necessitating careful management (Hasan et al. 2019; Serrano et al. 2024).
- *Phosphorus-Loaded Biochar:* The incorporation of phosphorus-loaded biochar has shown promising results in immobilizing Cd, Pb, and Cr in contaminated soils. This approach not only stabilizes contaminants but also improves soil nutrient profiles, enhancing plant health and remediation capacity (Serrano et al. 2024).
- *Iron Sulfates:* These compounds have been used effectively to immobilize arsenic in contaminated soils, supporting the growth of phytostabilizing species like *Salix purpurea* (Simiele et al. 2020).

10.2.2 Mechanisms of Enhancement

10.2.2.1 Soil Quality Improvement

Amendments such as biochar and compost enhance soil pH, organic matter content, and microbial activity, creating a conducive environment for plant growth. Improved soil quality supports the establishment of vegetation on degraded lands, which is essential for successful phytoremediation (Lebrun et al. 2018; Simiele et al. 2020).

10.2.2.2 Increased Metal Uptake and Stabilization

The addition of amendments facilitates both phytoextraction and phytostabilization. While biochar immobilizes metals in the rhizosphere, compost and chelating agents enhance their bioavailability for uptake into plant tissues (Gasco et al. 2019; Manori et al. 2021).

10.2.2.3 Enhanced Microbial Interactions

Amendments foster microbial diversity and activity, which play a vital role in nutrient cycling and contaminant degradation. For example, biochar-amended soils exhibit increased populations of beneficial bacteria, further supporting soil remediation (Ogundiran et al. 2018; Zhang et al. 2019).

10.3 DIRECT IMMOBILIZATION

Direct immobilization represents a cornerstone in the efforts to remediate contaminated soils, particularly those affected by heavy metals such as lead (Pb), cadmium (Cd), arsenic (As), and mercury (Hg). These pollutants, often resulting from industrial activities, mining operations, and agricultural practices, pose persistent threats due to their toxicity and long-term environmental stability (Ahmad et al. 2014; Mukherjee et al. 2022). Unlike extraction or transformation techniques, direct immobilization seeks to neutralize contaminants within the soil matrix, significantly reducing their bioavailability and leaching potential without disturbing the ecosystem's balance (Chen et al. 2011; Mukherjee et al. 2022).

In this context, biochar—a porous, carbon-rich material derived from the pyrolysis of biomass—has gained considerable attention. Its structural characteristics, including high surface area and abundant functional groups, position it as an effective agent for immobilizing contaminants, particularly in conjunction with biological crusts that stabilize the soil surface (Ahmad et al. 2014; Navas et al. 2023).

Several immobilization strategies have been explored for soil decontamination, each with distinct advantages and challenges. While biochar is increasingly recognized for its sustainability and cost-effectiveness, other materials like nanomaterials and inorganic amendments also contribute to remediation efforts. Table 10.1 provides a comparative overview of these approaches.

As outlined in Table 10.1, biochar offers a unique balance between effectiveness and sustainability, making it an attractive choice for agricultural soil decontamination. However, its potential limitations, such as long-term saturation, underscore the need for advancements in regeneration technologies. Compared to nanomaterials, biochar is more accessible and environmentally friendly, although it may lack the precision offered by advanced nanotechnologies.

Table 10.1 Comparison of immobilization strategies for soil decontamination, highlighting their advantages, limitations, and specific use cases

Immobilization Strategy	Advantages	Limitations	Example of Use
Biochar	High adsorption capacity; sustainable; cost-effective.	Long-term saturation; limited regeneration.	Agricultural soils are contaminated with Pb and Cd.
Nanomaterials	High efficiency; applicable at low concentrations.	Expensive; potential toxic effects.	Electronics industry.
Inorganic Amendments	Widely available; easy to apply.	Long-term effects are unknown.	Mining soils contaminated with As.

This section explores the principles, mechanisms, and applications of biochar in direct immobilization, emphasizing its synergistic role when combined with biological crusts to enhance soil decontamination processes.

10.3.1 Mechanisms of Direct Immobilization Using Biochar

Biochar's efficacy in direct immobilization is attributed to several interrelated mechanisms that limit the mobility of contaminants within the soil. The mechanisms through which biochar interacts with various soil contaminants are diverse, leveraging its structural and chemical properties to neutralize pollutants effectively. Table 10.2 provides an overview of how biochar functions across different contaminant types, highlighting specific immobilization mechanisms and their practical applications. This comparison underscores the versatility of biochar as a remediation tool, from reducing heavy metal bioavailability to supporting the degradation of hydrocarbons in petroleum-affected soils (Chen et al. 2011; Song et al. 2024).

Table 10.2 Illustrates the immobilization mechanisms employed by biochar to address different contaminant types, with specific examples showcasing its effectiveness in various soil contamination scenarios

Contaminant	Immobilization Mechanism	Example of Effectiveness
Lead (Pb)	Adsorption and precipitation.	Reduced bioavailability in agricultural soils.
Arsenic (As)	Specific adsorption through oxygenated groups.	Removal from mining soils.
Hydrocarbons	Adsorption within biochar pores.	Degradation in petroleum-contaminated soils.

10.3.1.1 Adsorption and Surface Interactions

The porous structure of biochar provides an extensive surface area for the adsorption of heavy metals and organic pollutants. Functional groups such as hydroxyl (–OH), carboxyl (–COOH), and carbonyl (C=O) groups interact with conta-minants via physical adsorption, ion exchange, or covalent bonding, creating stable complexes that are less bioavailable (Ahmad et al. 2014; Mukherjee et al. 2022).

10.3.1.2 Surface Precipitation

Biochar facilitates the precipitation of contaminants by acting as a nucleation site. Metals such as arsenic and lead can form insoluble compounds (e.g., arsenates and lead phosphates) on the biochar surface, further reducing their solubility and mobility (Chen et al. 2011).

10.3.1.3 Ion Exchange Processes

Biochar's negative surface charge enables it to exchange cations, such as Pb^{2+} and Cd^{2+}, with its inherent positively charged functional groups. This exchange stabilizes contaminants in forms that are less likely to leach into groundwater (Ahmad et al. 2014).

10.3.1.4 pH Alteration and Speciation

The alkaline nature of biochar can increase soil pH, affecting the speciation of metals. For example, at higher pH levels, metals like chromium and lead form less soluble hydroxides, which are immobilized in the soil matrix (Ghosh et al. 2017).

10.3.1.5 Interactions with Biological Crusts

In ecosystems where biological crusts are present, biochar provides structural support that enhances the crust's ability to bind pollutants. The symbiotic relationship between biochar and microbial communities may facilitate the biodegradation of organic contaminants while simultaneously immobilizing metals (Song et al. 2024).

10.3.2 Benefits and Limitations of Biochar in Direct Immobilization

10.3.2.1 Benefits of Biochar in Direct Immobilization

- *Economic Viability:* Biochar is produced from low-cost biomass sources such as agricultural residues, forestry byproducts, and urban organic waste, making it an economically feasible material for large-scale soil remediation projects. Studies indicate that biochar-based interventions can reduce overall remediation costs by 20–40% compared to conventional soil amendments (Mukherjee et al. 2022; Sanchez et al. 2024).
- *Environmental Sustainability*: Biochar is a carbon-negative material that sequesters carbon during its production through pyrolysis. This

sequestration contributes to climate change mitigation while enhancing soil health. Additionally, biochar improves soil structure, increases water retention, and reduces erosion, offering long-term environmental benefits (Chen et al. 2011; Ahmad et al. 2014).

- *Long-Term Stability and Persistence:* Biochar exhibits exceptional resistance to microbial degradation and environmental weathering, making it a reliable solution for long-term contaminant immobilization. Studies show that biochar can maintain over 90% of its structural integrity for decades under diverse soil conditions (Ahmad et al. 2014).
- *Versatility Across Soil Types*: Biochar effectively buffers pH and enhances nutrient availability, making it suitable for both acidic and alkaline soils. This adaptability enables its application across various contaminated environments, including agricultural lands and heavily polluted industrial sites (Mukherjee et al. 2022).

10.3.2.2 Limitations of Biochar in Direct Immobilization

- *Saturation of Adsorption Sites:* Over time, biochar's active sites become saturated with contaminants, reducing its efficiency. Regeneration methods such as chemical washing or thermal reactivation are often energy-intensive and may degrade the biochar's structure, limiting its reuse (Ghosh et al. 2017).
- *Competitive Adsorption in Organic-Rich Soils:* High organic matter content in soils can compete with contaminants for adsorption sites on biochar, reducing its immobilization efficiency. Organic matter can also alter the speciation of heavy metals, affecting their interaction with biochar (Song et al. 2016).
- *Variability Due to Feedstock and Pyrolysis Conditions:* The properties of biochar, such as porosity and functional groups, vary significantly based on feedstock type and pyrolysis parameters. This variability can lead to inconsistent performance, with some biochars exhibiting limited effectiveness against specific contaminants (Ahmad et al. 2014; Mukherjee et al. 2022).
- *Potential Environmental Risks:* Improperly produced biochar may release harmful compounds such as polycyclic aromatic hydrocarbons (PAHs) or residual heavy metals. These risks necessitate stringent quality control during production to ensure safety and efficacy (Chen et al. 2011).
- *Addressing Limitations and Enhancing Benefits:* Advances in biochar technology are addressing these challenges. Techniques such as chemical and biological activation enhance adsorption capacity, while improvements in production processes aim to reduce variability. Research into cost-effective regeneration methods, as well as the integration of biochar with advanced remediation technologies like nanomaterials and microbial consortia, offers promising pathways to overcome current limitations and maximize biochar's potential in soil decontamination.

10.3.3 Applications in Soil Decontamination

The application of biochar in soil decontamination has proven to be a versatile and effective approach for mitigating pollution across a variety of environmental contexts. Leveraging its structural and chemical properties, biochar addresses the immobilization of heavy metals, stabilization of contaminants, and enhancement of soil health. This section delves into its use in agricultural, industrial, and mining soils, as well as the benefits of integrated approaches, such as combining biochar with biological crusts, for improved outcomes in soil remediation (Jiang et al. 2022).

10.3.3.1 Agricultural Soils: Protecting Food Systems and Enhancing Productivity

Agricultural soils are often at risk of heavy metal contamination due to the application of phosphate fertilizers, irrigation with polluted water, and industrial emissions. Metals such as cadmium (Cd) and lead (Pb) pose significant threats to food safety and human health. Biochar has demonstrated remarkable potential in mitigating these risks by reducing metal bioavailability and enhancing soil quality.

One notable example is the use of rice husk biochar in paddy fields, where cadmium mobility was reduced by 70%, leading to safer rice production (Ghosh et al. 2017). The porous structure of biochar adsorbs metal ions, while its alkaline nature increases soil pH, precipitating metals into less soluble forms. Additionally, biochar enhances water retention and nutrient availability, which is critical for sustaining crop yields in contaminated soils.

Furthermore, biochar derived from agricultural waste aligns with circular economy principles, converting residues into valuable inputs for soil remediation. This dual benefit supports sustainable farming practices while addressing contamination challenges.

10.3.3.2 Industrial and Mining Soils: Tackling Toxic Legacy Pollutants

Industrial and mining activities contribute significantly to soil pollution, particularly through the release of arsenic (As), mercury (Hg), and other toxic elements. The proximity of these sites to natural ecosystems amplifies the urgency for effective decontamination strategies.

Biochar produced from woody biomass has been shown to immobilize arsenic effectively in mining-impacted soils. Chen et al. (2011) reported reductions in arsenic leaching rates exceeding 80%, highlighting biochar's potential to mitigate groundwater contamination risks. The mechanisms of action include specific adsorption facilitated by oxygen-containing functional groups and precipitation of arsenic compounds on the biochar surface.

In industrial settings, biochar's high surface area and tunable properties make it adaptable to different contaminants. For example, biochar treated with iron oxides has demonstrated enhanced adsorption of chromium (Cr) from industrial effluents, further broadening its applicability in soil remediation.

10.3.3.3 Combined Approaches: Synergistic Effects with Biological Crusts

Combining biochar with biological crusts offers an innovative solution to complex soil contamination challenges. Biological crusts—microbial communities comprising bacteria, fungi, and algae—stabilize soil surfaces and degrade organic pollutants. When integrated with biochar, these crusts benefit from its structural support and enhanced adsorption properties, creating a synergistic system for contaminant immobilization.

Song et al. (2016) demonstrated that microbial communities thrive on biochar surfaces, utilizing their porosity and functional groups to enhance pollutant degradation. Simultaneously, biochar reduces the mobility of heavy metals, allowing biological processes to focus on breaking down organic contaminants. This dual-action system has shown particular promise in remediating soils contaminated with hydrocarbons and mixed pollutants, offering a comprehensive approach to decontamination.

Moreover, the integration of biochar with biological crusts aligns with ecosystem restoration goals by promoting soil stability, reducing erosion, and supporting native microbial diversity. This approach not only addresses contamination but also enhances the ecological resilience of the treated area.

10.3.4 Synergies with Biological Crusts

The synergistic interaction between biochar and biological crusts represents a pioneering strategy in the field of soil decontamination. Biological crusts, comprising cyanobacteria, lichens, mosses, and other microorganisms, play a critical role in stabilizing soil surfaces, enhancing water retention, and mediating biogeochemical cycles (Büdel et al. 2014; Navas-Romero et al. 2023). When combined with biochar, these crusts exhibit an enhanced capacity to immobilize contaminants, providing both structural integrity and active pollutant removal through adsorption and microbial degradation processes.

Biochar serves as a physical support for the establishment and proliferation of biological crusts. Its porous structure offers a habitat for microbial communities, promoting the growth of cyanobacteria and other key organisms in crust formation (Chen et al. 2011; Palansooriya et al. 2019; Chang et al. 2022). Additionally, the chemical properties of biochar, such as the presence of oxygenated functional groups and its ability to modulate pH, create a favorable microenvironment for microbial activity, enhancing the crust's functionality in contaminant immobilization (Ahmad et al. 2014; Navas-Romero et al. 2021).

10.3.4.1 Enhanced Metal Adsorption and Stability

The integration of biochar and biological crusts has demonstrated significant improvements in the retention and immobilization of heavy metals, including lead (Pb) and arsenic (As). Studies have shown that biochar-modified soils with established crusts exhibit lower leaching rates of contaminants compared to soils treated with biochar alone (Chang et al. 2022; Mukherjee et al. 2022). This dual approach combines the adsorption capabilities of biochar with the biostabilization properties of crusts, effectively reducing the bioavailability of toxic metals and mitigating their environmental impact.

10.3.4.2 Organic Pollutant Degradation

In soils contaminated with hydrocarbons and other organic pollutants, biochar acts as a catalyst for microbial degradation. Biological crusts, enriched by the structural support of biochar, facilitate the breakdown of complex organic compounds into less harmful intermediates. This process is particularly effective in petroleum-affected soils, where microbial consortia within the crust utilize biochar as both a nutrient reservoir and a scaffold for enzymatic activity (Song et al. 2016; Navas-Romero et al. 2021).

10.3.4.3 Applications in Arid and Semi-Arid Environments

The utility of biochar and biological crusts is particularly relevant in arid and semi-arid regions, where soil degradation and contamination are prevalent challenges. The introduction of biochar enhances the water retention capacity of these soils, creating a stable foundation for the establishment of biological crusts. Research by Chang et al. (2022) highlights the potential of this integrated approach to restore functionality in degraded ecosystems, emphasizing its role in improving soil health and contaminants' immobilization under extreme environmental conditions.

10.4 MODIFICATION OF BIOCHAR FOR IMPROVING HEAVY METAL ADSORPTION

Soil contamination by heavy metals and persistent pollutants is a critical environmental issue, adversely affecting ecological stability and human health. These pollutants, often originating from industrial, agricultural, and urban activities, resist natural degradation, leading to bioaccumulation and long-term environmental risks (Chen et al. 2011; Ahmad et al. 2014). Effective soil decontamination strategies must not only immobilize contaminants but also promote soil recovery. In this context, biochar—a porous carbonaceous material derived from the pyrolysis of biomass—has emerged as a transformative solution due to its multifunctionality, cost-effectiveness, and environmental sustainability (Navas-Romero et al. 2021; Mukherjee et al. 2022).

Biochar's inherent properties, including a high surface area, abundant functional groups, and structural stability, make it a promising candidate for immobilizing heavy metals and organic pollutants (Chen et al. 2011). However, unmodified biochar often suffers from variability in performance due to differences in feedstock and pyrolysis conditions, limiting its application across diverse soil types (Ghosh et al. 2017). Recognizing these challenges, research has focused on strategic modifications—both physical and chemical—to enhance biochar's adsorption capacity and consistency (Ahmad et al. 2014; Song et al. 2016).

Physical modifications, such as steam or CO_2 activation, improve biochar's porosity and surface area, creating more adsorption sites for contaminants (Rashid et al. 2022). Chemical treatments, including acid functionalization and metal oxide doping, increase the reactivity and specificity of biochar, enabling it to target specific contaminants like lead (Pb), cadmium (Cd), and arsenic (As) (Mukherjee et al. 2022). These tailored modifications significantly enhance biochar's capacity to immobilize a wide range of heavy metals and organic pollutants, while also expanding its applicability in diverse soil contamination scenarios (Chen et al. 2011; Ahmad et al. 2014).

In addition to its intrinsic capabilities, modified biochar demonstrates synergistic effects when combined with biological crusts, microbial communities composed of cyanobacteria, lichens, and fungi. These crusts stabilize soil surfaces and contribute to biogeochemical cycling, while biochar provides a supportive substrate for microbial colonization and activity (Navas-Romero et al. 2021). This interaction enhances both pollutant immobilization and soil restoration, offering a sustainable and integrated solution for contaminated soils (Song et al. 2016; Chang et al. 2022).

This chapter delves into the principles and mechanisms underlying the modification of biochar for improving heavy metal adsorption, emphasizing its integration with biological crusts for enhanced decontamination. By tailoring biochar's properties to optimize its interaction with contaminants and leveraging its synergy with biological processes, this approach addresses the multifaceted challenges of soil contamination. These advancements underscore the critical role of innovation in bridging the gap between laboratory research and real-world environmental applications, particularly in regions with fragile ecosystems and high contamination levels (Ahmad et al. 2014; Navas-Romero et al. 2021).

10.4.1 Physical Activation of Biochar

Physical activation enhances biochar's structural properties, optimizing it for the adsorption of heavy metals and integration with biological crusts. These processes focus on increasing porosity, surface area, and mechanical resilience, which are critical for improving biochar's functionality in contaminated soils without altering its inherent chemical characteristics.

10.4.1.1 Porosity Enhancement

One of the primary goals of physical activation is to maximize biochar's porosity and surface area, key factors in its adsorption performance. Methods such as steam activation, CO_2 treatment, and high-temperature pyrolysis have demonstrated significant success in creating a network of micropores and mesopores critical for capturing heavy metals.

For instance, steam activation at temperatures above 800°C can enhance surface areas by over 300%, providing a greater number of adsorption sites (Ahmad et al. 2014). Studies show that lead (Pb) and cadmium (Cd) exhibit adsorption capacities up to 120 mg/g on steam-activated biochars, outperforming their unmodified counterparts (Mukherjee et al. 2022). High-temperature pyrolysis, while less specific than chemical methods, produces consistent porosity profiles suitable for large-scale applications (Ghosh et al. 2017).

10.4.1.2 Structural Stability

Physical activation not only enhances porosity but also strengthens the structural integrity of biochar, enabling its long-term application in various soil environments. Stability is particularly crucial for integration with biological crusts, as the activated biochar must maintain its mechanical properties under natural and contaminated conditions. Enhanced stability supports microbial communities, such as cyanobacteria and fungi, which thrive on biochar surfaces, aiding soil stabilization and pollutant immobilization in regions vulnerable to erosion and contamination (Navas-Romero et al. 2021).

10.4.1.3 Influence on Adsorption Kinetics

Activated biochar's improved surface structure facilitates rapid adsorption kinetics. Faster uptake rates result from increased accessibility of active sites and optimized pore distribution. Chen et al. (2011) reported that physically activated biochars reduced equilibrium times for lead adsorption by nearly 50%, making them highly effective for rapid remediation. This feature is particularly beneficial for industrial settings where time efficiency is critical.

10.4.2 Chemical Modification of Biochar

Chemical modification strategies enhance the surface chemistry of biochar, introducing functional groups or doping with additional elements to improve its affinity and selectivity for contaminants. These modifications enable biochar to immobilize a broader spectrum of heavy metals and other pollutants more effectively.

10.4.2.1 Acid and Base Treatments

- *Acidic Treatments:* Treatments using strong acids such as HNO_3 or H_2SO_4 are widely employed to increase the concentration of oxygen-containing functional groups on biochar's surface. These groups, such as carboxyl (–COOH) and hydroxyl (–OH), significantly improve biochar's ability to adsorb heavy metals through mechanisms like coordination and ion exchange. For example, nitric acid-treated biochar demonstrated a cadmium removal efficiency exceeding 90% in contaminated soils, offering significant advantages over unmodified biochar (Ahmad et al. 2014). These groups also enhance biochar's hydrophilicity, facilitating better interaction with contaminants in aqueous environments (Mukherjee et al. 2022).
- *Base Treatments:* Alkaline activators like NaOH and KOH modify biochar by increasing its surface alkalinity, which is particularly effective for adsorbing metals like chromium (Cr) and arsenic (As). Elevated pH

levels promote the precipitation of metals into hydroxides, reducing their solubility and mobility. Ghosh et al. (2017) demonstrated that KOH-activated biochar reduced arsenic leaching rates in mining soils by 85%.

10.4.2.2 Metal Oxide Doping

Doping biochar with metal oxides such as Fe_2O_3, MnO_2, or TiO_2 enhances its adsorption performance by introducing additional active sites and catalytic properties. Iron-doped biochar, for instance, effectively immobilizes arsenic through adsorption and reduction mechanisms, transforming it into less toxic forms (Song et al. 2024). These doped biochars also exhibit synergistic effects with biological crusts, fostering microbial activity that complements pollutant immobilization (Chang et al. 2022).

Research has highlighted that MnO_2-doped biochar excels in adsorbing lead and cadmium ions due to its oxidative properties, while TiO_2-doped biochar offers superior photodegradation capabilities, expanding its utility beyond heavy metals to organic pollutants (Chen et al. 2011).

10.4.2.3 Functionalization with Amine Groups

Functionalizing biochar with nitrogen-containing compounds such as ammonia or urea introduces amine ($-NH_2$) groups, which enhance its cation exchange capacity (CEC). This functionalization is particularly effective for adsorbing anionic pollutants like nitrate (NO_3^-) and phosphate (PO_4^{3-}) in soils co-contaminated with heavy metals.

Amine-functionalized biochar has been shown to remove over 95% of nitrates in agricultural soils while simultaneously immobilizing lead and cadmium (Mukherjee et al. 2022). This dual functionality makes it an invaluable tool for integrative soil remediation approaches, addressing nutrient pollution and heavy metal contamination concurrently.

10.4.3 Synergy Between Biochar Modifications and Biological Crusts

The integration of modified biochar and biological crusts offers a transformative approach to soil decontamination, leveraging the structural and chemical advancements of biochar to complement the ecological functionality of crust-forming organisms. This synergy enhances the stabilization of contaminated soils, augments pollutant immobilization, and fosters ecosystem resilience. Both physical and chemical modifications to biochar—such as porosity enhancement, functional group incorporation, and metal oxide doping—align with the biological and physicochemical needs of crusts, thereby amplifying their remediation potential.

10.4.3.1 Enhanced Metal Retention

The combination of biochar's modifications and biological crusts creates an optimized system for the retention and immobilization of heavy metals such as lead (Pb), cadmium (Cd), and arsenic (As). Modified biochar serves as a durable substrate, providing structural support and additional adsorption sites that complement the natural bio-remediating properties of biological crusts.

- *Mechanism of Enhanced Retention:* Physically activated biochar, characterized by high porosity and extensive surface area, offers abundant adsorption sites for metal ions. These sites are further augmented by chemical treatments, such as acid-induced carboxyl (–COOH) and hydroxyl (–OH) functional groups, which bind metals through ionic and covalent interactions (Ahmad et al. 2014; Ghosh et al. 2017). Once immobilized, heavy metals are stabilized within the matrix of the biological crust, where microbial communities enhance retention by transforming metals into less bioavailable forms via precipitation or redox processes (Navas-Romero et al. 2021).
- *Field Evidence:* Research conducted in mining-affected soils demonstrated that biochar treated with iron oxide increased arsenic retention by 85% when combined with biological crusts, reducing leaching rates and groundwater contamination risks (Ghosh et al. 2017). These findings underline the role of biochar-crust systems in addressing heavy metal contamination in industrial and agricultural landscapes.

10.4.3.2 Dual Action on Organic Pollutants

In mixed contamination scenarios, such as soils impacted by hydrocarbons and heavy metals, the integration of biochar modifications and biological crusts provides a synergistic solution. While biochar excels in adsorbing hydrophobic organic pollutants, biological crusts actively degrade them through enzymatic activity, achieving comprehensive remediation.

- *Biochar's Role:* Chemical modification techniques, such as nitrogen doping or the addition of amine functional groups ($-NH_2$), improve biochar's affinity for organic contaminants like polycyclic aromatic hydrocarbons (PAHs) and petroleum hydrocarbons. These functionalized biochars trap pollutants in their pores, reducing their mobility and bioavailability (Mukherjee et al. 2022).
- *Crusts' Contribution:* Biological crusts utilize microbial enzymes to break down complex organic compounds into less harmful byproducts. Cyanobacteria, lichens, and fungi within the crust form a dynamic biogeochemical network that enhances degradation rates (Navas-Romero et al. 2021). Enzymes such as laccases and peroxidases target the aromatic structures of hydrocarbons, while symbiotic interactions within the crust enhance efficiency in nutrient cycling and pollutant degradation (Büdel et al. 2014).
- *Case Studies and Applications:* A study on soils contaminated with diesel hydrocarbons revealed that integrating nitrogen-doped biochar with cyanobacteria-based crusts achieved a 70% reduction in hydrocarbon concentration over six months. This dual-action system not only degraded

organic pollutants but also immobilized cadmium co-contaminants, highlighting its effectiveness in remediating mixed pollution (Chen et al. 2011; Song et al. 2024).

10.4.3.3 Benefits of the Biochar-Crust Synergy

The integration of modified biochar and biological crusts provides several ecological and practical benefits:

- *Stabilization of Eroded Soils:* Biochar's structural resilience supports crust formation in degraded soils, preventing erosion and promoting ecosystem recovery (Ahmad et al. 2014).
- *Enhanced Biodiversity:* The microhabitats created by biochar modifications foster microbial and plant diversity, contributing to the restoration of contaminated ecosystems (Chang et al. 2022).
- *Sustainability and Cost-Effectiveness:* Utilizing agricultural waste-derived biochar and naturally occurring crusts offers an economically viable and environmentally sustainable remediation strategy (Mukherjee et al. 2022).
- *Climate Mitigation Co-Benefits:* Biochar's carbon-negative properties align with global climate goals, while crusts enhance soil carbon sequestration (Song et al. 2024).

10.5 BIOLOGICAL ACTIVATION OF BIOCHAR FOR ENHANCED HEAVY METAL ADSORPTION

Biochar, a carbon-rich material derived from the pyrolysis of organic biomass under limited oxygen conditions, has garnered significant attention in recent years due to its multifunctional applications in environmental remediation, particularly for the adsorption of heavy metals from contaminated soils and water systems. Its inherent physicochemical properties—such as high surface area, extensive porosity, and the presence of diverse functional groups (e.g., carboxyl, hydroxyl, and phenolic groups)—enable biochar to act as a potent sorbent for toxic metals like cadmium (Cd), lead (Pb), chromium (Cr), and arsenic (As) (Ahmad et al. 2014; Lehmann et al. 2015). Additionally, biochar's stability and cost-effectiveness make it an attractive alternative to traditional remediation techniques.

Despite its potential, the adsorption efficiency of unmodified biochar is often limited by its lack of specific binding sites and the variability in its surface chemistry, which depends on the feedstock and pyrolysis conditions (Inyang et al. 2012). This limitation can result in reduced affinity for certain metal ions or inadequate capacity to immobilize metals under field conditions. Consequently, researchers have explored various modification strategies to enhance biochar's adsorption capabilities. Among these, biological activation has emerged as a particularly promising approach.

Biological activation involves the inoculation of biochar with microorganisms or their metabolic byproducts to improve its physicochemical properties. This process not only enhances the surface functionalization of biochar through the introduction of new binding sites but also modifies its structural attributes, such as porosity and surface area (Shen et al. 2020; Wu et al. 2022). Additionally, microorganisms involved in the activation process can produce organic acids, extracellular polymeric substances (EPS), and other metabolites that facilitate the chelation and immobilization of heavy metals, thereby amplifying biochar's adsorption efficiency (Liang et al. 2021). This dual-action approach, leveraging both biochar's inherent properties and microbial contributions, offers a sustainable and effective solution for heavy metal remediation.

The application of biologically activated biochar in contaminated soils presents numerous advantages, including enhanced adsorption capacity, reduced metal bioavailability, and improved soil health through microbial activity. However, challenges remain, such as the scalability of the activation process, potential microbial degradation of biochar, and the variability in performance under different environmental conditions (Ahmad et al. 2014; Kumar et al. 2019). This section explores the mechanisms, methodologies, benefits, and challenges associated with the biological activation of biochar, focusing on its role in the remediation of heavy metal-contaminated soils.

10.5.1 Mechanisms of Biological Activation

The enhancement of biochar's adsorption capabilities through biological activation is primarily attributed to the interactions between microorganisms and the biochar matrix. These interactions result in significant modifications to the biochar's surface chemistry and structure, facilitating improved heavy metal binding. The key mechanisms include:

10.5.1.1 Microbial Colonization and Biofilm Formation

Microorganisms such as bacteria and fungi are capable of colonizing the biochar surface, forming biofilms that introduce additional functional groups (e.g., carboxyl, hydroxyl) with high affinity for heavy metals. Extracellular polymeric substances (EPS) produced by these microorganisms play a crucial role in enhancing metal binding. For instance, colonization of biochar by *Pseudomonas putida* has been shown to improve Pb(II) ion adsorption, as the EPS introduces negatively charged sites that increase metal retention capacity (Ahmad et al. 2014; Shen et al. 2020). Biofilms not only increase the functional group density but also protect the biochar surface from fouling, maintaining its adsorption efficiency over time.

10.5.1.2 Production of Organic Acids and Metabolites

Many microorganisms secrete organic acids (e.g., citric and oxalic acids) and metabolites capable of chelating heavy metals. These compounds form stable complexes that are subsequently adsorbed onto the biochar surface, effectively immobilizing the metals and reducing their bioavailability and toxicity. For example, Bacillus subtilis and Aspergillus niger have been reported to produce oxalic acid, which enhances the adsorption of Cd(II) and Pb(II) through the formation of metal-organic complexes (Wu et al. 2022). This process not only enhances metal retention but also contributes to the long-term stabilization of contaminants in soils.

10.5.1.3 Enhancement of Surface Area and Porosity

Microbial activity can lead to the degradation of organic components within biochar, creating new pores and increasing the overall surface area available for adsorption. The enhanced porosity provides additional active sites for the adsorption of heavy metals. For instance, biochar inoculated with *Aspergillus niger* exhibited a 25% increase in porosity, which significantly improved its capacity to adsorb Cd(II) ions (Liang et al. 2021). This structural modification is critical for applications in highly contaminated environments where adsorption capacity needs to be maximized.

10.5.1.4 Redox Reactions Mediated by Microorganisms

Certain bacteria possess enzymatic systems capable of reducing toxic metal ions to less toxic or insoluble forms, a process that can enhance the immobilization of metals within the biochar matrix. For example, *Shewanella oneidensis* reduces Cr(VI) to Cr(III), which precipitates as $Cr(OH)_3$, effectively sequestering the metal within the biochar (Kumar et al. 2019). Similarly, sulfate-reducing bacteria (SRB) can precipitate heavy metals such as Pb and Cd as sulfides, which are highly stable under reducing conditions, further enhancing the long-term immobilization of contaminants.

10.5.2 Methods of Biological Activation

The biological activation of biochar involves employing microorganisms to enhance its physicochemical properties and adsorption capabilities for heavy metal remediation. Several strategies have been developed, each with specific procedures, advantages, and applications. Below are the main methods, detailed with step-by-step procedures and supported by the literature.

10.5.2.1 Direct Inoculation with Microbial Cultures

Description: This method involves introducing specific microbial strains known for their metal-binding capabilities directly onto biochar. The process promotes microbial colonization and the production of extracellular polymeric substances (EPS), which enhance biochar's ability to adsorb heavy metals.

Procedure: Selection of Microbial Strain: Choose microorganisms with proven heavy metal binding properties, such as *Bacillus subtilis*, *Pseudomonas putida*, or *Aspergillus niger*. Preparation of Biochar: Clean and sterilize the biochar to remove contaminants and ensure a suitable surface for microbial attachment. Inoculation: Suspend the biochar in a liquid culture medium containing the selected microbial strain at an optimal concentration (e.g., 10^7 CFU/mL).

Incubation: Incubate the mixture under controlled conditions (e.g., 30–37°C, neutral pH) for a specific period (24–72 hours), allowing the microorganisms to colonize the biochar surface. Harvest and Drying: Separate the biochar from the culture medium, rinse gently to remove unattached cells, and dry under low temperatures to preserve microbial structures.

Example: *Bacillus subtilis*-inoculated biochar has demonstrated increased Pb(II) adsorption due to the production of EPS, which introduces carboxyl and hydroxyl groups onto the biochar surface (Ahmad et al. 2014; Shen et al. 2020).

10.5.2.2 Co-Composting with Organic Matter

Description: This method integrates biochar with organic waste materials during composting. The process fosters a diverse microbial community and enhances nutrient content, cation exchange capacity (CEC), and porosity of biochar.

Procedure: Preparation of Mixture: Combine biochar with organic waste materials, such as agricultural residues, manure, or kitchen waste, in a ratio of 1:3 to 1:5 (biochar: organic matter). Composting Setup: Pile the mixture in a composting bin or windrow, ensuring adequate aeration by turning it periodically. Moisture Management: Maintain moisture levels at 50–60% by adding water as needed. Temperature Monitoring: Allow the compost to undergo thermophilic and mesophilic phases (50–70°C) for 2–3 months, during which microbial activity modifies the biochar surface. Curing and Screening: Let the compost mature, then screen to separate the enriched biochar.

Example: Co-composted biochar has shown increased Cd(II) and Zn(II) adsorption compared to raw biochar, due to the enhanced microbial colonization and functionalization during the composting process (Liang et al. 2021; Wu et al. 2022).

10.5.2.3 Immobilization in Bioreactors

Description: Biochar is introduced into bioreactors where it interacts with microbial consortia under optimized environmental conditions. This method allows precise control over variables such as pH, temperature, and nutrient availability.

Procedure: Reactor Setup: Place biochar into a bioreactor system, such as a fixed-bed or fluidized-bed reactor. Introduction of Microbial Consortia: Inoculate the reactor with specific microbial consortia, such as sulfate-reducing bacteria (SRB) or metal-reducing bacteria like *Shewanella oneidensis*. Optimization of Conditions: Adjust parameters like pH (neutral to slightly alkaline), temperature (25–40°C), and nutrient levels to maximize microbial activity. Operation: Allow the system to run continuously or in batch mode for 1–2 weeks, facilitating microbial colonization and biochar modification. Harvesting: Retrieve the biochar, rinse to remove excess biomass, and dry under controlled conditions.

Example: Bioreactor-immobilized biochar has been shown to effectively remove Pb(II) and Cd(II) ions, leveraging the metabolic activities of SRB to precipitate metals such as sulfides (Kumar et al. 2019; Shen et al. 2020).

10.5.2.4 Pre-Treatment with Nutrient Solutions

Description: Pre-soaking biochar in nutrient-rich solutions before microbial inoculation enhances microbial growth and subsequent biochar modification, improving adsorption efficiency.
Procedure: (a) Preparation of Nutrient Solution: Prepare a nutrient-rich solution containing sources of carbon (e.g., glucose), nitrogen (e.g., ammonium nitrate), and phosphorus (e.g., potassium phosphate). (b) Soaking: Immerse the biochar in the nutrient solution for 24–48 hours, allowing it to absorb nutrients and enhance its compatibility with microbial growth. (c) Inoculation: Introduce microbial cultures such as *Pseudomonas aeruginosa* into the nutrient-pretreated biochar. (d) Incubation: Incubate under controlled conditions (e.g., 30°C, 24–72 hours) to ensure microbial colonization and the production of biofilm and EPS. (e) Drying and Activation: Dry the biochar gently to retain microbial biofilms while maintaining their structural integrity.

Example: Nutrient-pretreated biochar inoculated with *Pseudomonas aeruginosa* exhibited enhanced Cu(II) adsorption due to improved microbial colonization and increased active sites (Wu et al. 2022).

10.5.3 Benefits of Biological Activation

The biological activation of biochar offers several advantages in the context of heavy metal remediation:

10.5.3.1 Increased Adsorption Capacity

The introduction of microbial functional groups and metabolites enhances the number of active sites on biochar, leading to higher adsorption capacities for various heavy metals. For example, biochar activated with *Bacillus subtilis* exhibited a 50% increase in Pb(II) adsorption compared to non-activated biochar.

10.5.3.2 Improved Selectivity

Certain microorganisms can produce specific binding agents that preferentially adsorb particular heavy metals, allowing for targeted remediation strategies. Studies have shown that biochar inoculated with *Pseudomonas putida* selectively adsorbs Pb(II) ions in the presence of other metal ions.

10.5.3.3 Enhanced Stability of Adsorbed Metals

Microbial metabolites, such as extracellular polymeric substances (EPS), form stable complexes with heavy metals, reducing their mobility and bioavailability in the soil. This long-term stabilization minimizes environmental risks and prevents secondary contamination. For example, biochar activated with fungal strains like *Aspergillus niger* demonstrated superior stabilization of cadmium (Cd) in contaminated agricultural soils over six months (El-Naggar et al. 2022; Mekonnen et al. 2024).

10.5.3.4 Improved Soil Properties

Biologically activated biochar not only immobilizes heavy metals but also enhances soil quality by increasing microbial diversity, organic matter content, and nutrient cycling. This dual functionality makes it an ideal amendment for remediating degraded soils. In one study, the application of biochar activated with *Trichoderma harzianum* improved soil fertility and increased crop yield by 20% in a lead (Pb)-contaminated field (Mekonnen et al. 2024).

10.5.3.5 Cost-Effectiveness and Sustainability

Biological activation is an eco-friendly alternative to chemical modification methods, as it leverages naturally occurring microorganisms and their byproducts. It is also economically viable for large-scale applications, particularly when combined with low-cost organic materials during co-composting (El-Naggar et al. 2022; Mekonnen et al. 2024).

10.5.4 Heavy Metals: Most Studied and Specific Activation Strategies

10.5.4.1 Chromium (Cr)

- *Key Findings:* Chromium remediation has focused on reducing toxic Cr(VI) to the less toxic and less mobile Cr(III). Microbial strains like *Shewanella oneidensis* have been shown to mediate this redox transformation effectively when paired with biochar.
- *Recommended Strategy:* Direct inoculation with redox-active bacteria is particularly effective for Cr(VI) reduction. In one study, biochar inoculated with *Shewanella* reduced Cr(VI) concentrations by 90% within 48 hours in highly contaminated soils (Mekonnen et al. 2024).

10.5.4.2 Cadmium (Cd)

- *Key Findings:* Cadmium adsorption is enhanced through the production of microbial EPS and organic acids, which form stable complexes with Cd ions.
- *Recommended Strategy:* Co-composting biochar with organic waste materials and cadmium-tolerant microbial consortia, such as *Pseudomonas fluorescens* and *Aspergillus niger*, has proven effective, increasing Cd immobilization by up to 80% (El-Naggar et al. 2022).

10.5.4.3 Lead (Pb)

- *Key Findings:* Lead remediation benefits from microbial processes that introduce functional groups, such as carboxyl and hydroxyl, on the biochar surface, enhancing adsorption capacity.

- *Recommended Strategy:* Pre-treatment of biochar with nutrient solutions followed by inoculation with lead-tolerant strains like *Bacillus subtilis* is a cost-effective method for Pb immobilization in contaminated soils (El-Naggar et al. 2022, Mekonnen et al. 2024).

10.5.4.4 Zinc (Zn) and Copper (Cu)

- *Key Findings:* These metals are effectively adsorbed by biochar inoculated with fungal mycelia due to the production of organic acids that chelate metal ions.
- *Recommended Strategy:* Immobilization in bioreactors with fungal consortia, such as *Trichoderma* species, has been demonstrated to achieve high adsorption efficiencies for Zn and Cu (Mekonnen et al. 2024).

10.6 MICROBIOME AND BIOCHAR FOR REMEDIATION OF CONTAMINATED SOIL WITH PETROLEUM HYDROCARBONS

Petroleum hydrocarbon (PHC) contamination is a significant global environmental issue, stemming primarily from oil spills, industrial activities, and improper waste disposal. These compounds, consisting of alkanes, aromatics, and polycyclic aromatic hydrocarbons (PAHs), exhibit recalcitrance in the environment, posing risks to ecosystems and human health (Haque et al. 2022; Li et al. 2023).

Integrating biochar with microbiomes has emerged as an innovative approach to enhance the remediation of PHC-contaminated soils. Biochar, produced through the pyrolysis of organic material, serves as a structural and functional enhancer. It improves soil physicochemical properties, facilitates microbial colonization, and supports the degradation of PHCs by increasing microbial activity and soil enzymatic functions (Liu et al. 2023; Minnikova et al. 2023). Studies have demonstrated that the combination of biochar and microbial consortia can significantly reduce PHC concentrations, leveraging biochar's sorption capacity and microbial metabolic capabilities (Sarma et al. 2019; Haque et al. 2022).

Biological soil crusts (BSCs) offer an additional mechanism for stabilizing and restoring contaminated soils. These crusts, comprising cyanobacteria, algae, fungi, and lichens, can improve soil aggregation and nutrient cycling. When combined with biochar, BSCs enhance microbial diversity and support pollutant degradation (Lababpour 2016; Wang et al. 2022). Research highlights the potential of biochar to enrich microbial communities by providing a stable habitat and promoting the proliferation of hydrocarbon-degrading bacteria such as *Bacillus* and *Paenibacillus* (Minnikova et al. 2023).

Recent advancements show that integrating biochar with plant-microbiome systems can further accelerate PHC degradation. For instance, biochar-amended soils treated with microbial inoculants and plants such as ryegrass exhibit enhanced enzymatic activity and pollutant breakdown (Kudakwashe et al. 2022; Liu et al. 2023). Moreover, co-application strategies, involving biochar and microbial consortia engineered for specific PHC degradation pathways, have shown promise in addressing both organic pollutants and associated heavy metals (Sarma et al. 2019).

In conclusion, the synergistic use of biochar and microbiomes represents a sustainable and cost-effective solution for the remediation of PHC-contaminated soils. Future research should focus on optimizing biochar production techniques and tailoring microbial consortia to specific contamination scenarios, ensuring broader applicability across diverse environmental conditions.

10.6.1 The Role of Microbiomes in PHC Degradation

Microbial communities play a critical role in the bioremediation of soils contaminated with petroleum hydrocarbons (PHCs). These microorganisms, including aerobic and anaerobic bacteria such as *Pseudomonas*, *Bacillus*, and *Paenibacillus*, have evolved metabolic pathways to degrade hydrocarbons into less harmful compounds like carbon dioxide (CO_2) and water, making them essential for environmental recovery efforts (Haque et al. 2022; Kudakwashe et al. 2022; Minnikova et al. 2023).

Key mechanisms utilized by these microbes include:

10.6.1.1 Enzymatic Degradation

Hydrocarbon-degrading bacteria produce specialized enzymes, such as monooxygenases and dioxygenases, which initiate the oxidation of PHCs. These enzymes are particularly effective in breaking down polycyclic aromatic hydrocarbons (PAHs), which are among the most recalcitrant contaminants. For example, monooxygenases catalyze the introduction of oxygen into hydrocarbon molecules, thereby increasing their solubility and susceptibility to further microbial action (Kudakwashe et al. 2022; Minnikova et al. 2023).

10.6.1.2 Biosurfactant Production

Many hydrocarbon-degrading microbes, such as *Pseudomonas* and *Bacillus* species, produce biosurfactants like glycolipids and lipopeptides. These compounds enhance the bioavailability of hydrophobic hydrocarbons by reducing surface tension and facilitating microbial access to contaminants (Sarma et al. 2019; Kudakwashe et al. 2022). Biosurfactants also play a crucial role in emulsifying hydrocarbons, thus accelerating their degradation in soil environments (Kudakwashe et al. 2022).

10.6.1.3 Community Dynamics

The synergistic interactions within microbial communities are essential for efficient hydrocarbon degradation. Certain microbes degrade simpler hydrocarbons, creating byproducts that serve as substrates for other species. This interdependence enhances the overall degradation capacity of the microbial consortia (Kudakwashe et al. 2022; Liu et al. 2023). Studies have shown that microbial diversity and richness are positively correlated with the degradation rates of PHCs, as diverse communities are better equipped to metabolize a range of hydrocarbon compounds (Wang et al. 2022).

Despite these capabilities, microbial degradation is often constrained by environmental factors such as nutrient availability, soil pH, and the inherent toxicity of hydrocarbons. The presence of heavy metals, salinity, or extreme pH levels can inhibit microbial activity (Sarma et al. 2019; Kudakwashe et al. 2022). Furthermore, soil structure and low oxygen availability in compacted or waterlogged soils can reduce microbial efficiency (Wang et al. 2022; Minnikova et al. 2023).

10.6.2 Biochar as a Catalyst in Soil Remediation

Biochar, a carbon-rich material produced through the pyrolysis of organic biomass, has gained significant attention as a catalyst in soil remediation efforts. Its unique physical and chemical properties make it an effective agent for enhancing the degradation of petroleum hydrocarbons (PHCs) in contaminated soils.

10.6.2.1 Sorption Properties

Biochar's high surface area and porosity allow it to adsorb hydrophobic hydrocarbons effectively, reducing their bioavailability and toxicity. This sorption capability immobilizes contaminants, thereby preventing their spread to groundwater or surrounding ecosystems. Simultaneously, it promotes microbial access to hydrocarbons by concentrating them in biochar's porous network, facilitating enzymatic degradation (Kudakwashe et al. 2022; Liu et al. 2023). The strong affinity of biochar for non-polar compounds such as alkanes and PAHs stems from its carbon-rich structure, which interacts with hydrocarbons via hydrophobic interactions and $\pi-\pi$ stacking (Sarma et al. 2019; Kudakwashe et al. 2022).

10.6.2.2 Improved Soil Properties

In addition to its sorption properties, biochar improves key soil characteristics, making it more conducive to microbial activity and plant growth. It enhances soil aeration, water retention, and nutrient availability, which are critical for the proliferation of hydrocarbon-degrading microbial communities. By stabilizing soil pH and increasing cation exchange capacity, biochar creates an environment that supports enzymatic activity and microbial metabolic processes (Kudakwashe et al. 2022; Minnikova et al. 2023). Studies have shown that biochar amendments lead to better water infiltration and retention, which is particularly valuable in arid and semi-arid regions prone to petroleum contamination (Lababpour 2016; Wang et al. 2022).

10.6.2.3 Microbial Habitat

Biochar's highly porous structure serves as a stable and protective habitat for microbial colonization. Hydrocarbon-degrading microbes, such as *Pseudomonas*, *Bacillus*, and *Paenibacillus*, benefit from the physical shelter and nutrient reservoirs provided by biochar (Haque et al. 2022; Wu et al. 2022). This protective environment enables microbial communities to persist and thrive, even in soils contaminated with toxic hydrocarbons (Minnikova et al. 2023). Biochar also supports the establishment of biofilms, which further enhance microbial resilience and degradation efficiency (Lababpour 2016).

10.6.2.4 Enhancement of Enzymatic Activities

Biochar has been shown to significantly increase the activity of soil enzymes such as dehydrogenase, lipase, and urease, which play crucial roles in the breakdown of hydrocarbons. Dehydrogenase, for instance, is involved in microbial respiration and indicates overall microbial activity, while lipase catalyzes the hydrolysis of lipid-based hydrocarbons (Kudakwashe et al. 2022; Liu et al. 2023). The increased enzymatic activity observed in biochar-amended soils is attributed to improved microbial proliferation and the stabilization of enzyme structures within the biochar matrix (Kudakwashe et al. 2022).

10.6.2.5 Interaction with Heavy Metals and Co-Contaminants

In many PHC-contaminated sites, heavy metals co-exist as additional pollutants. Biochar's ability to bind heavy metals reduces their toxicity and mitigates their inhibitory effects on microbial activity. This dual function enables biochar to facilitate PHC degradation while concurrently addressing heavy metal contamination (Sarma et al. 2019; Minnikova et al. 2023). The interaction between biochar and co-contaminants enhances soil recovery by neutralizing multiple pollutants simultaneously.

10.6.2.6 Biochar Modification for Enhanced Remediation

Recent advancements in biochar production and modification have further expanded its application in remediation. Functionalized biochars, enriched with metal oxides or engineered for specific surface properties, have demonstrated improved sorption capacities and catalytic efficiencies (Kudakwashe et al. 2022). For instance, biochars impregnated with iron or manganese oxides have been shown to enhance redox reactions, accelerating the degradation of recalcitrant hydrocarbons such as PAHs (Kudakwashe et al. 2022).

10.6.3 Synergistic Use of Biochar and Microbiomes

The combination of biochar and microbiomes offers enhanced remediation outcomes. Key benefits include:

- *Enhanced Microbial Diversity:* Biochar-enriched soils show higher microbial diversity and richness, especially in hydrocarbon-degrading bacteria (Wang et al. 2022; Minnikova et al. 2023).
- *Targeted Biodegradation:* Engineered microbial consortia, when combined with biochar, exhibit improved efficiency in degrading complex hydrocarbon mixtures (Sarma et al. 2019).

- *Rhizodegradation:* Integrating biochar with plant-root systems (e.g., ryegrass) further accelerates PHC degradation by leveraging plant-microbe interactions (Kudakwashe et al. 2022; Liu et al. 2023).

Field studies have demonstrated that biochar, when applied with bacterial strains such as *Paenibacillus* and *Bacillus*, can reduce hydrocarbon concentrations by up to 70% in contaminated soils (Minnikova et al. 2023).

10.6.4 Biological Soil Crusts (BSCs) as a Complementary Approach

Biological soil crusts (BSCs) are complex communities of cyanobacteria, fungi, algae, and other microbes that form cohesive layers on the soil surface. These crusts play a crucial role in maintaining soil stability, reducing erosion, and enhancing soil fertility, especially in arid and semi-arid ecosystems where vegetation cover is limited. Incorporating biochar into BSC systems has been shown to amplify their ecological benefits and improve their potential in the remediation of petroleum hydrocarbon (PHC)-contaminated soils.

10.6.4.1 Enhanced Soil Stabilization

BSCs inherently reduce soil erosion by aggregating soil particles through biological and biochemical processes. The addition of biochar to BSC systems enhances this function by improving the structural integrity of the soil matrix. Biochar's porous structure binds soil particles more effectively, while its high carbon content stabilizes organic matter, reducing the susceptibility of soils to wind and water erosion (Lababpour 2016; Wang et al. 2022). In PHC-contaminated sites, biochar also serves as a sorbent for hydrocarbons, immobilizing these pollutants and preventing their migration to groundwater or adjacent ecosystems (Lababpour 2016; Kudakwashe et al. 2022). In field studies conducted in arid regions, biochar-amended BSCs significantly reduced soil erosion rates while increasing the retention of hydrocarbons within the upper soil layers. This dual function of stabilization and sorption makes biochar-BSC systems particularly effective in areas prone to desertification and contamination (Lababpour 2016).

10.6.4.2 Support Microbial Communities

The synergy between biochar and BSCs creates a conducive environment for diverse microbial communities, which are essential for hydrocarbon degradation. Cyanobacteria, a dominant component of BSCs, contribute to hydrocarbon breakdown through oxygenic photosynthesis, which increases the availability of oxygen for aerobic degradation pathways (Kudakwashe et al. 2022; Wang et al. 2022). Biochar, on the other hand, provides a stable habitat for hydrocarbon-degrading bacteria and fungi, promoting their persistence even under harsh environmental conditions (Kudakwashe et al. 2022; Minnikova et al. 2023). Research has shown that biochar-enriched BSCs support higher microbial diversity and richness compared to non-amended crusts. This diversity enhances the efficiency of PHC degradation by enabling different species to metabolize a broader range of hydrocarbon compounds (Lababpour 2016). Moreover, biochar's ability to adsorb heavy metals mitigates their toxic effects on microbial communities, further enhancing the functionality of BSC systems (Sarma et al. 2019; Kudakwashe et al. 2022).

10.6.4.3 Improved Water Retention and Nutrient Cycling

One of the significant challenges in arid and semi-arid regions is the low water retention capacity of soils. Biochar improves the water-holding capacity of soils, ensuring that BSCs remain hydrated for extended periods. This is critical for maintaining microbial activity and facilitating the enzymatic breakdown of hydrocarbons (Lababpour 2016; Wang et al. 2022). Additionally, biochar enhances nutrient cycling within BSC systems by providing a reservoir for essential nutrients such as nitrogen and phosphorus. This not only supports microbial and plant growth but also accelerates the natural recovery of contaminated soils (Lababpour 2016; Kudakwashe et al. 2022).

10.6.4.4 Applicability in Arid and Semi-Arid Regions

BSC-biochar systems have shown particular efficacy in arid and semi-arid regions where traditional remediation methods are less viable. These systems are cost-effective, require minimal maintenance, and can be deployed over large areas with relative ease (Minnikova et al. 2023; Lababpour 2016). By leveraging the natural processes of BSC formation and the catalytic properties of biochar, these systems offer a sustainable solution for restoring degraded soils in water-scarce environments. In a study conducted in desertified areas, biochar-amended BSCs significantly improved soil stability, reduced hydrocarbon bioavailability, and enhanced microbial activity. The findings suggest that integrating biochar into BSC systems can extend the applicability of biological remediation to environments previously considered unsuitable for such approaches (Kudakwashe et al. 2022; Wang et al. 2022).

10.6.4.5 Potential for Long-Term Ecosystem Restoration

The combination of biochar and BSCs not only addresses immediate contamination issues but also contributes to the long-term restoration of soil ecosystems. By stabilizing soil structure, enhancing microbial diversity, and improving water and nutrient dynamics, biochar-enriched BSC systems lay the foundation for the recovery of vegetation and the return of ecosystem services (Lababpour 2016; Kudakwashe et al. 2022).

10.6.5 Mechanisms of Action in PHC Remediation

The remediation of petroleum hydrocarbons (PHCs) in contaminated soils is enhanced significantly by the synergistic interaction between biochar and microbial consortia. These mechanisms not only mitigate the toxicity of hydrocarbons but also promote their complete degradation through biological and chemical pathways. Below are the key mechanisms involved:

10.6.5.1 Immobilization and Bioavailability Reduction

Biochar plays a pivotal role in the immobilization of hydrocarbons, which reduces their bioavailability and toxicity in the soil. Its porous structure and high surface area allow for the adsorption of hydrophobic hydrocarbons, effectively concentrating these compounds in localized areas where microbes can degrade them more efficiently (Haque et al. 2022; Liu et al. 2023).

This sorption process also prevents the leaching of hydrocarbons into groundwater and minimizes the spread of contamination. Furthermore, biochar's ability to bind heavy metals and other co-contaminants reduces their inhibitory effects on microbial communities, creating a more conducive environment for microbial activity (Lababpour 2016; Kudakwashe et al. 2022).

10.6.5.2 Enzymatic Enhancement

The presence of biochar in contaminated soils has been shown to enhance the activity of hydrocarbon-degrading enzymes such as monooxygenases, dioxygenases, dehydrogenases, and lipases. These enzymes are critical in breaking down complex hydrocarbons into simpler intermediates that are easier to metabolize (Kudakwashe et al. 2022; Minnikova et al. 2023).

For example:

- Monooxygenases and Dioxygenases initiate the oxidation of aromatic hydrocarbons, including PAHs, introducing oxygen atoms into their structure and making them more accessible for further degradation (Wang et al. 2022).
- Lipases target lipid-based hydrocarbons, accelerating their hydrolysis into fatty acids and glycerol, which are metabolizable by microbial communities (Lababpour 2016).
- Dehydrogenases, indicators of microbial respiration, facilitate the degradation of hydrocarbons by catalyzing oxidation-reduction reactions in microbial metabolic pathways (Liu et al. 2023).

Biochar supports enzymatic activity by providing a stable microenvironment and buffering against pH fluctuations that could otherwise denature these enzymes.

10.6.5.3 Biochemical Pathways

Microbial consortia use hydrocarbons as a carbon and energy source through metabolic pathways that break them down into less toxic compounds such as carbon dioxide, water, and biomass. The key biochemical pathways include:

- *Aerobic Degradation:* Microorganisms utilize oxygen to oxidize hydrocarbons, producing intermediates like alcohols, aldehydes, and acids, which are further broken down into CO_2 and water (Kudakwashe et al. 2022; Wang et al. 2022).

- *Anaerobic Degradation:* In oxygen-deprived environments, microbes metabolize hydrocarbons through reductive processes, involving nitrate, sulfate, or iron as electron acceptors. This pathway is particularly important in waterlogged or compacted soils (Kudakwashe et al. 2022).

Biochar enhances these pathways by improving oxygen diffusion in aerobic conditions and providing surface area for electron transfer in anaerobic conditions (Lababpour 2016). Additionally, biochar reduces competition among microbial species for hydrocarbons by immobilizing contaminants, allowing diverse metabolic pathways to coexist effectively.

10.6.5.4 Reduction of Toxicity

Hydrocarbons, particularly PAHs and other recalcitrant compounds, can exert toxic effects on soil microbial communities, inhibiting their activity and growth. Biochar mitigates these effects by adsorbing toxic compounds, thus shielding microbes from their harmful impacts (Minnikova et al. 2023). This mechanism not only preserves microbial functionality but also promotes the establishment of diverse and resilient microbial consortia (Wang et al. 2022).

- *Interaction with Soil Microorganisms:* Biochar facilitates close interactions between microbial communities and hydrocarbons by acting as a scaffold for biofilm formation. Biofilms enhance the resilience of microbes against environmental stressors and improve their efficiency in degrading hydrocarbons (Lababpour 2016; Wang et al. 2022). Furthermore, biochar fosters the exchange of genetic material among microbes, promoting the spread of hydrocarbon-degrading capabilities through horizontal gene transfer (Kudakwashe et al. 2022).
- *Activation of Plant-Microbe Synergies:* In rhizosphere-associated remediation systems, biochar amplifies the interaction between plants and microbes, enhancing rhizodegradation processes. Plants release root exudates that serve as carbon sources for microbes, while biochar immobilizes hydrocarbons and provides a favorable habitat for microbial proliferation (Kudakwashe et al. 2022; Liu et al. 2023).

These mechanisms collectively highlight the multifaceted role of biochar and microbiomes in PHC remediation, making them a powerful and sustainable tool for soil restoration. Future research should aim to optimize these interactions and explore their application in diverse environmental settings.

The utilization of biochar as a tool for soil decontamination presents a transformative approach to addressing the pervasive challenges posed by heavy metal contamination and organic pollutants. This chapter has highlighted biochar's multifaceted capabilities, from immobilizing toxic metals and enhancing soil structure to fostering microbial activity and synergizing with biological soil crusts (BSCs). These properties make biochar an indispensable resource for advancing sustainable soil remediation practices.

Biochar's ability to immobilize heavy metals, facilitated by its high surface area, functional groups, and structural stability, significantly reduces the bioavailability and environmental mobility of contaminants. When modified chemically or biologically, biochar's adsorption efficiency is enhanced, allowing for targeted remediation strategies tailored to specific pollutants. The integration of biochar with biological soil crusts further exemplifies the material's versatility, creating systems that not only immobilize contaminants but also promote ecological resilience and long-term soil recovery.

The combination of biochar with innovative remediation techniques, such as microbial inoculation and tailored amendments, underscores the potential for integrated and hybrid solutions. These approaches address the limitations of traditional methods while ensuring cost-effectiveness, scalability, and environmental compatibility. The role of biochar in supporting microbial diversity and enzymatic degradation processes highlights its capacity to restore ecosystem functionality, particularly in highly degraded or contaminated sites.

Future research should aim to standardize biochar production, optimize modification techniques, and evaluate long-term field applications to ensure consistent and sustainable outcomes. Moreover, the development of tailored microbial consortia and hybrid remediation systems can enhance biochar's applicability across diverse environmental settings. Collaborative efforts among scientists, policymakers, and industry stakeholders are essential to advance the adoption of biochar-based technologies and integrate them into global soil management frameworks.

In conclusion, biochar serves not only as a catalyst for decontamination but also as a cornerstone for promoting soil health and ecological restoration. Its application represents a critical step toward mitigating environmental pollution while fostering sustainable land management practices.

NOMENCLATURE

HMs	:	Heavy Metals
PAHs	:	Polycyclic Aromatic Hydrocarbons
EPS	:	Extracellular Polymeric Substances
CEC	:	Cation Exchange Capacity
SRB	:	Sulfate-Reducing Bacteria
BSCs	:	Biological Soil Crusts
PHCs	:	Petroleum Hydrocarbons
EDTA	:	Ethylenediaminetetraacetic Acid

REFERENCES

Ahmad, M., A.U. Rajapaksha, J.E. Lim, M. Zhang, N. Bolan, D. Mohan, et al. 2014. Biochar as a sorbent for contaminant management in soil and water: A review. Chemosphere. 99: 19–33. https://doi.org/10.1016/j.chemosphere.2013.10.071.

Ali, H., E. Khan and I. Ilahi. 2019. Environmental chemistry and ecotoxicology of hazardous heavy metals: environmental persistence, toxicity, and bioaccumulation. Journal of Chemistry. 6730305. https://doi.org/10.1155/2019/6730305.

Aziz, K.H.H., F.S. Mustafa, K.M. Omer, S. Hama, R.F. Hamarawf and K.O. Rahman. 2023. Heavy metal pollution in the aquatic environment: efficient and low-cost removal approaches to eliminate their toxicity: A review. RSC Advances. 13(26): 17595–17610. https://doi.org/10.1039/d3ra00723e.

Bautista-Olivas, A.L., F. Cruz-Bautista, C.R. Álvarez-Chávez, A.G. Zavala-Reyna, L.A. Sánchez-Landero and J. Alvarado-Ibarra. 2017. Concentration of heavy metals in condensed atmospheric water vapor at three Mexican localities. Atmósfera. 30(3): 209–220. https://doi.org/10.20937/ATM.2017.30.03.02.

Bharti, R. and R. Sharma. 2022. Effect of heavy metals: An overview. Materials Today: Proceedings. 51: 880–885. https://doi.org/10.1016/j.matpr.2021.06.278.

Briffa, J., E. Sinagra and R. Blundell. 2020. Heavy metal pollution in the environment and their toxicological effects on humans. Heliyon. 6(9): e04691. https://doi.org/10.1016/j.heliyon.2020.e04691

Büdel, B., C. Colesie, T.A. Green, M. Grube, R. Lázaro Suau, K. Loewen-Schneider, et al. 2014. Improved appreciation of the functioning and importance of biological soil crusts in Europe: The Soil Crust International Project (SCIN). Biodiversity and Conservation. 23: 1639–1658. https://doi.org/10.1007/s10531-014-0645-2.

Chang, Y., W. Liu, Y. Mao, T. Yang and Y. Chen. 2022. Biochar addition alters C:N:P stoichiometry in moss crust-soil continuum in Gurbantünggüt desert. Plants. 11(6): 814. https://doi.org/10.3390/plants11060814.

Chen, B., Z. Chen and S. Lv. 2011. A novel magnetic biochar efficiently sorbs organic pollutants and phosphate. Bioresource Technology. 102(2): 716–723. https://doi.org/10.1016/j.biortech.2010.08.067.

Cimboláková, I., I. Uher, K.V. Laktičová, M. Vargová, T. Kimáková and I. Papajová. 2019. Heavy metals and the environment. *In*: Ivan, U. (Ed.). Environmental Factors Affecting Human Health. 1–31. IntechOpen.

Edo, G.I., P.O. Samuel, G.O. Oloni, G.O. Ezekiel, V.O. Ikpekoro, P. Obasohan, et al. 2024. Environmental persistence, bioaccumulation, and ecotoxicology of heavy metals. Chemistry and Ecology. 40(3): 322–349. https://doi.org/10.1080/02757540.2024.2306839.

El-Naggar, A., A. Mosa, N. Ahmed, N.K. Niazi, B. Yousaf, B. Sarkar, et al. 2022. Modified and pristine biochars for remediation of chromium contamination in soil and aquatic systems. Chemosphere. 303: 134942. https://doi.org/10.1016/j.chemosphere.2022.134942.

Gasco, G., M.L. Alvarez, J. Paz-Ferreiro and A. Mendez. 2019. Combining phytoextraction by *Brassica napus* and biochar amendment for the remediation of a mining soil in Riotinto (Spain). Chemosphere. 231: 562–570. https://doi.org/10.1016/j.chemosphere.2019.05.168.

Ghosh, P., I.S. Thakur and A. Kaushik. 2017. Bioassays for toxicological risk assessment of landfill leachate: A review. Ecotoxicology and Environmental Safety. 141: 259–270. https://doi.org/10.1016/j.ecoenv.2017.03.023.

Haque, S., N. Srivastava, D.B. Pal, M.F. Alkhanani, A.H. Almalki, M.Y. Areeshi, et al. 2022. Functional microbiome strategies for the bioremediation of petroleum-hydrocarbon and heavy metal contaminated soils: A review. Science of the Total Environment. 833: 155222. https://doi.org/10.1016/j.scitotenv.2022.155222.

Hasan, M.M., M.N. Uddin, I. Ara-Sharmeen, H.F. Alharby, Y. Alzahrani, K.R. Hakeem, et al. 2019. Assisting phytoremediation of heavy metals using chemical amendments. Plants. 8(9): 295. https://doi.org/10.3390/plants8090295.

Inyang, M., B. Gao, Y. Yao, Y. Xue, A. Zimmerman, A. Mosa, et al. 2012. A review of biochar as a low-cost adsorbent for aqueous heavy metal removal. Critical Reviews in Environmental Science and Technology. 42(5): 410–453. https://doi.org/10.1080/10643389.2011.627016.

Jadaa, W. and H. Mohammed. 2023. Heavy metals–definition, natural and anthropogenic sources of releasing into ecosystems, toxicity, and removal methods–an overview study. Journal of Ecological Engineering. 24(6): 249–271. https://doi.org/10.12911/22998993/162955ISSN.

Jaishankar, M., T. Tseten, N. Anbalagan, B.B. Mathew and K.N. Beeregowda. 2014. Toxicity, mechanism, and health effects of some heavy metals. Interdisciplinary Toxicology. 7(2): 60–72. https://doi.org/10.2478/intox-2014-0009.

Jiang, M., K. Wang, Y. Wang, Q. Zhao and W. Wang. 2022. Technologies for the cobalt-contaminated soil remediation: a review. Science of the Total Environment. 813: 151908. https://doi.org/10.1016/j.scitotenv.2021.151908.

Khalid, N., M. Aqeel, A. Noman, S.M. Khan and N. Akhter. 2021. Interactions and effects of microplastics with heavy metals in aquatic and terrestrial environments. Environmental Pollution. 290: 118104. https://doi.org/10.1016/j.envpol.2021.118104.

Kudakwashe, M.E.K.I., L.I.U. Qiang, W.U. Shuai and Y.U.A.N. Yanfei. 2022. Plant- and microbe-assisted biochar amendment technology for petroleum hydrocarbon remediation in saline-sodic soils: A review. Pedosphere. 32(1): 211–221. https://doi.org/10.1016/S1002-0160(21)60041-3.

Kumar, R., A. Kumari and I. Bahadur. 2019. Bioreduction of chromium using biochar-supported microorganisms: A novel approach to remediation of contaminated sites. Journal of Environmental Management. 242: 10–19. https://doi.org/10.1016/j.jenvman.2019.03.105.

Lababpour, A. 2016. Potentials of the microalgae inoculant in restoration of biological soil crusts to combat desertification. International Journal of Environmental Science and Technology. 13: 2521–2532. https://doi.org/10.1007/s13762-016-1074-4.

Lebrun, M., F. Miard, N. Hattab-Hambli, S. Bourgerie and D. Morabito. 2018. Assisted phytoremediation of a multi-contaminated industrial soil using biochar and garden soil amendments associated with Salix alba or Salix viminalis: abilities to stabilize As, Pb, and Cu. Water, Air, and Soil Pollution. 229: 1–13. https://doi.org/10.1007/s11270-018-3816-z.

Lehmann, J. and S. Joseph (Eds). 2015. Biochar for Environmental Management: Science, Technology, and Implementation (2nd ed.). Routledge. https://doi.org/10.4324/9780203762264.

Lenart-Boroń, A. and P. Boroń. 2014. The effect of industrial heavy metal pollution on microbial abundance and diversity in soils—A review. *In*: Environmental Risk Assessment of Soil Contamination. IntechOpen. https://doi.org/10.5772/57406.

Li, C., C. Cui, J. Zhang, J. Shen, B. He, Y. Long, et al. 2023. Biodegradation of petroleum hydrocarbons based pollutants in contaminated soil by exogenous effective microorganisms and indigenous microbiome. Ecotoxicology and Environmental Safety. 253: 114673. https://doi.org/10.1016/j.ecoenv.2023.114673.

Liang, C., J. Chen and Y. Zhou. 2021. Microbial-enhanced biochar for the remediation of cadmium-contaminated soil: Mechanisms and effectiveness. Journal of Hazardous Materials. 403: 124037. https://doi.org/10.1016/j.jhazmat.2020.124037.

Liu, Z., Z. Li, S. Chen and W. Zhou. 2023. Enhanced phytoremediation of petroleum-contaminated soil by biochar and urea. Journal of Hazardous Materials. 453: 131404. https://doi.org/10.1016/j.jhazmat.2023.131404.

Manori, S., V. Shah, V. Soni, K. Dutta and A. Daverey. 2021. Phytoremediation of cadmium-contaminated soil by Bidens pilosa L.: impact of pine needle biochar amendment. Environmental Science and Pollution Research. 1–13. https://doi.org/10.1007/s11356-021-12953-3.

Mekonnen, B.A., T.A. Aragaw and M.B. Genet. 2024. Bioremediation of petroleum hydrocarbon contaminated soil: A review on principles, degradation mechanisms, and advancements. Frontiers in Environmental Science. 12: 1354422. https://doi.org/10.3389/fenvs.2024.1354422.

Minnikova, T., S. Kolesnikov, N. Minin, A. Gorovtsov, N. Vasilchenko and V. Chistyakov. 2023. The influence of remediation with *Bacillus* and *Paenibacillus* strains and biochar on the biological activity of petroleum-hydrocarbon-contaminated haplic chernozem. Agriculture. 13(3): 719. https://doi.org/10.3390/agriculture13030719.

Mitra, S., A.J. Chakraborty, A.M. Tareq, T.B. Emran, F. Nainu, A. Khusro, et al. 2022. Impact of heavy metals on the environment and human health: Novel therapeutic insights to counter the toxicity. Journal of King Saud University-Science. 34(3): 101865. https://doi.org/10.1016/j.jksus.2022.101865.

Mukherjee, A., A.R. Zimmerman and W. Harris. 2022. Surface chemistry variations among a series of laboratory-produced biochars. Geoderma. 163(3–4): 247–255. https://doi.org/10.1016/j.geoderma.2011.04.021.

Naqash, N., S. Prakash, D. Kapoor and R. Singh. 2020. Interaction of freshwater microplastics with biota and heavy metals: A review. Environmental Chemistry Letters. 18(6): 1813–1824. https://doi.org/10.1007/s10311-020-01044-3.

Navas-Romero, A.L., E. Martínez-Carretero and M.A. Herrera-Moratta. 2021. Restauración de costras biológicas del suelo: Pasado, presente y futuro. Multequina. 30(2): 25–47.

Navas-Romero, A.L., M.A. Herrera-Moratta, E. Martinez-Carretero, B. Vento and M.C. Fernández-Belmonte. 2023. Insights into variations of seasonal and daily soil temperatures under the effect of biocrusts in central-western Argentina. Acta Botánica Mexicana. (130). https://doi.org/10.21829/abm130.2023.2212.

Ogundiran, M.B., N.S. Mekwunyei and S.A. Adejumo. 2018. Compost and biochar-assisted phytoremediation potentials of Moringa oleifera for remediation of lead-contaminated soil. Journal of Environmental Chemical Engineering. 6(2): 2206–2213. https://doi.org/10.1016/j.jece.2018.03.025.

Palansooriya, K.N., J.T.F. Wong, Y. Hashimoto, L. Huang, J. Rinklebe, S.X. Chang, et al. 2019. Response of microbial communities to biochar-amended soils: A critical review. Biochar. 1: 3–22. https://doi.org/10.1007/s42773-019-00009-2.

Priya, M.M. and S.R. Shankar. 2020. Effect of heavy metals on growth and development of cultivated plants with reference to cadmium, chromium, and lead—A review. Journal of Stress Physiology and Biochemistry. 16(3): 84–102.

Rahman, Z. and V.P Singh. 2019. The relative impact of toxic heavy metals (THMs) (arsenic (As), cadmium (Cd), chromium (Cr) (VI), mercury (Hg), and lead (Pb)) on the total environment: An overview. Environmental Monitoring and Assessment. 191: 1–21. https://doi.org/10.1007/s10661-019-7528-7.

Rashid, M.S., G. Liu, B. Yousaf, Y. Hamid, A. Rehman, M. Arif, et al. 2022. A critical review on biochar-assisted free radicals mediated redox reactions influencing transformation of potentially toxic metals: Occurrence, formation, and environmental applications. Environmental Pollution. 315: 120335. https://doi.org/10.1016/j.envpol.2022.120335.

Rehman, K., F. Fatima, I. Waheed and M.S.H. Akash. 2018. Prevalence of exposure of heavy metals and their impact on health consequences. Journal of Cellular Biochemistry. 119(1): 157–184. https://doi.org/10.1002/jcb.26234.

Rubiya Syed, R.S., D.K. Dhriti Kapoor and A.A. Bhat. 2018. Heavy metal toxicity in plants: A review. Plant Archives. 18(2): 1229–1238.

Sanchez, E., R.B. Zabaleta Nievas, A.L. Navas-Romero, R.A. Torres Sciancalepore and G.G. Fouga. 2024. Assessment of Pistachio Shell-Based Biochar Application in the Sustainable Amendment of Soil and Its Performance in Enhancing Bell Pepper (*Capsicum annuum* L.) Growth. Sustainability. 16(11): 1–22. https://doi.org/10.3390/su16114429.

Sarker, A., M.A. Al Masud, D.M. Deepo, K. Das, R. Nandi, M.W.R. Ansary, et al. 2023. Biological and green remediation of heavy metal contaminated water and soils: A state-of-the-art review. Chemosphere. 332: 138861. https://doi.org/10.1016/j.chemosphere.2023.138861.

Sarma, H., S. Sonowal and M.N.V. Prasad. 2019. Plant-microbiome assisted and biochar-amended remediation of heavy metals and polyaromatic compounds—A microcosmic study. Ecotoxicology and Environmental Safety. 176: 288–299. https://doi.org/10.1016/j.ecoenv.2019.03.081.

Serrano, M.F., J.E. López, N. Henao and J.F. Saldarriaga. 2024. Phosphorus-loaded biochar-assisted phytoremediation to immobilize cadmium, chromium, and lead in soils. ACS Omega. 9(3): 3574–3587. https://doi.org/10.1021/acsomega.3c07433.

Sharifuzzaman, S.M., H. Rahman, S.M. Ashekuzzaman, M.M. Islam, S.R. Chowdhury and M.S. Hossain. 2016. Heavy metals accumulation in coastal sediments. pp. 21–42. *In*: H. Hasegawa, I. Rahman and M. Rahman (eds). Environmental Remediation Technologies for Metal-Contaminated Soils. Springer, Tokyo. https://doi.org/10.1007/978-4-431-55759-3_2.

Shen, Z., Y. Zhang, F. Jin, D. Hou, J. Shi and O. McMillan. 2020. The role of biofilms on biochar in the immobilization of heavy metals in contaminated sediments. Science of The Total Environment. 742: 140598. https://doi.org/10.1016/j.scitotenv.2020.140598.

Simiele, M., M. Lebrun, F. Miard, D. Trupiano, P. Poupart, O. Forestier, et al. 2020. Assisted phytoremediation of a former mine soil using biochar and iron sulphate: Effects on As soil immobilization and accumulation in three Salicaceae species. Science of The Total Environment. 710: 136203. https://doi.org/10.1016/j.scitotenv.2019.136203.

Song, X., G. Pan, C. Zhang, L. Zhang and H. Wang. 2016. Effects of biochar application on fluxes of three biogenic greenhouse gases: a meta-analysis. Ecosystem Health and Sustainability. 2(2): e01202. https://doi.org/10.1002/ehs2.1202.

Song, H., S.F. Chen, G. Si, K. Bhatt, S.H. Chen and W.J. Chen. 2024. Removal of environmental pollutants using biochar: Current status and emerging opportunities. Environmental Geochemistry and Health. 46(10): 384. https://doi.org/10.1007/s10653-024-02142-9.

Wang, J., R. Huang, L. Zhu, H. Guan, L. Lin, H. Fang, et al. 2022. The effects of biochar on microbial community composition in and beneath biological soil crusts in a *Pinus massoniana* lamb. plantation. Forests. 13(7): 1141. https://doi.org/10.3390/f13071141.

Wu, X., S.J. Cobbina, G. Mao, H. Xu, Z. Zhang and L. Yang. 2016. A review of toxicity and mechanisms of individual and mixtures of heavy metals in the environment. Environmental Science and Pollution Research. 23: 8244–8259. https://doi.org/10.1007/s11356-016-6333-x.

Wu, H., Z. Yan, J. Li, Q. Zhou and Y. Zhao. 2022. Enhancing heavy metal immobilization in soil by biochar modified with microbial metabolic products. Environmental Pollution. 311: 119964. https://doi.org/10.1016/j.envpol.2022.119964.

Zhang, M., J. Wang, S.H. Bai, Y. Zhang, Y. Teng and Z. Xu. 2019. Assisted phytoremediation of a co-contaminated soil with biochar amendment: Contaminant removals and bacterial community properties. Geoderma. 348: 115–123. https://doi.org/10.1016/j.geoderma.2019.04.031.

Chapter 11

Biochar Uses for Gas Cleaning

José Soria*[1,a], Germán Mazza[1], Daniela Asensio[1,b] and Rodrigo Torres-Sciancalepore[1,c]

[1]Instituto de Investigación y Desarrollo en Ingeniería de Procesos, Biotecnología y Energías Alternativas, PROBIEN (CONICET-Universidad Nacional del Comahue), Calle Buenos Aires 1400, Neuquén 8300, Argentina

11.1 INTRODUCTION

Rapid industrialization and widespread use of fossil fuels have significantly increased air pollution, posing serious environmental and human health hazards (Srinadh et al. 2024). Common gaseous pollutants emitted by industries include sulfur dioxide (SO_2), hydrogen sulfide (H_2S), carbon dioxide (CO_2), elemental mercury (Hg^0), volatile organic compounds (VOCs), and ammonia (NH_3). Several technologies, such as flue gas desulfurization, wet oxidation, scrubbing, biological oxidation, and adsorption are used to reduce these emissions (Abdulrasheed et al. 2018; Liu and Wang 2019; Peluso et al. 2019). Adsorption stands out for its efficiency and selectivity, particularly at lower concentrations, using customizable adsorbents with remarkable selectivity and high surface area (Du et al. 2020; Vikrant et al. 2017). Activated carbon (AC), carbon nanotubes (CNT), carbon molecular sieves, zeolites, and metal-organic frameworks (MOFs) are commonly employed as solid sorbents in adsorption devices (Cha et al. 2016; Liu et al. 2020), though their high costs and integration challenges limit broader industrial use. Therefore, cost-effective and efficient alternatives are necessary.

Biochar, a carbon-rich solid produced through thermochemical conversion of carbonaceous biomass under oxygen-limited or oxygen-free conditions, has

*Corresponding author: José Soria (E-mail: jose.soria@probien.gob.ar)

DOI: 10.1201/9781032726694-11

emerged as a promising material for air pollution control due to its affordability (90–1,200 \$/t) (Fdez-Sanromán et al. 2020), high adsorption capacity, large surface area, and enriched surface functional groups (Sun et al. 2017). Initially employed in soil amendment and water treatment, biochar is now extensively explored for removing industrial gaseous pollutants (Gwenzi et al. 2021). Recently, biochar modification and functionalization developments using physical, chemical, and thermal activation methods have significantly improved the adsorption potential (Chhimwal et al. 2022). However, its effectiveness can vary depending on pollutant characteristics, gas stream composition, and environmental conditions, as well as the complex interactions among various gaseous pollutants (Zhao et al. 2022). Understanding these interactions and removal mechanisms is essential to optimizing biochar's effectiveness.

This chapter reviews biochar preparation, activation, and functionalization methods, discusses mechanisms involved in pollutant removal, evaluates factors affecting biochar performance, examines regeneration methods, and identifies research gaps, aiming to guide future studies for enhanced air pollution management using biochar.

11.2 PREPARATION AND MODIFICATION OF BIOCHAR FOR THE REMOVAL OF AIR POLLUTANTS

Biochar's effectiveness in capturing contaminants depends significantly on its physicochemical properties, which are greatly influenced by preparation, activation, and functionalization methods. These processes alter biochar's physical and chemical attributes, allowing for tailored pollutant removal. The following sections explore different biochar preparation, activation, and functionalization techniques to enhance its efficiency in air pollution control.

11.2.1 Biochar Preparation

Biochar can be produced from many biomass feedstocks, including agricultural wastes, food waste, sludge, animal dung, invasive plant species, and aquatic sources like seaweed (Wang and Wang 2019). Common synthesis methods include pyrolysis, gasification, hydrothermal carbonization (HTC), and microwave-assisted pyrolysis, each of which significantly impacts the biochar characteristics and effectiveness in adsorbing gaseous pollutants (Cha et al. 2016).

Pyrolysis involves the thermal degradation of a carbonaceous material under an inert atmosphere. Slow pyrolysis, characterized by a lower heating rate and extended residence time, favors the production of biochar (Weber and Quicker 2018). Pyrolysis temperature (300–500°C) strongly affects biochar properties; higher temperatures generally increase specific surface area (SSA) and porosity but may reduce functional groups beneficial for adsorption (Usevičiūtė and Baltrėnaitė-Gedienė 2021).

Gasification partially combusts biomass with controlled oxygen or steam (600–1,200°C), primarily yielding syngas, yet the resultant biochar can exhibit specific adsorption properties (Igalavithana et al. 2020). Hydrothermal carbonization employs moderate temperatures (180–250°C) and high pressures (2–20 MPa), producing biochar with unique surface groups and acidity beneficial for specific adsorption uses (Patra et al. 2021; Zhang et al. 2019).

Microwave-assisted pyrolysis uses electromagnetic waves to uniformly heat biomass, achieving a more efficient and controlled carbonization process. This method can provide biochar with enhanced porosity and structural consistency, suitable for contaminant adsorption (Wahi et al. 2017). The uniform heating provided by microwaves eliminates the usual temperature variations observed in conventional heating methods, resulting in a biochar with a more consistent structure (Wahi et al. 2017).

The choice of production method should be determined by the desired biochar characteristics, which are affected by variables such as the biomass type, the production technique, and the processing conditions that determine biochar suitability for gaseous contaminant removal (Patra et al. 2021).

11.2.2 Biochar Activation

Various activation techniques improve biochar's adsorption capacity by enhancing surface area, pore structure, and surface chemistry. Physical activation methods increase the exposed surface area and porosity of biochar. Commonly used physical activation techniques include exposure to an oxidizing medium (steam, carbon dioxide, ozone, or air), size reduction through ball milling, activation through radiation (microwave heating, plasma radiation, and electrochemical heating), and ultrasonic agitation.

Activation with CO_2 usually results in the formation of micropores, which are essential for adsorbing small gas molecules. On the other hand, steam activation increases porosity by removing volatile compounds at elevated temperatures, forming a more developed pore structure. This technique enhances the biochar's capacity to remove different gases, such as CO_2 and SO_2 (Wen et al. 2023). Plasma activation uses ionized gases to modify the surface chemistry efficiently while minimizing structural damage (Xu et al. 2018a). By employing different gas conditions, plasma modification can be customized to produce particular surface functional groups, thereby improving the efficacy of biochar in removing specified contaminants, such as Hg^0 (Xu et al. 2018a). Mechanical treatments like ball milling increase SSA and reduce particle size, further improving biochar's adsorption capacity (Shen and Zhang 2019).

Chemical activation typically involves treating biochar with acids, alkalis, or metal salts to increase the SSA, pore volume, and the number of active sites. Acidic agents like H_3PO_4 can increase microporosity and introduce acidic functional groups, improving biochar's capacity to adsorb basic gases (Guo and Lua 2003). In contrast, alkaline treatment using KOH or NaOH enhances biochar's basicity and specific surface area, making it especially effective for adsorbing acidic gases like CO_2 and SO_2 (Wang and Wang 2019). Despite higher efficiency, chemical activation involves higher costs and environmental considerations (Feliz Florian et al. 2024).

11.2.3 Functionalization of Biochar

Functionalization involves adding active components or chemicals to biochar to enhance specific adsorption capabilities. The affinity of the biochar for different gaseous contaminants can be enhanced by integrating metal oxides, amines, and other active substances onto the biochar surface.

Incorporating metal oxides (Fe, Mn, Ce, and Cu) significantly boosts gas adsorption through catalytic sites, facilitating the adsorption and oxidation of gas molecules, especially chemicals like NO_x, SO_2, and elemental mercury (Hg^0) (Chen et al. 2017). The redox properties of Fe and Mn oxides, in particular, enable the conversion of toxic gases to less harmful compounds, enhancing biochar's overall adsorption performance and durability (Wang and Wang 2019).

Amination by impregnating or grafting organic amines (e.g., polyethyleneimine, monoethanolamine) onto biochar surfaces increases alkalinity, enhancing the adsorption of acidic gases like CO_2 and SO_2 (Samanta et al. 2012). For example, the selectivity and capacity of biochar for CO_2 adsorption are greatly improved as the amine groups form strong interactions with CO_2 molecules, forming carbamate compounds.

Functionalization by plasma treatment in chlorine-rich environments or hydrochloric acid impregnation significantly improves mercury adsorption efficiency (Luo et al. 2019). For instance, biochar functionalized with chlorine effectively converts elemental mercury (Hg^0) into mercuric chloride ($HgCl_2$) or other mercury and chlorine complexes, enhancing their adsorption efficiency.

11.3 MECHANISMS OF CONTAMINANT REMOVAL

The removal of gaseous pollutants by biochar depends on several mechanisms, mainly affected by the interactions between pollutant molecules and biochar's physical and chemical characteristics. The adsorption mechanisms include physical adsorption, pore filling, chemical adsorption, catalysis, partitioning, and specific interactions such as π–π binding, electrostatic interactions, hydrogen bonding, and hydrophobic interactions, all of which play a vital role in efficiently adsorbing biochar. In this sense, Figure 11.1 depicts the main removal mechanisms of different gaseous contaminants.

11.3.1 Physical Adsorption and Pore Filling

Physical adsorption (physisorption) is the main pathway by which biochar removes gaseous contaminants. The effectiveness of physisorption is mainly determined by the physical characteristics of biochar, including specific surface area, porosity, and pore sizes. The process is characterized by weak, non-covalent forces—like Van der Waals forces—between the gas molecules and the biochar surface. Pore filling, on the other hand, refers to the process in which gas molecules physically occupy the volume of the pores within the biochar structure. It is typically driven by the small size of the gas molecules and the presence of micropores (<2 nm) or mesopores (2–50 nm) in the biochar.

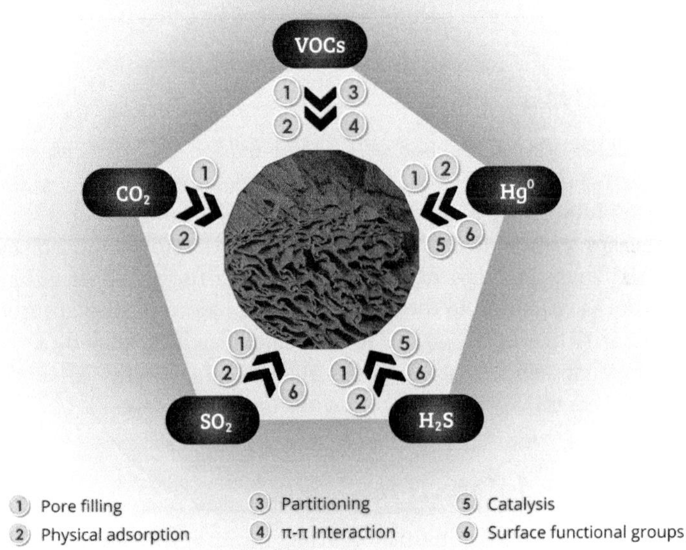

Figure 11.1 Main gaseous contaminant removal mechanisms.

Micropores are the primary sites for physisorption, in which gas molecules are adsorbed through volume filling rather than forming a surface layer formation (Wen et al. 2023). The high adsorption potential of these micropores is important for capturing gases, making biochar with a well-developed microporous structure particularly effective for gas adsorption, particularly CO_2 and other small gas molecules.

In addition to enhancing its physisorption capacity, biochar's large SSA provides additional sites where gas molecules can interact. Thus, biochars with both high SSA and micropore volumes are very efficient in adsorbing gases such as CO_2 and SO_2. These biochar properties are also advantageous for enhancing VOC adsorption through physical adsorption and pore filling, where VOCs can become constrained within the pores (Sadegh et al. 2024).

11.3.2 Chemical Adsorption

Chemical adsorption (chemisorption) involves stronger interactions, forming chemical bonds between gas molecules and active sites on biochar, typically resulting in irreversible adsorption. Functionalizing biochar with specific surface groups (e.g., sulfur, carbonyl, oxygen-containing groups) enhances its selectivity and reactivity for pollutants.

A highly efficient method for trapping acidic gases such as CO_2, SO_2, and H_2S is functionalizing biochar with basic groups, such as amines ($-NH_2$). Strong chemical interactions are formed between the basic sites on the biochar surface and the acidic gases, improving the biochar's adsorption capacity and selectivity (Zhao et al. 2022). Similarly, basic gases like NH_3 can be effectively removed from biochar surfaces using acidic functional groups like –COOH, $-SO_3H$, and –OH since acid-base interactions promote effective gas adsorption and retention.

11.3.3 Catalysis

The functionalization of biochar with metal oxides or other catalytic materials greatly contributes to effectively eliminating specific gaseous contaminants via catalysis. Catalytic processes occurring on the surface of biochar can transform hazardous gases such as H_2S and Hg^0 into less toxic molecules. For example, the compound H_2S can undergo oxidation to form sulfuric acid or sulfate, whereas Hg^0 can undergo oxidation to form Hg^{2+}, which is more readily absorbed and held by the biochar (Zhao et al. 2022). Adding metal oxides, including Fe, Mn, and Cu, to the biochar surface increases the quantity of catalytic active sites, thereby improving the overall efficacy of removing these pollutants.

11.3.4 Specific Interactions

These interactions are distinguished from more general forces like Van der Waals forces or physical adsorption because they arise due to specific chemical and molecular characteristics of the adsorbent (biochar) and the adsorbate (gas molecules). The adsorption of aromatic VOCs such as benzene is particularly affected by specific interactions, mainly $\pi-\pi$ interactions, which take place between the π-electrons of the aromatic rings in VOC molecules and the aromatic structures found on the biochar surface. For instance, biochar obtained from biomass feedstocks with high carbon content can contain many aromatic rings, thus improving the efficiency of capturing chemicals such as benzene, toluene, and other aromatic VOCs (Amalina et al. 2023).

Partitioning is a major mechanism in the adsorption of VOCs onto biochar, especially when the biochar retains significant amounts of non-carbonized organic matter (NOM). VOCs dissolve into the NOM fraction rather than adhering to the biochar surface, with hydrophobic interactions driving the process. This mechanism is more prevalent in biochars produced at lower pyrolysis temperatures since they contain more organic matter like cellulose and lignin. These components create a more flexible, hydrophobic environment conducive to partitioning non-polar or weakly polar VOCs.

11.4 FACTORS INFLUENCING THE REMOVAL OF AIR POLLUTANTS

The effectiveness of biochar in capturing air pollutants depends on various factors, including biochar preparation methods, its physicochemical properties, and environmental conditions during adsorption. A comprehensive understanding of these factors is essential to optimize biochar performance in air pollution control.

11.4.1 Preparation Conditions of Biochar

Biochar's physical and chemical characteristics, such as the SSA, pore structure, and surface functional groups (SFGs), are closely related to the preparation conditions, including the feedstock type, pyrolysis conditions, and activation methods. These properties directly impact biochar's adsorption performance.

The type of feedstock used in biochar synthesis determines the resulting pore structure and surface composition. Plant-derived biochars typically exhibit an increased SSA and better-developed pore structures than biochars produced from animal waste or sludge. The primary factor contributing to this phenomenon is the elevated levels of lignin and cellulose in plants, resulting in the development of a biochar matrix with higher porosity (Zhao et al. 2022). Generally, biochar obtained from the pyrolysis of feedstocks rich in cellulose contains a higher amount of micropores. In contrast, those generated from feedstocks rich in lignin tend to have larger pore diameters (Song et al. 2022).

Pyrolysis parameters such as temperature, heating rate, and residence time significantly influence the properties of biochar. Higher pyrolysis temperatures often enhance the SSA and microporosity of biochar by promoting the release of volatile compounds and the formation of a more refined carbonized structure. Extended exposure to elevated temperatures can destroy essential functional groups required for chemical adsorption (Sadegh et al. 2024). The heating rate governs the development of the pore structure, with lower rates promoting microporosity and higher rates leading to the buildup of macropores (Fu et al. 2012).

The choice of the activation agent (like CO_2, steam, or chemical agents) and the specific conditions for activation (such as temperature and residence time) can significantly influence the properties of the biochar. Activation can enhance microporosity, promoting the adsorption of small gas molecules such as CO_2 and H_2S (Nicolae et al. 2022). Nevertheless, adverse activation conditions, such as high temperature or extended activation time, might worsen the structure of biochar and diminish its efficacy (Iberahim et al. 2019).

11.4.2 Physicochemical Properties of Biochar

The physicochemical properties of biochar significantly influence its adsorption performance. Among these properties, specific surface area, pore structure, and surface functional groups are especially critical. A higher specific surface area typically corresponds to greater adsorption capacity, as it provides more active sites for pollutant adsorption.

The pore structure, particularly micropores, strongly influences the adsorption of small gas molecules. The highly effective adsorption performance of micropores is attributed to their capacity to confine gas molecules within limited areas, improving physisorption (Sadegh et al. 2024). Furthermore, mesopores and macropores are also important, since they allow the movement of gas molecules into the microporous areas where adsorption takes place (Chen et al. 2017).

Surface functional groups, such as carboxyl, hydroxyl, and amine groups, also significantly impact biochar adsorption. These groups interact with contaminants through mechanisms like hydrogen bonding, acid-base interactions, and covalent bonding (Srinadh et al. 2024). For example, nitrogen-containing groups enhance acidic gas adsorption (e.g., CO_2) via acid-base interactions, whereas oxygen-containing groups can facilitate the oxidation of pollutants like SO_2 and Hg^0 (Wen et al. 2023).

11.4.3 Environmental Conditions

The adsorption process is subject to several environmental conditions, including adsorption temperature, humidity, and coexisting gases. The adsorption efficiency of biochar is temperature-dependent. In general, lower temperatures promote physisorption by reducing the kinetic energy of gas molecules, which enables them to adsorb more easily to the surface of biochar. In contrast, elevated temperatures can enhance chemisorption by supplying the necessary energy for chemical interactions to occur between the biochar and the pollutants. However, excessively high temperatures can deteriorate the pore structure and active sites of biochar, thus decreasing its adsorption capacity.

Humidity levels can affect the adsorption process, especially for contaminants that are soluble in water. High relative humidity increases the likelihood of water molecules and contaminants competing for adsorption sites, diminishing biochar's total adsorption capacity. Furthermore, the presence of a water film on the surface of biochar can hinder the diffusion of contaminants into the pores, reducing the effectiveness of adsorption (Amalina et al. 2023).

Interactions with other gases in the surroundings can impact the adsorption of particular pollutants. For example, oxygen might increase the adsorption of Hg^0 by facilitating its oxidation to Hg^{2+}, which is more readily adsorbed by biochar. In contrast, gases like SO_2 might compete with other contaminants for adsorption sites, reducing biochar's efficiency in environments containing several pollutants (Wen et al. 2023).

11.5 THERMODYNAMICS, KINETICS, AND ISOTHERM MODELS

A comprehensive understanding of the thermodynamics, kinetics, and isotherm models that govern the adsorption process on biochar is necessary to optimize its use in removing air pollutants. These studies offer valuable information regarding the viability, effectiveness, and mechanisms of the adsorption process, enabling the development of more robust biochar-based systems.

11.5.1 Adsorption Kinetics

Adsorption kinetics refers to the study of the rate at which contaminants are adsorbed onto the biochar surface. It provides a valuable understanding of the mechanisms that govern the adsorption process. The study of adsorption kinetics involves two main methods: fitting experimental data to kinetic models such as pseudo–first-order and pseudo–second-order models, and thermogravimetric analysis (Gwenzi et al. 2021).

The pseudo–first-order kinetic model postulates that the adsorption rate is directly proportional to the number of unoccupied sites on the surface of the adsorbent. This characteristic renders it appropriate for the description of physisorption processes. Previous studies have demonstrated that this model is highly compatible with the adsorption of CO_2 at ambient temperature, indicating that the adsorption of CO_2 on biochar mostly takes place via physisorption (Bamdad 2019). Nevertheless, the pseudo–second-order model, which proposes that the adsorption rate is directly related to the square of the number of unoccupied sites, is better suited for chemisorption processes. For example, studies on the adsorption of Hg^0 and volatile organic compounds like polar acetone have shown that the pseudo-second-order model provided a better fit, indicating that the process occurs mainly by chemisorption (Johari et al. 2016; Kumar et al. 2020).

The Michaelis–Menten model has been used to characterize the rate at which H_2S is adsorbed in plug-flow reactors packed with biochar. The adsorption process is well-fitted by this model, which accurately describes the maximum removal capacity and the saturation constant (Shang et al. 2013). The intraparticle diffusion model, introduced by Weber and Morris (1963), is a commonly employed model for characterizing adsorption processes in which the rate is governed by the diffusion of gas molecules within the biochar pores (Khuong and Nguyen 2022). This model explains the adsorption phases, such as the initial rapid uptake followed by slower intraparticle diffusion. Complex adsorption systems that include both physical and chemical interactions are commonly characterized by the Avrami model. This model accounts for the influence of several kinetic processes, rendering it especially advantageous for functionalized biochars in which both physisorption and chemisorption may have significant effects (Zhang et al. 2022).

11.5.2 Thermodynamics

A thermodynamic study allows for determining essential parameters like the variation in Gibbs free energy (ΔG), enthalpy change (ΔH), and entropy change (ΔS), which provide valuable information about the spontaneity, feasibility, and nature of the adsorption process.

A negative ΔG value indicates that the adsorption process is spontaneous and thermodynamically favorable. For example, studies on CO_2 adsorption on biochar have consistently demonstrated negative ΔG values at different temperatures, therefore validating the spontaneous nature of the process (Bamdad et al. 2019). The ΔH value determines whether adsorption is characterized by exothermic or endothermic behavior. Most gases, such as CO_2 and VOCs, undergo exothermic adsorption on biochar, as shown by negative ΔH values. These findings show that adsorption is typically more favorable at lower temperatures (Gwenzi et al. 2021).

The entropy change quantifies the level of disorder at the interface between the adsorbent and adsorbate. It can serve as an indicator of the adsorbate's affinity for the surface of the biochar. Negative ΔS values denote a decrease in randomness and a strong interaction between the gas molecules and the biochar, while positive values imply an increase in randomness (Khuong and Nguyen 2022). The presence of negative ΔS values in the adsorption of CO_2 on biochar suggests a highly ordered state at the surface of the adsorbent, which is in line with physisorption interaction (Bamdad et al. 2019).

11.5.3 Isotherm Modeling

Isotherm models describe how pollutants interact with biochar surfaces at equilibrium, offering valuable insights into the specific capacity and nature of the adsorption process. The Langmuir isotherm model assumes that adsorption occurs in a monolayer on a surface with a limited number of identical sites, without any interaction between the adsorbed molecules. This model helps describe adsorption on uniform surfaces, where all sites exhibit the same affinity for the adsorbate. It has been successfully employed to characterize the adsorption of gases such as SO_2 on biochar, demonstrating a monolayer adsorption mechanism (Atanes et al. 2012).

Alternatively, the Freundlich isotherm model is an empirical equation that describes the adsorption process on surfaces with sites of different affinities. This model accounts for multilayer adsorption, in which the adsorption capacity increases with concentration and does not reach saturation. It is commonly used to represent the adsorption of gases such as CO_2 and VOCs on biochar, suggesting the presence of a heterogeneous surface with multiple adsorption sites (Zhang et al. 2022; Pan et al. 2023).

Alternative isotherm models, such as the Temkin, Sips, and Dubinin–Astakhov isotherms, are used to characterize complex adsorption phenomena. The Temkin isotherm specifically accounts for the influence of indirect interactions between the adsorbate and adsorbent. It posits that the heat of adsorption diminishes linearly as the coverage increases. Adsorption on surfaces displaying both uniform and non-uniform properties can be effectively described using the Sips isotherm, which integrates the characteristics of the Langmuir and Freundlich models. The application of this model to the adsorption of CO_2 on biochar has proven effective in accurately representing the adsorption process across a broad pressure range (Siemak and Michalkiewicz 2023).

11.6 BIOCHAR APPLICATION IN GASEOUS POLLUTION CONTROL

The efficiency of biochar in adsorbing gaseous contaminants depends on the preparation methods, activation, and functionalization processes, as well as specific environmental conditions such as temperature and humidity during adsorption. The following sections discuss the biochar removal performance of SO_2, CO_2, H_2S, Hg^0, and VOCs, highlighting the influence of preparation conditions, activation, functionalization, and adsorption environment.

11.6.1 Removal Performance of SO_2

Sulfur dioxide (SO_2) is a major environmental contaminant that is mainly produced by the burning of sulfur-containing biomass and fossil fuels, as well as by volcanic eruptions and the oxidation of organic materials (Iberahim et al. 2022). SO_2 has harmful impacts, such as the generation of acid rain and sulfuric acid smoke, which can decrease the pH levels of soil and water, causing adverse consequences for plant and animal life (Iberahim et al. 2018). Considering these environmental impacts, eliminating SO_2 from industrial emissions with biochar is an important topic of study.

11.6.1.1 Influence of Biochar Preparation Conditions

The performance of biochar in removing SO_2 is significantly affected by its preparation conditions, including the pyrolysis temperature, the heating rate, and the type of activation process. In general, pristine biochar has a limited capacity to adsorb SO_2, frequently below 20 mg/g (Shao et al. 2018). Nevertheless, adjusting the pyrolysis conditions can significantly improve the adsorptive characteristics of biochar. For example, Iberahim et al. (2018) reported that the capacity of biochar to adsorb SO_2 rose as the pyrolysis temperature increased, peaking at 450°C before decreasing at higher temperatures owing to pore collapse and the destruction of surface functional groups.

Activation methods, such as CO_2 activation, can enhance the adsorption capacity of biochar by creating a more porous structure and increasing the specific surface area. Sumathi et al. (2009) found a maximum SO_2 removal capacity when biochar was activated with CO_2 at 1,100°C. This capacity was attributed to the creation of micropores that preferentially adsorb SO_2 due to their small kinetic diameter of 0.364 nm. In a similar vein, Zhang et al. (2021) reported that biochar activated at 800°C with CO_2 presented the highest removal capacity of 167.5 mg/g. Moreover, Atanes et al. (2012) showed that chemical activation of biochar with KOH presented a higher SO_2 uptake value when compared to that of CO_2-activated biochar (95.2 vs. 61.0, respectively). The authors explained that this difference was due to a higher porosity produced by the chemical activation method.

11.6.1.2 Impacts of Functionalization

Functionalizing biochar, particularly with nitrogen-containing compounds, can greatly enhance its efficacy in removing SO_2. Improved physical adsorption of nitrogen-doped biochar is attributed to increased Van der Waals and electrostatic interactions. In contrast, chemical adsorption is enhanced by the creation of stable complexes between SO_2 and nitrogen-containing functional groups (Shao et al. 2018; Qu et al. 2018).

A study conducted by Shao et al. (2018) showed that the addition of nitrogen functionalization to corncob biochar modified using CO_2 and methyldiethanolamine (MDEA) resulted in an enhanced micropore structure and the introduction of C–N and N–H groups. This, in turn, led to a significant adsorption capacity of 156.22 mg/g for SO_2. This capacity was attained by combining physisorption and chemisorption, emphasizing the significance of both porous structure and the presence of functional groups in maximizing the removal of SO_2. Similar results were obtained by Zhang et al. (2021), in which the CO_2-activated and NH_3-modified soybean straw biochar presented an uptake of 175.9 mg SO_2/g.

11.6.1.3 Adsorption Conditions

The adsorption temperature is an important parameter for achieving optimal efficiency of SO_2 removal with biochar. Studies have shown that the impact of temperature on adsorption can be either favorable or unfavorable, contingent upon the equilibrium between physical adsorption (which diminishes with temperature) and chemical adsorption (which may increase with temperature). For instance, whereas physisorption is more efficient at lower temperatures, elevated temperatures can increase the rate of chemisorption between SO_2 and active sites on biochar. For example, Zhang et al. (2021) compared the SO_2 uptake at adsorption temperatures of 30 and 120°C for CO_2-activated and NH_3-modified biochars. They observed no significant differences in the uptake between the two biochars at 30°C. However, the NH_3-modified biochar showed a better SO_2 uptake at 120°C. This difference was related to the presence of nitrogen functional groups on the surface of biochar that promote the chemical adsorption mechanism.

Additionally, humidity has a considerable impact on SO_2 adsorption. Elevated atmospheric moisture levels can improve SO_2 adsorption by creating a liquid membrane on the surface of biochar. This membrane enables the transformation of SO_2 into sulfuric acid and its subsequent adsorption. However, excessive humidity might reduce the partial pressure of SO_2, making it more difficult for SO_2 molecules to compete with water vapor for adsorption sites (Iberahim et al. 2019).

Furthermore, the presence of other gases, such as oxygen (O_2) and nitric oxide (NO), may influence the adsorption of SO_2. Oxygen can promote the oxidation of SO_2 to SO_3, which is then converted into H_2SO_4 on the biochar surface, increasing its adsorption capacity. Nitric oxide has been shown to have a synergistic effect with SO_2, enhancing the capacity of SO_2 adsorption by promoting the production of more easily adsorbed intermediates (Zhang et al. 2020a).

11.6.2 Removal Performance of CO_2

Elevated carbon dioxide (CO_2) levels in the atmosphere, caused mainly by burning fossil fuels and land use changes, play a major role in increasing global warming and climate change. Thus, technology development for carbon dioxide (CO_2) capture and storage has gained significant importance (Dissanayake et al. 2020). Biochar's potential for CO_2 removal stems from its remarkable capacity to efficiently adsorb CO_2, particularly when subjected to targeted modifications or activation.

11.6.2.1 Influence of Biochar Preparation Conditions

The preparation conditions influence the CO_2 adsorption capacity of biochar. Thermal treatment, specifically pyrolysis, strongly influences the development of biochar's porous structure and surface characteristics. Research findings indicate that raising the pyrolysis temperature to a range of 600–900°C typically improves the adsorption of CO_2 by promoting the formation of microporosity and surface activity in the biochar. For instance, biochars generated at elevated temperatures frequently exhibit micropores ranging from 0.4 to 0.9 nm in size. These micropores are especially conducive to CO_2 adsorption due to their kinetic diameter of 0.33 nm (González et al. 2013).

Microwave pyrolysis has been shown to generate biochar with a greater CO_2 adsorption capacity at lower temperatures, such as 321°C, compared to traditional pyrolysis. This is attributed to a larger specific surface area (Huang et al. 2015). Moreover, pre-treatment procedures such as hydrothermal carbonization can reduce the oxygen functionality in biochar, improving its hydrophobic characteristics and increasing its efficiency in adsorbing non-polar CO_2 molecules (Dissanayake et al. 2020).

Activation of biochar is a key step in enhancing the CO_2 adsorption capacity. Physical activation methods, such as CO_2 or steam activation, enhance microporosity, an important attribute for CO_2 capture. A study by Zhang et al. (2014) showed that biochar activated with CO_2 at 800°C exhibited a notable enhancement in adsorption capacity due to the formation of micropores. However, extremely high activation temperatures can lead to the merging of micropores into larger mesopores, thereby decreasing the adsorption efficacy (Goel et al. 2021).

Another efficient way to increase the CO_2 adsorption capacity of biochar is through chemical activation, especially with KOH. Activation with KOH results in an increased presence of micropores and basic oxygen functional groups, which possess advantageous properties for the adsorption of CO_2. For example, Igalavithana et al. (2020) found that the CO_2 removal capacity of food waste/wood biochar activated with KOH rose from 114 mg/g to 141 mg/g. Activation with K_2CO_3 is sometimes preferred over KOH due to its lower environmental impact, while producing a biochar with a similar micropore volume and surface area, leading to effective CO_2 chemisorption through the formation of $KHCO_3$ (Yue et al. 2018).

11.6.2.2 Impacts of Functionalization

Functionalizing biochar with nitrogen-containing compounds or metals can greatly improve its CO_2 removal capacity. Ammonia or amine modification introduces nitrogen functional groups, such as amines and pyridine-like functionalities, which enhance biochar's surface polarity and basicity. In particular, at low CO_2 partial pressures, these modifications improve CO_2 adsorption through chemical interactions. For instance, ammonium-hydroxide-modified biochar exhibited a 130% enhancement in CO_2 removal capacity compared to unmodified biochar, with adsorption uptake up to 162.8 mg/g (Bamdad et al. 2018a).

Metal modification is also a highly efficient approach. Including metals such as Mg, Al, or Fe into the biochar structure can improve the adsorption of CO_2 by both physical and chemical means. For instance, biochar loaded with magnesium has demonstrated higher capacities for adsorbing CO_2 compared to pristine biochar. Studies have reported CO_2 adsorption rates of up to 82 mg/g, attributed to the combined effects of physical and chemical interactions (Lahijani et al. 2018). Adding metal oxides or hydroxides enhances the biochar's alkalinity, improving the capacity to capture CO_2.

11.6.2.3 Adsorption Conditions

The effectiveness of biochar in adsorbing CO_2 is influenced by factors like temperature and composition of the gas mixture. Carbon dioxide adsorption capacity generally declines as temperature rises because of the exothermic nature of physical adsorption. Nevertheless, biochars featuring increased chemisorption capacities, such as those altered with metals or nitrogen groups, can sustain or even accelerate the uptake of CO_2 at elevated temperatures. Optimal CO_2 adsorption at 80°C was observed in MgO-loaded biochar, effectively balancing both physical and chemical adsorption mechanisms (Liu et al. 2013).

Other gases, such as N_2, O_2, and moisture, also impact CO_2 adsorption. Specifically, moisture can improve the adsorption of CO_2 by creating a liquid membrane over the biochar's surface, allowing chemisorption by producing carbonates. Nonetheless, elevated moisture levels might result in the obstruction of pores, diminishing the effectiveness of adsorption (Murge et al. 2018). The optimal selectivity of biochar for CO_2 over other gases, such as N_2 and O_2, is important for practical applications in flue gas environments, which include several gas components. The biochar exhibits enhanced selectivity in adsorbing CO_2 due to its high quadrupole moment and polarizability, particularly when it is functionalized with nitrogen or metals (Parshetti et al. 2015).

11.6.3 Removal Performance of H_2S

H_2S is a highly toxic gas produced during biogas generation, fossil fuel combustion, and landfill processes. It may cause equipment corrosion, catalyst poisoning, and potential health hazards (Ahmad et al. 2021). Thus, biochar can act as an effective H_2S adsorbent due to its large specific surface area, porosity, and functional groups that can capture and convert H_2S.

11.6.3.1 Influence of Biochar Preparation Conditions

Pyrolysis temperature, residence time, and activation method are among the most important parameters for modifying biochar's specific surface area, porous structure, and chemical properties, which in turn influence the H_2S adsorption efficiency of biochar. For example, biochar derived from rice husk at 500°C was reported to have a maximum H_2S adsorption of 382.70 mg/g (Shang et al. 2016). The carbonization process increases the specific surface area and pore volume, resulting in an enhancement in adsorption. A higher porosity provides more adsorption sites, thus leading to a significant increase of 21% in adsorption capacity (Ahmad et al. 2021).

The pH of the biochar also determines the uptake capacity. As an acidic gas, H_2S is more readily adsorbed by alkaline-pH biochar. Biochar produced at higher pyrolysis temperatures tends to be more alkaline as volatile components are removed, increasing the biochar's adsorption efficiency (Ahmad et al. 2021). Sun et al. (2017) reported that the pH of rice husk biochar rose from 9 to 10 when the biomass was pyrolyzed from 400 to 500°C, improving H_2S removal efficiency.

11.6.3.2 Impacts of Functionalization

Biochar functionalization with metal ions or nitrogen can enhance H_2S uptake capacity. The use of $FeCl_3$ to modify corn stover and maple biochars increased H_2S removal by 22 times and 1.23 times, respectively, with the maximum capture efficiency of modified maple being 269 mg/g (Choudhury and Lansing 2020). The enhancement can be ascribed to the more extensive surface area, increased positive charge, and catalytic H_2S oxidation of metal oxides.

Nitrogen doping also serves as an effective functionalization method, increasing the surface basicity and introducing nitrogen-based functional groups that improve biochar's capacity to adsorb and oxidize H_2S. Qin et al. (2020) reported an H_2S uptake for nitrogen-doped biochar of 332 mg/g due to its increased surface area, functional groups, and synergistic effects from microbial growth and biodegradation.

11.6.3.3 Adsorption Conditions

Adsorption parameters such as moisture level, oxygen, and other gas availability influence biochar's effectiveness in H_2S elimination. H_2S adsorption increases with moisture since it favors dissociation into H^+ and HS^- ions that are more readily adsorbed by biochar (Ahmad et al. 2021). For example, Xu et al. (2014) found that pig manure biochar adsorbed 65.4 mg/g at a moisture level of 25 wt% compared to 59.6 mg/g when adsorption was carried out under dry conditions.

Oxygen can either enhance or inhibit the removal of H_2S, depending on the concentration in the surrounding environment. Under an oxygen-limited atmosphere, H_2S is adsorbed into biochar's micropores, transforming it to elemental sulfur (S^0), which may be further oxidized to sulfate (SO_4^{2-}) in an oxygen-rich environment (Xu et al. 2014). However, high oxygen levels can create sulfur dioxide (SO_2), intensifying the biochar's surface acidity and reducing H_2S removal efficiency (Ahmad et al. 2021).

11.6.4 Removal Performance of Hg^0

Mercury, a highly toxic and volatile metal, presents serious environmental and health hazards. It is released into the air from different sources, such as coal-fired boilers, mining operations, and natural events like volcanic eruptions (Johari et al. 2016). Among the different states of mercury, elemental mercury (Hg^0) is difficult to remove because of its volatility, low solubility in water, and tendency to remain in the air (Oliveira et al. 2017).

11.6.4.1 Influence of Biochar Preparation Conditions

The biochar preparation conditions, especially the pyrolysis temperature, affect the Hg^0 adsorption performance. Higher pyrolysis temperatures generally enhance biochar's specific surface area and micropore volume, improving its physical adsorption capacity for Hg^0 (Johari et al. 2016). Biochar pyrolyzed at about 800°C had a more porous structure and higher Hg^0 removal efficiency than that prepared at lower pyrolysis temperatures (Klasson et al. 2014). In addition, biochar produced from animal manure, which contains higher concentrations of sulfur, also removes Hg^0 more efficiently due to the interaction with the active sulfur groups included in the biochar (Oliveira et al. 2017).

The specific surface area and pore structure are also relevant properties for Hg^0 capture. Micropores (<2 nm) show excellent Hg^0 adsorption potential because of the size compatibility with mercury atoms, which have an approximate radius of 0.15 nm (Shen et al. 2019). Carbonization temperatures between 500 and 700°C generate biochars with functional groups (C=O) that interact with Hg^0 through chemisorption. However, excessive temperatures (>800°C) can degrade these functional groups, lowering the Hg^0 removal efficiency (Xu et al. 2018b).

11.6.4.2 Impacts of Functionalization

Biochar functionalization, especially with halogens and metal oxides, greatly increases Hg^0 removal capacity. Halogen impregnation (e.g., KI, KBr, and KCl) introduces covalent-bonded halogens to the biochar surface, forming active sites that transform Hg^0 to mercury halides (HgX and HgX_2), with X = Cl, Br, or I (Xu et al. 2018a). These modifications are more effective in the order I > Br > Cl because of halogen size and reactivity (Liu et al. 2018). Li et al. (2016) performed adsorption experiments with ammonium-halides-modified biochar and observed a similar trend, where NH_4I-modified biochar presented the highest uptake (91.0%) when compared to the NH_4Cl and NH_4Br-modified biochar, with uptake values of 78.2% and 85.5%, respectively.

Another effective strategy for enhancing the Hg^0 removal capacity of biochar is through metal oxide modification. The biochars modified with metal oxides (CuO, MnO_2, and Fe_2O_3) increase the Hg^0 removal by catalytic oxidation and charge transfer reactions (Xu et al. 2018a). For instance, Yi et al. (2018) reported that biochar impregnated with Cu–Mn oxide can promote electron transfer from Mn^{4+} to Mn^{3+}, oxidizing the adsorbed Hg^0 to HgO. In addition, metal oxides interacting with halides on biochar surfaces might provide numerous active sites, improving the efficiency of Hg^0 capture.

11.6.4.3 Adsorption Conditions

Adsorption conditions—temperature, flue gas composition, and moisture content—actively influence the efficiency of biochar for capturing Hg^0. Hg^0 removal by biochar is typically effective at temperatures from 120 to 180°C, where a rise in temperature within this range is favorable for chemisorption to occur (Li et al. 2017). For example, biochar modified with $FeCl_3$ showed a 95% removal efficiency in this temperature interval (Altaf et al. 2021). A similar trend was observed by Li et al. (2017), who tested the influence of temperature on Hg^0 uptake for different biochars modified with various halides. The authors observed that for temperatures between 60 and 180°C, the adsorption capacity increased from 631.7 to 3,089.4 μg/g for the Br-modified biochar and from 5,337.6 to 7,752.0 μg/g for the I-modified biochar.

Flue gas composition can also affect Hg^0 removal. The presence of O_2 and NO enhances Hg^0 oxidation, which improves its adsorption on biochar surfaces. O_2 facilitates the replenishment of chemisorbed oxygen and supports electron transfer, aiding the conversion of Hg^0 to oxidized forms like HgO or $Hg(NO_3)_2$, which are easier to adsorb. Moreover, NO_2, a product of NO oxidation, further promotes Hg^0 oxidation. However, excessive NO concentrations may consume surface oxygen and hinder adsorption (Wen et al. 2023). On the other hand, SO_2 exhibits a more complex influence on Hg^0 removal. SO_2 can have both positive and negative effects, depending on its concentration. At lower concentrations (<700 ppm), SO_2 may enhance Hg^0 removal by promoting the formation of $HgSO_4$. However, at higher concentrations, SO_2 competes with Hg^0 for active sites and forms stable metal sulfates that poison the biochar, reducing its capacity for Hg^0 capture (Liu et al. 2018).

Moisture in flue gas may enhance or inhibit Hg^0 uptake. At low concentrations, water vapor contributes to active hydroxyl radicals (OH) on the biochar, promoting Hg^0 oxidation. However, at higher moisture levels (~3–5%), water competes with Hg^0 for adsorption sites, forming a water film that prevents Hg^0 from interacting with the active sites on the biochar (Yang et al. 2017).

11.6.5 Removal Performance of VOCs

VOCs comprise a range of organic compounds that are typically released as gases via different sources, including industrial activities, fossil fuel combustion, and several natural processes. The toxic nature and potential to cause the formation of secondary pollutants like ozone and fine particulate matter make VOCs a significant environmental concern that requires suitable removal treatments (Sadegh et al. 2024).

11.6.5.1 Influence of Biochar Preparation Conditions

The preparation conditions of biochar, such as the pyrolysis temperature, activation techniques employed, and feedstock type, affect the biochar's capacity to remove VOCs. These conditions influence the biochar's surface area, porosity, and functional groups. Pyrolysis temperature directly impacts the surface area, pore structure, and the content of non-carbonized organic matter (NOM). Increasing pyrolysis temperature generally raises the SSA and micropore volume of the biochar, but the VOC adsorption capacity can be reduced due to the loss of NOM. Biochars prepared under lower pyrolysis temperatures (for instance, 300°C) contain a higher NOM content, promoting the partitioning mechanism, resulting in a greater VOC adsorption uptake. It was reported that the acetone adsorption capacity for biochar prepared at 300°C was 14.05 mg/g. In comparison, the 600°C biochar had a lower adsorption capacity of 11.49 mg/g, demonstrating the importance of NOM in VOC removal (Zhang et al. 2017). Considering the overall adsorption process, it can be noted that a mixture of adsorption-partition mechanisms takes place, with partitioning governing in low-temperature biochars with higher NOM content and adsorption controlling in high-temperature biochars with extensive surface areas and less NOM content.

The feedstock choice also considerably impacts the VOC adsorption capacity of biochar. Different feedstocks lead to biochars with different SSA, ash content, and functional groups, affecting VOC removal capacity. According to Pongkua et al. (2018), bagasse-derived biochar presented a maximum adsorption of methyl tert-butyl ether (MTBE) of 8.26 mg/g compared to bone-derived biochar, which had MTBE adsorption of 3.51 mg/g due to its large ash content. This behavior can be explained by the fact that bagasse biochar functional groups, such as carbonyls and esters, interacted more effectively with MTBE than phosphate-rich bone biochar.

Biochar activation by agents such as H_3PO_4 and K_2CO_3 is important for increasing porosity and surface functional groups. Chemical activation significantly raises the SSA and total pore volume (TPV), improving the VOC adsorption potential. For instance, H_3PO_4-modified hickory wood biochar led to a total SSA increase to 815.11 m^2/g, achieving a removal potential of acetone of 147.77 mg/g and cyclohexane at 159.66 mg/g (Zhang et al. 2019). The improved surface area and pore structure enhanced physical adsorption, which is the main mechanism for VOC removal for non-polar and weakly polar VOCs, such as toluene and cyclohexane. Furthermore, CO_2 activation also enhances biochar porosity. Analysis of acetone and cyclohexane removal showed that CO_2-activated biochar is effective due to improved mesopore and micropore structures (Zhang et al. 2020b). Kinetic studies indicated that physical adsorption was the controlling mechanism in VOC removal, especially in biochars with larger surface areas and pore volumes.

11.6.5.2 Impacts of Functionalization

Biochar functionalization can significantly improve the adsorption performance of VOCs. For example, nitrogen-doping biochar can enhance π-electron polarization, which generates π-electron enrichment sites that promote aromatic VOC-like toluene adsorption through π–π interactions (Amalina et al. 2023). In this sense, Cheng et al. (2023) reported that N-doped biochar achieved a toluene adsorption capacity of 223 mg/g, while the non-doped biochar exhibited a lower uptake of 123 mg/g under the same conditions.

Furthermore, functionalization with metal oxides (such as CuO and MnO_2) improves biochar's adsorption capacity for polar and nonpolar VOCs since metal oxides act as active sites for chemisorption, facilitating electron transfer and reinforcing the interaction between biochar and the VOC molecules. Additionally, metal oxides increase the surface area and pore volume of biochar. For instance, CuO-modified biochar significantly improved toluene adsorption, reaching an adsorption capacity of 501.8 mg/g at 60°C compared to the uptake value of 51.9 mg/g obtained with pristine biochar (Zhang et al. 2023)

11.6.5.3 Adsorption Conditions

Temperature and the coexistence of multiple VOCs are adsorption conditions affecting biochar's VOC removal performance. VOC adsorption is generally exothermic, and higher adsorption temperatures reduce adsorption capacities. For example, increasing the temperature from 20 to 40°C decreased biochar's acetone adsorption capacity by 73.1–98.2% (Zhang et al. 2017). A temperature rise also reduced the adsorption of benzene and xylenes since VOCs with higher kinetic energy showed fewer interactions with the biochar's surface (Opatokun et al. 2017; Rajabi et al. 2021).

When multiple VOCs coexist in a system, competitive adsorption may occur, where VOCs with stronger intermolecular forces and larger molecular sizes can displace other adsorbed VOCs on the biochar surface. Zhao et al. (2018) reported that chlorobenzene gradually replaced toluene on biochar when adsorption saturation was approached. This behavior was explained by the chlorobenzene's lower volatility and stronger adsorption potential. Rajabi et al. (2021) pointed out that the adsorption performance in multi-component systems is usually similar to the sum of single VOC adsorption uptakes, indicating that most active sites on biochar are used.

11.7 REGENERATION AND DISPOSAL OF SPENT BIOCHAR

The sustainable use of biochar in environmental applications strongly depends on the regeneration and disposal of spent biochar. Regeneration extends the life cycle of biochar, reducing the need for continuous production, while proper disposal ensures that spent biochar does not contribute to environmental contamination.

11.7.1 Regeneration of Spent Biochar

Regeneration of spent biochar is of the utmost importance for reducing production costs and recovering absorbed contaminants, enabling several cycles of reuse. Even though there are different techniques to regenerate biochar, thermal treatment is the most common method to achieve the desorption of the contaminants and recover the biochar's adsorption capacity. Furthermore, the effectiveness of biochar regeneration may be influenced by the specific type of adsorption mechanism involved. Thus, pollutants that are physically adsorbed require lower temperatures to desorb than those that are chemically adsorbed. Bamdad et al. (2018b) showed that CO_2 adsorbed by amine-modified wood chip biochar could be efficiently regenerated at room temperature and exhibited a mere 20% reduction in adsorption capacity after 10 cycles. Weak bonds, such as hydrogen bonds or reversible interactions between CO_2 and amine groups, may explain this behavior.

Additionally, biochar exhausted with VOCs, including toluene, ethanol, cyclohexane, chloroform, and acetone, can be effectively regenerated by heating to 60–115°C, achieving 80–90% desorption efficiency after five adsorption-desorption cycles (Xiang et al. 2020), indicating that physical adsorption is the main mechanism, as suggested by Zhang et al. (2017). However, for certain pollutants, a higher temperature level may be necessary to regenerate biochar. Stable sites retain species like Hg and are, therefore, more challenging to desorb, requiring higher desorption temperatures, typically above 200°C. Braghiroli et al. (2019) pointed out that regenerating wood biochar spent with SO_2 required a temperature of 600°C to maintain a removal efficiency of 78% after four cycles, which suggests that chemisorption controls the adsorption process. However, thermal regeneration may alter the physical structure of biochar, modifying the microporosity and surface area. Furthermore, overheating may degrade surface functional groups and active sites, reducing adsorption capacity over time. In some cases, it may enhance the adsorption properties of biochar. Braghiroli et al. (2019) reported that after six cycles of adsorption-desorption of SO_2, biochar showed an increase in micropore volume and surface area, which promoted further adsorption.

Hybrid regeneration techniques, combining high-temperature and low-temperature treatments, could improve the regeneration efficiency. For example, exhausted magnetic biochar was first heated at 400°C under nitrogen and then regenerated at 250°C under air to desorb Hg^0, which resulted in a removal efficiency of 78% after five cycles (Shan et al. 2019). Similar hybrid methods may provide energy-saving alternatives for biochar regeneration rather than a one-step high-temperature treatment.

Thermal swing adsorption (TSA) and pressure swing adsorption (PSA) are two pressure-based regeneration techniques that can be used to regenerate exhausted biochar. TSA uses heat to desorb air contaminants, whereas PSA disengages pollutants by lowering pressure. For instance, both approaches have effectively achieved CO_2 recovery with a minimal decrease in biochar sorption capacity throughout multiple cycles (Karimi et al. 2022). These regeneration processes are characterized by their simplicity and cost-effectiveness, largely due to the limited CO_2 interactions with biochar, enabling widespread industrial implementation.

11.7.2 Spent Biochar Disposal

After several regeneration cycles, biochar reaches a point where its adsorption capacity is reduced, making its final disposal difficult. Disposing of spent biochar must be done appropriately to mitigate environmental hazards and optimize resource recovery. Land application as a soil amendment is one of the first disposal pathways. Due to biochar's intrinsic properties, land application is considered a sustainable approach since it combines waste management with environmental benefits. For example, spent biochar with CO_2 can be used in soils as a carbon sequestration strategy since CO_2 is stable under normal atmospheric and soil conditions (Wen et al. 2023). However, biochar with heavy metals or other toxic elements should not be applied in agriculture to avoid contaminating the food chain (Gwenzi et al. 2017). If biochar is polluted with toxic substances like heavy metals (e.g., Hg^0), its disposal becomes more complex. One disposal option is landfilling; however, the leaching of harmful pollutants is a possibility that should be carefully analyzed and strictly controlled. In this situation, encapsulation in stable materials or mixing with other waste to reduce the likelihood of leaching represents a viable option.

Incineration may be considered as a possibility to dispose of spent biochar, especially when loaded with organic contaminants like VOCs. Incineration can combust organic compounds effectively; however, given the possibility of the re-emission of pollutants back into the atmosphere, unless suitable flue gas treatment measures are applied, this procedure's sustainability is threatened. Moreover, biochar combustion also nullifies biochar's capacity as a carbon storage medium, therefore diminishing the sustainability of this method.

Another route for biochar disposal is its incorporation into construction materials, where it can act as a filler material. Biochar's low density and thermal conductivity render it appropriate for insulation materials, carbon-sequestering concrete, and sustainable building materials (Gwenzi et al. 2021). Biochar rich in non-hazardous adsorbates, such as CO_2 or SO_2, can be beneficial in these materials without posing major environmental risks. Moreover, biochar incorporation into construction material can increase durability and reduce permeability, offering added value to its disposal.

An alternative disposal method is converting spent biochar to activated carbon. This procedure entails more heat and chemical treatment, resetting the biochar's surface area and functional groups. The activated carbon can be reapplied in adsorption processes, closing the loop of biochar disposal. However, the energy and chemical requirements in this process must be carefully evaluated to ensure its sustainability.

11.8 CONCLUSIONS AND FUTURE PERSPECTIVES

The use of biochar to adsorb gaseous pollutants is a promising, cost-effective, and renewable pathway to deal with industrial emissions. Previous research has underscored biochar's potential to efficiently capture several air pollutants, including VOCs, CO_2, SO_2, and heavy metals, through processes like physical adsorption, pore filling, chemical adsorption, and catalytic degradation, amongst others. The versatility of biochar, along with its potential for modification, allows its use across various contaminants, making it a valuable, environmentally friendly tool. Yet, despite these promising results, some constraints must be addressed for biochar's industrial-scale applications.

The major difficulty in progressing with biochar technology is the requirement for more comprehensive research at an industrial level. Current studies are generally restricted to lab-scale conditions, and a knowledge gap remains in the field of biochar performance under complex, realistic conditions. Also, the absence of standardized techniques for biochar preparation, customization, and use further complicates its integration. Consequently, pilot and demonstration projects should be established to bridge these gaps and enable the shift from laboratory research to industrial execution.

Another research focus should be on spent biochar's regeneration and disposal. Even though thermal regeneration can regain biochar's adsorption potential, it requires substantial energy and may reduce the uptake effectiveness in the following cycles. Alternative regeneration processes, like pressure transition adsorption or chemical treatments, should be explored to pinpoint more sustainable and cost-efficient solutions. Also, the safe disposal of spent biochar, particularly when it contains harmful pollutants, is a top priority. Practices for environmentally sustainable disposal, like application in soil amendment or as a filler in construction materials, need to be developed and comprehensively evaluated.

Rigorous life-cycle analysis (LCA) is pivotal to gauge the environmental implications of biochar, from production to final disposal stages. Such research will help to identify the most sustainable practices and eliminate any likely negative impacts associated with biochar use.

Future research should also investigate state-of-the-art engineered biochars with enhanced adsorption properties and custom-made characteristics for particular usages. This includes looking into nanotechnology and other specialized materials to modify biochar and increase its efficiency. Furthermore, biochar can be designed not just to adsorb pollutants but also to decompose them into less harmful components. The integration of biochar with other emerging technologies, like photocatalysis and microbial degradation, might create room for new methodologies to eliminate pollutants and recover resources.

In conclusion, biochar holds excellent potential for air pollution mitigation, yet realizing these advantages demands determined efforts to overcome the existing challenges. These issues encompass scaling research to match industry requirements, designing cost-efficient regeneration and disposal methods, as well as carrying out comprehensive environmental assessments.

REFERENCES

Abdulrasheed, A.A., A.A. Jalil, S. Triwahyono, M.A.A. Zaini, Y. Gambo and M. Ibrahim. 2018. Surface modification of activated carbon for adsorption of SO_2 and NO_X: A review of existing and emerging technologies. Renewable and Sustainable Energy Reviews. 94: 1067–1085.

Ahmad, W., S. Sethupathi, G. Kanadasan, L.C. Lau and R. Kanthasamy. 2021. A review on the removal of hydrogen sulfide from biogas by adsorption using sorbents derived from waste. Reviews in Chemical Engineering. 37: 407–431.

Altaf, A.R., H. Teng, M. Zheng, I. Ashraf, M. Arsalan, A.U. Rehman, et al. 2021. One-step synthesis of renewable magnetic tea-biochar derived from waste tea leaves for the removal of Hg^0 from coal-syngas. Journal of Environmental Chemical Engineering. 9: 105313.

Amalina, F., S. Krishnan, A.W. Zularisam and M. Nasrullah. 2023. Biochar and sustainable environmental development towards adsorptive removal of pollutants: Modern advancements and future insight. Process Safety and Environmental Protection. 173: 715–728.

Atanes, E., A. Nieto-Márquez, A. Cambra, M.C. Ruiz-Pérez and F. Fernández-Martínez. 2012. Adsorption of SO_2 onto waste cork powder-derived activated carbons. Chemical Engineering Journal. 211–212: 60–67.

Bamdad, H., K. Hawboldt and S. Macquarrie. 2018a. Nitrogen functionalized biochar as a renewable adsorbent for efficient CO_2 removal. Energy and Fuels. 32: 11742–11748.

Bamdad, H., K. Hawboldt and S. MacQuarrie. 2018b. A review on common adsorbents for acid gases removal: Focus on biochar. Renewable and Sustainable Energy Reviews. 81: 1705–1720.

Bamdad, H. 2019. A theoretical and experimental study on biochar as an adsorbent for removal of acid gases (CO_2 and H_2S). Ph.D. Thesis, Memorial University of Newfoundland, Canada.

Bamdad, H., K. Hawboldt, S. MacQuarrie and S. Papari. 2019. Application of biochar for acid gas removal: experimental and statistical analysis using CO_2. Environmental Science and Pollution Research. 26: 10902–10915.

Braghiroli, F.L., H. Bouafif and A. Koubaa. 2019. Enhanced SO_2 adsorption and desorption on chemically and physically activated biochar made from wood residues. Industrial Crops and Products. 138.

Cha, J.S., S.H. Park, S.C. Jung, C. Ryu, J.K. Jeon, M.C. Shin, et al. 2016. Production and utilization of biochar: A review. Journal of Industrial and Engineering Chemistry. 40: 1–15.

Chen, Y., X. Zhang, W. Chen, H. Yang and H. Chen. 2017. The structure evolution of biochar from biomass pyrolysis and its correlation with gas pollutant adsorption performance. Bioresource Technology. 246: 101–109.

Cheng, T., Y. Bian, J. Li, X. Ma, L. Yang, L. Zhou, et al. 2023. Nitrogen-doped porous biochar for selective adsorption of toluene under humid conditions. Fuel. 334: 126452.

Chhimwal, M., D. Pandey and R.K. Srivastava. 2022. Pristine Biochar and Engineered Biochar: Differences and application. pp. 3–19. *In*: S. Ramola, D. Mohan, O. Masek, A. Méndez and T. Tsubota (eds). Engineered Biochar. Springer Nature Singapore, Singapore.

Choudhury, A. and S. Lansing. 2020. Biochar addition with Fe impregnation to reduce H_2S production from anaerobic digestion. Bioresource Technology. 306: 123121.

Dissanayake, P.D., S. You, A.D. Igalavithana, Y. Xia, A. Bhatnagar, S. Gupta, et al. 2020. Biochar-based adsorbents for carbon dioxide capture: A critical review. Renewable and Sustainable Energy Reviews. 119: 109582.

Du, Y., Z. Fan, T. Guo, J. Xu, Z. Han, Y. Pan, et al. 2020. Characteristics of as-prepared biochar derived from catalytic pyrolysis within moderate-temperature ionic liquid for CO_2 uptake. The Canadian Journal of Chemical Engineering. 98: 690–704.

Fdez-Sanromán, A., M. Pazos, E. Rosales and M.A. Sanromán. 2020. Unravelling the environmental application of biochar as low-cost biosorbent: A review. Applied Sciences. 10: 1–22.

Feliz Florian, G., M. Ragoubi, N. Leblanc, B. Taouk and L. Abdelouahed. 2024. Biochar production and its potential application for biocomposite materials: A comprehensive review. Journal of Composites Science. 8: 220.

Fu, P., S. Hu, J. Xiang, L. Sun, S. Su and J. Wang. 2012. Evaluation of the porous structure development of chars from pyrolysis of rice straw: Effects of pyrolysis temperature and heating rate. Journal of Analytical and Applied Pyrolysis. 98: 177–183.

Goel, C., S. Mohan and P. Dinesha. 2021. CO_2 capture by adsorption on biomass-derived activated char: A review. Science of the Total Environment. 798: 149296.

González, A.S., M.G. Plaza, F. Rubiera and C. Pevida. 2013. Sustainable biomass-based carbon adsorbents for post-combustion CO_2 capture. Chemical Engineering Journal. 230: 456–465.

Guo, J. and A.C. Lua. 2003. Adsorption of sulphur dioxide onto activated carbon prepared from oil-palm shells with and without pre-impregnation. Separation and Purification Technology. 30: 265–273.

Gwenzi, W., N. Chaukura, C. Noubactep and F.N.D. Mukome. 2017. Biochar-based water treatment systems as a potential low-cost and sustainable technology for clean water provision. Journal of Environmental Management. 197: 732–749.

Gwenzi, W., N. Chaukura, T. Wenga and M. Mtisi. 2021. Biochars as media for air pollution control systems: Contaminant removal, applications and future research directions. Science of the Total Environment. 753: 142249.

Huang, Y.F., P. Te Chiueh, C.H. Shih, S.L. Lo, L. Sun, Y. Zhong, et al. 2015. Microwave pyrolysis of rice straw to produce biochar as an adsorbent for CO_2 capture. Energy. 84: 75–82.

Iberahim, N., S. Sethupathi and M.J.K. Bashir. 2018. Optimization of palm oil mill sludge biochar preparation for sulfur dioxide removal. Environmental Science and Pollution Research. 25: 25702–25714.

Iberahim, N., S. Sethupathi, C.L. Goh, M.J.K. Bashir and W. Ahmad. 2019. Optimization of activated palm oil sludge biochar preparation for sulphur dioxide adsorption. Journal of Environmental Management. 248: 109302.

Iberahim, N., S. Sethupathi, M.J.K. Bashir, R. Kanthasamy and T. Ahmad. 2022. Evaluation of oil palm fiber biochar and activated biochar for sulphur dioxide adsorption. Science of the Total Environment. 805: 150421.

Igalavithana, A.D., S.W. Choi, P.D. Dissanayake, J. Shang, C.H. Wang, X. Yang, et al. 2020. Gasification biochar from biowaste (food waste and wood waste) for effective CO_2 adsorption. Journal of Hazardous Materials. 391: 121147.

Johari, K., N. Saman, S.T. Song, S.C. Cheu, H. Kong and H. Mat. 2016. Development of coconut pith chars towards high elemental mercury adsorption performance - Effect of pyrolysis temperatures. Chemosphere. 156: 56–68.

Karimi, M., M. Shirzad, J.A.C. Silva and A.E. Rodrigues. 2022. Biomass/Biochar carbon materials for CO_2 capture and sequestration by cyclic adsorption processes: A review and prospects for future directions. Journal of CO_2 Utilization. 57: 101890.

Khuong, D.A. and H.N. Nguyen. 2022. Engineered biochar as gas adsorbent. pp. 237–258. *In*: Ramola, S., D. Mohan, O. Masek, A. Méndez, and T. Tsubota (eds). Engineered Biochar. Springer, Singapore.

Klasson, K.T., L.L. Boihem, M. Uchimiya and I.M. Lima. 2014. Influence of biochar pyrolysis temperature and post-treatment on the uptake of mercury from flue gas. Fuel Processing Technology. 123: 27–33.

Kumar, A., E. Singh, A. Khapre, N. Bordoloi and S. Kumar. 2020. Sorption of volatile organic compounds on non-activated biochar. Bioresource Technology. 297: 122469.

Lahijani, P., M. Mohammadi and A.R. Mohamed. 2018. Metal incorporated biochar as a potential adsorbent for high capacity CO_2 capture at ambient condition. Journal of CO_2 Utilization. 26: 281–293.

Li, G., S. Wang, Q. Wu, F. Wang and B. Shen. 2016. Mercury sorption study of halides modified bio-chars derived from cotton straw. Chemical Engineering Journal. 302: 305–313.

Li, G., S. Wang, Q. Wu, F. Wang, D. Ding and B. Shen. 2017. Mechanism identification of temperature influence on mercury adsorption capacity of different halides modified bio-chars. Chemical Engineering Journal. 315: 251–261.

Liu, W.J., H. Jiang, K. Tian, Y.W. Ding and H.Q. Yu. 2013. Mesoporous carbon stabilized MgO nanoparticles synthesized by pyrolysis of $MgCl_2$ preloaded waste biomass for highly efficient CO_2 capture. Environmental Science & Technology. 47: 9397–9403.

Liu, Z., W. Yang, W. Xu and Y. Liu. 2018. Removal of elemental mercury by bio-chars derived from seaweed impregnated with potassium iodine. Chemical Engineering Journal. 339: 468–478.

Liu, Y. and Y. Wang. 2019. Removal of gaseous hydrogen sulfide by a photo-Fenton wet oxidation scrubbing system. Energy and Fuels. 33: 10812–10819.

Liu, D., C. Li, J. Wu and Y. Liu. 2020. Novel carbon-based sorbents for elemental mercury removal from gas streams: A review. Chemical Engineering Journal. 391: 123514.

Luo, J., M. Jin, L. Ye, Y. Cao, Y. Yan, R. Du, et al. 2019. Removal of gaseous elemental mercury by hydrogen chloride non-thermal plasma modified biochar. Journal of Hazardous Materials. 377: 132–141.

Murge, P., S. Dinda and S. Roy. 2018. Adsorbent from Rice Husk for CO_2 Capture: Synthesis, Characterization, and Optimization of Parameters. Energy and Fuels. 32: 10786–10795.

Nicolae, S.A., J. Louis-Therese, S. Gaspard, P.Á. Szilágyi and M.M. Titirici. 2022. Biomass derived carbon materials: Synthesis and application towards CO_2 and H_2S adsorption. Nano Select. 3: 165–177.

Oliveira, F.R., A.K. Patel, D.P. Jaisi, S. Adhikari, H. Lu and S.K. Khanal. 2017. Environmental application of biochar: Current status and perspectives. Bioresource Technology. 246: 110–122.

Opatokun, S.A., A. Prabhu, A. Al Shoaibi, C. Srinivasakannan and V. Strezov. 2017. Food wastes derived adsorbents for carbon dioxide and benzene gas sorption. Chemosphere. 168: 326–332.

Pan, X., N. Zhang, L. Yang, C. He, X. Ma, X. Liu, et al. 2023. Preparation of a novel straw-sludge activated biochar and its adsorption mechanisms for removal of VOCs. ACS Omega. 8: 39329–39344.

Parshetti, G.K., S. Chowdhury and R. Balasubramanian. 2015. Biomass derived low-cost microporous adsorbents for efficient CO_2 capture. Fuel. 148: 246–254.

Patra, B.R., A. Mukherjee, S. Nanda and A.K. Dalai. 2021. Biochar production, activation and adsorptive applications: A review. Environmental Chemistry Letters. 19: 2237–2259.

Peluso, A., N. Gargiulo, P. Aprea, F. Pepe and D. Caputo. 2019. Nanoporous materials as H_2S adsorbents for biogas purification: A review. Separation & Purification Reviews. 48: 78–89.

Pongkua, W., R. Dolphen and P. Thiravetyan. 2018. Effect of functional groups of biochars and their ash content on gaseous methyl tert-butyl ether removal. Colloids and Surfaces A: Physicochemical and Engineering Aspects. 558: 531–537.

Qin, L., X. Huang, Q. Xue, L. Liu and Y. Wan. 2020. In-situ biodegradation of harmful pollutants in landfill by sludge modified biochar used as biocover. Environmental Pollution. 258: 113710.

Qu, Z., F. Sun, X. Liu, J. Gao, Z. Qie and G. Zhao. 2018. The effect of nitrogen-containing functional groups on SO_2 adsorption on carbon surface: Enhanced physical adsorption interactions. Surface Science. 677: 78–82.

Rajabi, H., M.H. Mosleh, P. Mandal, A. Lea-Langton and M. Sedighi. 2021. Sorption behaviour of xylene isomers on biochar from a range of feedstock. Chemosphere. 268: 129310.

Sadegh, F., N. Sadegh, W. Wongniramaikul, R. Apiratikul and A. Choodum. 2024. Adsorption of volatile organic compounds on biochar: A review. Process Safety and Environmental Protection. 182: 559–578.

Samanta, A., A. Zhao, G.K.H. Shimizu, P. Sarkar and R. Gupta. 2012. Post-combustion CO_2 capture using solid sorbents: A review. Industrial & Engineering Chemistry Research. 51: 1438–1463.

Shan, Y., W. Yang, Y. Li, Y. Liu and J. Pan. 2019. Preparation of microwave-activated magnetic bio-char adsorbent and study on removal of elemental mercury from flue gas. Science of the Total Environment. 697: 134049.

Shang, G., G. Shen, L. Liu, Q. Chen and Z. Xu. 2013. Kinetics and mechanisms of hydrogen sulfide adsorption by biochars. Bioresource Technology. 133: 495–499.

Shang, G., L. Liu, P. Chen, G. Shen and Q. Li. 2016. Kinetics and the mass transfer mechanism of hydrogen sulfide removal by biochar derived from rice hull. Journal of the Air & Waste Management Association. 66: 439–445.

Shao, J., J. Zhang, X. Zhang, Y. Feng, H. Zhang, S. Zhang, et al. 2018. Enhance SO_2 adsorption performance of biochar modified by CO_2 activation and amine impregnation. Fuel. 224: 138–146.

Shen, B., Z. Liu, H. Xu and F. Wang. 2019. Enhancing the absorption of elemental mercury using hydrogen peroxide modified bamboo carbons. Fuel. 235: 878–885.

Shen, Y. and N. Zhang. 2019. Facile synthesis of porous carbons from silica-rich rice husk char for volatile organic compounds (VOCs) sorption. Bioresource Technology. 282: 294–300.

Siemak, J. and B. Michalkiewicz. 2023. Adsorption equilibrium of CO_2 on microporous activated carbon produced from avocado stone using H_2SO_4 as an activating agent. Sustainability. 15: 16881.

Song, G., F. Qin, J. Yu, L. Tang, Y. Pang, C. Zhang, et al. 2022. Tailoring biochar for persulfate-based environmental catalysis: Impact of biomass feedstocks. Journal of Hazardous Materials. 424: 127663.

Srinadh, R.V., R. Neelancherry and A. Verma. 2024. Biochar as a filter media for air pollution control systems. pp. 105–123. *In*: Agricultural Waste to Value-Added Products. Springer Nature Singapore, Singapore.

Sumathi, S., S. Bhatia, K.T. Lee and A.R. Mohamed. 2009. Optimization of microporous palm shell activated carbon production for flue gas desulphurization: Experimental and statistical studies. Bioresource Technology. 100: 1614–1621.

Sun, Y., G. Yang, L. Zhang and Z. Sun. 2017. Preparation of high performance H_2S removal biochar by direct fluidized bed carbonization using potato peel waste. Process Safety and Environmental Protection. 107: 281–288.

Usevičiūtė, L. and E. Baltrėnaitė-Gedienė. 2021. Dependence of pyrolysis temperature and lignocellulosic physical-chemical properties of biochar on its wettability. Biomass Convers. Biorefinery. 11: 2775–2793.

Vikrant, K., V. Kumar, K.H. Kim and D. Kukkar. 2017. Metal-organic frameworks (MOFs): Potential and challenges for capture and abatement of ammonia. Journal of Materials Chemistry A. 5: 22877–22896.

Wahi, R., N.F.Q. Zuhaidi, Y. Yusof, J. Jamel, D. Kanakaraju and Z. Ngaini. 2017. Chemically treated microwave-derived biochar: An overview. Biomass and Bioenergy. 107: 411–421.

Wang, J. and S. Wang. 2019. Preparation, modification and environmental application of biochar: A review. Journal of Cleaner Production. 227: 1002–1022.

Weber, W.J. and J.C. Morris. 1963. Kinetics of adsorption on carbon from solution. Journal of the Sanitary Engineering Division. 89: 31–59.

Weber, K. and P. Quicker. 2018. Properties of biochar. Fuel. 217: 240–261.

Wen, C., T. Liu, D. Wang, Y. Wang, H. Chen, G. Luo, et al. 2023. Biochar as the effective adsorbent to combustion gaseous pollutants: Preparation, activation, functionalization and the adsorption mechanisms. Progress in Energy and Combustion Science. 99: 101098.

Xiang, W., X. Zhang, K. Chen, J. Fang, F. He, X. Hu, et al. 2020. Enhanced adsorption performance and governing mechanisms of ball-milled biochar for the removal of volatile organic compounds (VOCs). Chemical Engineering Journal. 385: 123842.

Xu, X., X. Cao, L. Zhao and T. Sun. 2014. Comparison of sewage sludge- and pig manure-derived biochars for hydrogen sulfide removal. Chemosphere. 111: 296–303.

Xu, W., A. Hussain and Y. Liu. 2018a. A review on modification methods of adsorbents for elemental mercury from flue gas. Chemical Engineering Journal. 346: 692–711.

Xu, Y., F. Deng, Q. Pang, S. He, Y. Xu, G. Luo, et al. 2018b. Development of waste-derived sorbents from biomass and brominated flame retarded plastic for elemental mercury removal from coal-fired flue gas. Chemical Engineering Journal. 350: 911–919.

Yang, W., Y. Liu, Q. Wang and J. Pan. 2017. Removal of elemental mercury from flue gas using wheat straw chars modified by Mn-Ce mixed oxides with ultrasonic-assisted impregnation. Chemical Engineering Journal. 326: 169–181.

Yi, Y., C. Li, L. Zhao, X. Du, L. Gao, J. Chen, et al. 2018. The synthetic evaluation of CuO-MnOx-modified pinecone biochar for simultaneous removal formaldehyde and elemental mercury from simulated flue gas. Environmental Science and Pollution Research. 25: 4761–4775.

Yue, L., Q. Xia, L. Wang, L. Wang, H. DaCosta, J. Yang, et al. 2018. CO_2 adsorption at nitrogen-doped carbons prepared by K_2CO_3 activation of urea-modified coconut shell. Journal of Colloid and Interface Science. 511: 259–267.

Zhang, X., S. Zhang, H. Yang, Y. Feng, Y. Chen, X. Wang, et al. 2014. Nitrogen enriched biochar modified by high temperature CO_2-ammonia treatment: Characterization and adsorption of CO_2. Chemical Engineering Journal. 257: 20–27.

Zhang, X., B. Gao, Y. Zheng, X. Hu, A.E. Creamer, M.D. Annable, et al. 2017. Biochar for volatile organic compound (VOC) removal: Sorption performance and governing mechanisms. Bioresource Technology. 245: 606–614.

Zhang, X., B. Gao, J. Fang, W. Zou, L. Dong, C. Cao, et al. 2019. Chemically activated hydrochar as an effective adsorbent for volatile organic compounds (VOCs). Chemosphere. 218: 680–686.

Zhang, J., J. Shao, D. Huang, Y. Feng, X. Zhang, S. Zhang, et al. 2020a. Influence of different precursors on the characteristic of nitrogen-enriched biochar and SO_2 adsorption properties. Chemical Engineering Journal. 385: 123932.

Zhang, X., W. Xiang, B. Wang, J. Fang, W. Zou, F. He, et al. 2020b. Adsorption of acetone and cyclohexane onto CO_2 activated hydrochars. Chemosphere. 245: 125664.

Zhang, X., H. Zheng, G. Li, J. Gu, J. Shao, S. Zhang, et al. 2021. Ammoniated and activated microporous biochar for enhancement of SO_2 adsorption. Journal of Analytical and Applied Pyrolysis. 156: 105119.

Zhang, X., L. Cao, W. Xiang, Y. Xu and B. Gao. 2022. Preparation and evaluation of fine-tuned micropore biochar by lignin impregnation for CO_2 and VOCs adsorption. Separation and Purification Technology. 295: 121295.

Zhang, J., J. Shao, X. Zhang, G. Rao, G. Li, H. Yang, et al. 2023. Facile synthesis of Cu-BTC@biochar with controlled morphology for effective toluene adsorption at medium–high temperature. Chemical Engineering Journal. 452: 139003.

Zhao, X., X. Zeng, Y. Qin, X. Li, T. Zhu and X. Tang. 2018. An experimental and theoretical study of the adsorption removal of toluene and chlorobenzene on coconut shell derived carbon. Chemosphere. 206: 285–292.

Zhao, Z., B. Wang, B.K.G. Theng, X. Lee, X. Zhang, M. Chen, et al. 2022. Removal performance, mechanisms, and influencing factors of biochar for air pollutants: a critical review. Biochar. 4: 1–24.

Chapter 12

Biochar as an Additive to Composite Materials

María Paula Fabani* [a,b], Analia Alejandra Alvarez[c,d], Laura Noel García[d,e], Virginia Castro[a] and Maria Celia Roman[a]

[a]Instituto de Ingeniería Química, Fac. de Ingeniería, Grupo Vinculado al PROBIEN (Universidad Nacional de San Juan, CONICET, Universidad Nacional del Comahue), San Juan, Argentina

[b]Instituto de Biotecnología, Fac. de Ingeniería, Universidad Nac. de San Juan, Argentina

[c]Instituto Regional de Planeamiento y Hábitat, Universidad Nacional de San Juan, CONICET, J5402DCS San Juan, Argentina

[d]Instituto de Mecánica Aplicada, Universidad Nac. de San Juan, San Juan, Argentina

[e]Instituto de Ciencias Astronómicas, de la Tierra y del Espacio, Univ. Nacional de San Juan, CONICET, San Juan, Argentina

12.1 INTRODUCTION

Biochar as an additive in composite materials is a field of study of relevant interest because its unique properties can improve certain properties in the materials in which it is incorporated (Yang et al. 2025). Biochar, obtained mainly through the pyrolysis of biomasses such as wood, agricultural residues, or organic waste, has characteristics that make it interesting for use as an additive in the manufacture of this new type of material, such as improved mechanical strength, reduced environmental impact, and modification of physical properties (Feliz Florian et al. 2024;

*Corresponding author: María Paula Fabani (paufabani@unsj.edu.ar)

DOI: 10.1201/9781032726694-12

Rashid et al. 2024). Among its most notable characteristics in terms of its potential use as an aggregate in the formulation of new materials are the following:

Porosity: Biochar has a porous structure, which gives it a high specific surface area (Liang et al. 2023). This property makes it ideal for use as an additive in composite materials, as it can improve adhesion between the matrix and reinforcement phases.

Carbon: Due to its high carbon content, biochar can influence the thermal and mechanical properties of materials, acting as a reinforcement or as an agent that modifies the thermal and electrical conductivity of the matrix.

Surface Functionalization: Biochar can be chemically modified to increase its reactivity and compatibility with other materials, such as polymers or cement matrices. This can improve its integration into the mixture and the uniform distribution in the composite matrix (Dixit et al. 2021; Ahmad et al. 2023).

12.1.1 Advantages of Biochar as an Additive in Composite Materials

Matrix Reinforcement: When used in small quantities, biochar can act as a reinforcement for composite materials, such as plastics or types of cement. Its porous and lightweight structure can improve flexural, compressive, and impact strength properties (Feliz Florian et al. 2024).

Improved Thermal Resistance: Biochar has good thermal stability and can increase the resistance of composite materials to high temperatures. This is useful for applications where material integrity needs to be maintained in thermally demanding environments (Pudelko et al. 2021).

Absorption Properties: Due to its high specific surface area, biochar can absorb unwanted organic or chemical compounds, which can be useful for the treatment of composite materials intended for use in aggressive or polluting environments.

Reduced Carbon Footprint: Using biochar in composite materials allows the use of recycled material with a low environmental impact. This reduces the use of synthetic or non-renewable materials and also contributes to carbon sequestration, making products more sustainable (Liu et al. 2022).

12.1.2 Biochar Applications in Composite Materials

Composite Polymers: Biochar has been used as an additive in composite polymers to improve mechanical and thermal strength. Its incorporation can result in a material with lower weight and better performance in terms of tensile and compressive strength. In this case, biochar functions as a lightweight reinforcement that contributes to improved overall strength of the material (Aboughaly et al. 2023; Feliz Florian et al. 2024). Similarly, Pudelko et al. (2021) reported that waste-derived biochar can serve as an alternative filler in polymer-based biocomposites, significantly enhacing their mechanical, thermal, and morphological properties.

Cementitious Composites: In cementitious composites, biochar can partially replace cement or be added as an additive, improving characteristics such as durability, compressive strength, and water absorption. In addition, it has been observed that it can act as a concrete microstructure modifier, improving density and reducing porosity, making the final material more durable (Barbhuiya et al. 2024; Feliz Florian et al. 2024).

Hybrid Composites: Biochar is also being used in combination with other reinforcements, such as glass or carbon fibers, to improve the properties of composite materials. In this context, biochar can contribute to the greater sustainability of the final product, reducing the number of synthetic materials and improving their performance (Feliz Florian et al. 2024; Ghaedi et al. 2024).

12.1.3 Biochar Use Considerations

Amount Added: The proportion depends on the desired properties of the final product. For example, in some cases, a small amount may improve the properties without compromising the integrity of the material; higher amounts may decrease the mechanical strength or affect the cohesion of the material matrix (Zouari et al. 2022).

Compatibility with the Matrix: Compatibility with the matrix in which it is to be incorporated is essential. In the case of polymer composites, for example, it may be necessary to treat the biochar to improve its adhesion to the polymer and avoid separation or the formation of micro-fractures.

Processing: Biochar must be properly processed before incorporation into the composite mixture. This may include heat or chemical treatment to improve its reactivity properties, facilitate its distribution in the matrix, and optimize phase interaction.

12.1.4 Challenges and Limitations

Biochar Variability: The quality and properties of biochar can vary significantly depending on the biomass source and pyrolysis conditions, which can make it difficult to predict the behavior of biochar in composites and additional testing may be necessary to ensure consistency in product performance.

Interaction with Other Additives: Biochar may interact with other additives used in the composite blend, which could influence the performance of the final material. Compatibility with other reinforcements, additives, and the matrix itself must be carefully evaluated (Feliz Florian et al. 2024).

12.2 PHYSICOCHEMICAL CHARACTERISTICS OF BIOCHAR AS A FILLER OR CEMENT REPLACEMENT MATERIAL IN CEMENTITIOUS COMPOSITES

Biochar, derived from the pyrolysis of wood, agricultural residues, or organic waste, has attracted interest as an alternative material in construction, especially in the partial replacement of cement in cementitious composites. This is due to its unique properties and its ability to improve certain characteristics of building materials. Among the physicochemical characteristics that make biochar attractive for this type of application are (Ahmad et al. 2023; Liang et al. 2023; Feliz Florian et al. 2024; Rashid et al. 2024; Yang et al. 2025) (Table 12.1):

Table 12.1. Physicochemical characteristics that make biochar attractive as a filler or cement replacement material in cementitious composites

	Chemical Composition
Carbon (C)	Biochar is rich in C, which gives it adsorption properties and improves its chemical stability under certain conditions.
Ash content	Depending on the source biomass and the pyrolysis process, biochar may contain a variable amount of inorganic minerals, such as calcium (Ca), iron (Fe), and potassium (K), which influence its reactive properties.
Hydrogen (H) and oxygen (O)	Their presence in the biochar structure can influence its ability to interact with cement and other compounds.
	Physical Properties
Bulk density	Biochar has a low bulk density due to its porous structure. This can affect the total mass of cementitious composites in which it is used as a partial cement substitute, which could decrease the final density of the materials.
Specific surface area	Biochar has a high specific surface area, which makes it useful for water adsorption or interaction with other materials, such as additives and chemical compounds present in cementitious mixtures.
Porosity	The porous structure of biochar influences adhesion and reactivity in cementitious compounds, which can improve the microstructure of concrete if used properly.
Particle size	Biochar particles tend to be small and fine, which facilitates their incorporation into the cement mix and improves homogeneous distribution in the mix.
	Reactive Properties
Pozzolanic activity	Depending on its chemical composition, biochar may have some pozzolanic activity, similar to other materials such as fly ash or blast furnace slag. This means that it can react with the calcium hydroxides present in cement in the presence of water, forming additional compounds that contribute to the strength of the material.
Effect on cement hydration	Biochar can influence cement hydration reactions, depending on its chemical reactivity, which can enhance or retard the process depending on the quantity used and the quality of the biochar.

	Mechanical Properties
Compressive strength	Partial replacement of cement with biochar can influence the compressive strength of cementitious composites. In some research, it has been observed that a moderate amount of biochar improves the initial strength, while excessive amounts can reduce it due to a lack of cohesion and the presence of increased porosity.
Durability	The use of biochar in cementitious mixtures can improve the durability of materials by reducing permeability and increasing resistance to aggressive agents such as sulfates or chemical aggression. This is due to its porous structure and chemical characteristics, which can facilitate the formation of secondary products that reinforce the concrete matrix.
	Environmental Effects
CO_2 absorption	As biochar is a carbon-rich material, its use in construction could indirectly contribute to carbon capture and storage, which represents an additional environmental benefit. Moreover, the partial substitution of cement, a material with a high carbon footprint due to its manufacturing process, can contribute to the reduction of greenhouse gas emissions.
Low environmental impact	Biochar is a by-product of biomass, which means it can be a more sustainable alternative to traditional cement, especially if it comes from renewable sources and is produced through low environmental impact processes.
	Interactions with Other Materials
Compatibility with additives	Biochar may interact with other additives used in the cement mix, such as setting retarders or accelerators. These interactions can modify the workability of the mix and the setting time.
Improved thermal properties	The addition of biochar can improve the thermal properties of cementitious composites, such as thermal conductivity and resistance to extreme temperatures, due to its porous structure and heat absorption capacity.
	Interactions with Other Materials
Compatibility with additives	Biochar may interact with other additives used in the cement mix, such as setting retarders or accelerators. These interactions can modify the workability of the mix and the setting time.
Improved thermal properties	The addition of biochar can improve the thermal properties of cementitious composites, such as thermal conductivity and resistance to extreme temperatures, due to its porous structure and heat absorption capacity.

In summary, biochar offers a promising alternative as a partial cement replacement material in cementitious composites, with advantages such as improved porosity, reduced carbon footprint, and possibly improved durability. However, its effectiveness depends on the specific characteristics of biochar and its appropriate dosage in the mixes.

12.3 ENGINEERING PROPERTIES OF BIOCHAR-MODIFIED CEMENTITIOUS COMPOSITES

Cementous materials and their applications are associated with the increase in urbanization and improved standards of living. Their versatility and high strength make them one of the most used materials in the construction industry. However, its production has a high demand for Ordinary Portland Cement (OPC). In this regard, the industry represents 4–7% of global emissions of CO_2 (Danish et al. 2021) and is one of the largest consumers of natural resources (Gupta et al. 2018). That makes cementitious-based materials less sustainable than other construction materials. Therefore, one option is to replace or reduce the use of OPC in cementitious composites without compromising their mechanical properties and hardness (Danish et al. 2021; Wen et al. 2023). Another potential solution is the sequestration of carbon in cementitious materials (Gupta et al. 2018). In both cases, a feasible way to incorporate sustainable benefits and properties into these composites is to use biochar to replace OPC (Gupta et al. 2018; Danish et al. 2021).

Biochar results from biomass pyrolysis, a thermochemical process that eliminates the adverse effects of cellulose, hemicellulose, and lignocellulose on cement hydration and possesses unique physical and chemical properties that make it a potential cement additive or alternative (Wen et al. 2023). In this sense, the techniques to produce biochar depend on the type of biomass, available resources, and intended future applications. Therefore, it is necessary to consider its physiochemical properties to determine its usability as cement replacements or fillers in cementitious composites (Danish et al. 2021).

In this context, biochar seems to have a high potential for integration into the matrix of various building materials. In this direction, there is evidence that low-carbon cementitious composites can be prepared with biochar as filler or OPC replacement, allowing a composite to capture CO_2 (Danish et al. 2021; Rashid et al. 2024). According to the literature, using one kilogram of biochar in concrete can sequester roughly 2.5 kg of CO_2 emissions (Rashid et al. 2024). In this regard, adding biochar to cementitious composites is associated with less consumption of limestone and, consequently, with a 25% reduction in greenhouse gas emissions (Wen et al. 2023).

Thus, the goal of their characterization is to establish their chemical and physical properties based on their production conditions and types of biomass, their usability fields, and strategies to remove ecotoxicological parameters. Considering the targets to accomplish, fresh cementitious compounds present different proportions of cement, fine/coarse aggregates, and water. Hence, the mix design can incorporate fillers that, by hand, decrease the amount of OPC and, on the other hand, improve composite properties such as workability, self-compatibility, rheological, durability, or mechanical behavior (Danish et al. 2021). In other words, adding biochar to cement mortar and concrete mixtures modifies their initial properties, such as flowability, setting time, workability, compressive strength, flexural strength, tensile strength, durability, drying rate, and shrinkage. For that reason, the engineering properties of the composites depend on parameters such as the type of feedstock, quantity of biochar, the pre-treatment processes, and the materials and testing conditions (Rashid et al. 2024).

The chemical properties of biochar influence its content demand in cementitious composite, hydration rate, setting time, and pozzolanic reactivity. Carbon content and impurities determine the usability of biochar as a filler. Their reactivity influences their capability as an additive to improve the durability properties of biochar-modified cementitious composites. Another remarkable feature of biochar that influences the strength, durability, and serviceability of cement-based structures is its pH value. In this context, it is noteworthy that adding biochar generates a porous structure on cementitious composites that can be associated with a high thermal insulation potential. Furthermore, the physical properties of biochar affect surface area, density, porosity, morphology, and pore volume. Also, the particle size of biochar determines water absorption capacity, reduces the water/cement ratio, and contributes to the densification effect (Danish et al. 2021; Liang et al. 2023).

Concerning the influence on engineering properties of biochar use in cementitious composites, it is outstanding that the incorporation of fine or coarse biochar on mortar determines its yield stress and plastic viscosity. In this sense, including 0.25–2% of fine biochar increased yield stress and plastic viscosity of mortar, while coarse biochar is more effective in maintaining workability. Biochar's adverse effects on the rheology of cementitious materials could be overcome, among other aspects, by changing various parameters such as the kind of biomass or porosity of biochar. Related to the hydration kinetics of cementitious composites, it is notable that biochar facilitates the creation of nucleation clusters that result in the improvement of the hydration reaction. Moreover, adding biochar to cementitious composites reduces setting times and increases the air content of the mortar.

To summarize, using coarser biochar in cementitious composites increases the hydration degree resulting from the formation of macropores, but also, rheology, strength, and energy cost decrease. Using finer biochar, rheology, hydration, strength, and cost energy increase their values because the water absorption capacity of fine biochar particles is significantly lower than that of coarse biochar particles, and the water film that forms around it is thicker. When using a blender biochar conformer by equal proportions of coarse and finer biochar, rheology, hydration, and strength increase while cost and energy decrease by 27–37% (Wen et al. 2023).

Regarding hardness properties, the literature suggests that biochar reduces compressive strength attributable to the dilution effect, porous structure of biochar, or localized segregation of biochar from the matrix. However, some research indicates that adding 0.08 wt% coconut shell-derived biochar improves the compressive strength by 25% and fracture toughness of cement paste by 76% (Gupta et al. 2018). Thus, 1% biochar as a fine aggregate can increase the compressive strength of the composite by 10% (Danish et al. 2021). Other research explains that a 5% replacement of weight cement with biochar improved the 28-day compressive strength of mortar by about 10% (Gupta et al. 2018). On the other hand, it is necessary to consider that the strength of biochar-modified composites increases with age, which indicates that the incorporation of biochar does not affect the progression of the hydration reaction in a negative way. That results in the generation of more hydroxide in cement pastes with biochar. This means their addition improves hydration. In this sense, some evidence affirms that the compressive strength of mortar, especially at day 7, is significantly increased due to the addition of unsaturated biochar and biochar with adsorbed CO_2.

Nevertheless, thermogravimetry analysis and X-ray diffraction (XRD) confirmed that saturated biochar is associated with lower late strength due to carbonation by adsorbed CO_2. In addition, mortar flowability decreases when the percentage of biochar is greater than 5%. As the rate is nearly 15% by weight of cement, flow is reduced by 20% (Gupta et al. 2018).

According to the literature, replacing 1–2 wt% of cement with biochar improves elastic modulus, shrinkage properties, water absorption behavior, compressive, splitting tensile, and flexural strength of cementitious materials. This behavior is related to pore refinement, pozzolanic reactivity, and internal curing effect (Gupta et al. 2018; Danish et al. 2021).

At this point, in normal curing conditions, it is suggested that an optimal amount of biochar is near <1% as a promising approach regarding utilizing waste material. This rate reduces cement content in the mix design and allows for storing 50% of ambient CO_2 permanently in the cementitious matrix (Mishra et al. 2023). Thus, biochar as an additive is a sustainable solution to improve mortar properties because it is derived from bio-waste and is also an acceptable, cost-effective option, especially in developing countries, since its production and mixing in the mortar does not require sophisticated techniques or setup.

Based on the above, it concluded that biomass type and biochar carbonization conditions influence the features that biochar can confer on cementitious composites. Nevertheless, more research related to the effects associated with the use of biochar on the engineering properties of cementitious materials is required. In this sense, obtaining the durability properties of biochar-modified cementitious material requires more research about the chemical composition of biochar. Furthermore, Life Cycle Assessment (LCA) needed to be studied to determine the environmental impact and cost-effectiveness of using biochar to replace OPC.

12.4 SYNTHESIS OF BIOCHAR-BASED NANO-COMPOSITES

The synthesis of nanocomposites based on different types of biochar is an emerging area in advanced materials research due to the unique properties that can be derived from the combination of biochar and nanomaterials (Chausale et al. 2021; Verma and Singh 2023). Promising applications have been reported in fields such as adsorption of pollutants, catalysis, electrochemistry, and the improvement of mechanical properties in composite materials.

12.5 FUNCTIONALIZATION OF BIOCHAR

To improve the properties of biochar and allow its integration with nanometric materials, various modifications can be made, such as (Chausale et al. 2021; Zhang et al. 2022).

Oxidation: The introduction of functional groups such as carboxyl, hydroxyl, and carbonyl improves the interaction of biochar with other compounds.

Metal Doping: Metals such as iron (Fe), zinc (Zn), or copper (Cu) can be incorporated into the biochar structure, which increases its catalytic activity.

Coating: Biochar surfaces can be coated with materials such as metal oxides or polymers to improve their stability and functionality.

12.6 INCORPORATION OF NANOMATERIALS

Nanomaterials, such as carbon nanotubes, metal oxides (such as TiO_2 or ZnO), metal nanoparticles (silver, gold, iron), and semiconducting nanoparticles, can be integrated with biochar in a variety of ways (Chaubey et al. 2024):

Impregnation: Nanoparticles can be impregnated onto the surface of biochar by methods such as chemical precipitation or adsorption.

In Situ Synthesis: In some cases, nanoparticles are synthesized directly on the biochar during its preparation or by chemical reactions occurring in the modified biochar.

Covalent Bonding or Electrostatic Interactions: In some situations, interactions between functional groups in biochar and nanoparticles are key to ensuring the stabilization and dispersion of nanomaterials.

12.7 APPLICATIONS OF NANO-BIOCHAR AND BIOCHAR-BASED NANOCOMPOSITES

Nano-biochar (N–BC) has emerged as an adaptable material with multiple sustainable applications due to its high surface area, porosity, and surface charge. In environmental remediation, it is used to remove pollutants such as heavy metals, hydrocarbons, and inorganic compounds, as well as to improve agricultural soils by optimizing water and nutrient retention (Anupama and Khare 2021; Jiang et al. 2023; Chaubey et al. 2024). It also serves as an additive in fuels, reducing emissions, and as a catalyst in cleaner chemical processes. Its ability to immobilize enzymes and microorganisms expands its use in biotechnology and decontamination. These applications establish N–BC as a key material in the transition toward a circular and sustainable economy.

12.8 REMOVAL OF WATER CONTAMINANTS

N–BC has emerged as a promising tool for the remediation of contaminated water due to its ability to form stable colloidal solutions and its high surface area, porosity, and absolute surface charge, characteristics that make it highly effective for adsorbing various contaminants (Jiang et al. 2023). Among the pollutants N–BC can remove are metals such as Cd (II), Cu (II), Pb (II), Hg (II), Ni (II), As(V), and Cr (VI) (Chaubey et al. 2024), as well as herbicides, polychlorinated biphenyls, polycyclic aromatic hydrocarbons, and inorganic compounds, which establishes it as a versatile and highly effective tool for water remediation (Anupama and Khare 2021; Jiang et al. 2023). Moreover, its versatility is not limited to being an excellent adsorbent, as it also functions as a nanocarrier, significantly expanding its applications in environmental remediation (Anupama and Khare 2021).

12.9 SOIL REMEDIATION

N–BC offers significant potential for soil improvement and agriculture due to its unique properties, such as high specific surface area, increased cation exchange capacity, and a microporous structure that optimizes water and nutrient retention. Its application has been shown to increase soil moisture, reduce nutrient leaching, and strengthen soil aggregate stability, decreasing erosion and runoff losses (Chaubey et al. 2024). N–BC has been used as an effective filler to enhance soil properties and composite materials. In soils, N–BC increases water retention and reduces crack formation during drying and wetting cycles due to its porous structure. Additionally, amending the soil with 10% N–BC improves the growth of crops like corn, increasing plant height, stem diameter, and leaf area. Even low applications of N–BC (0.5–1%) yeild results comparable to higher doses of biochar for nitrate fixation and nitrogen loss control, making it a cost-effective option (Anupama and Khare 2021). Furthermore, N–BC improves fertilizer use efficiency by providing controlled nutrient release that aligns with plant needs during development. These benefits also include promoting plant growth, with significant increases in crop yields and seed germination. Compared to conventional biochar, N–BC achieves superior results with lower application rates, positioning it as an economical and sustainable tool for modern agriculture (Chaubey et al. 2024). In this context, rice husk-derived nanocomposites have proven effective in removing arsenic and other contaminants, significantly contributing to the rehabilitation of contaminated soils (Chausali et al. 2021).

12.10 REMOVAL OF ALLELOPATHIC COMPOUNDS

N–BC is an effective solution for the controlled release of allelopathic compounds, limiting their leaching and reducing their environmental impact. It has been shown to mitigate the negative effects on crops, such as rice, promoting growth and increasing biomass. Additionally, its use enhances the efficiency of bioherbicides, highlighting its potential as a tool in sustainable agriculture (Anupama and Khare 2021; Chausali et al. 2021).

12.11 ENZYME IMMOBILIZATION

Nano-biochar is an effective material for enzyme and microorganism immobilization due to its high porosity, abundance of functional groups (such as carboxyls), and adsorption capacity. Immobilizing the enzyme laccase on N–BC significantly improves its stability and catalytic activity, even under extreme pH and temperature conditions, and allows for its reuse across multiple cycles. This process also protects enzymes from denaturation and unraveling, increasing their structural rigidity and operability (Feliz Florian et al. 2024; Yang et al. 2025). Moreover, N–BC serves as a support for microorganisms such as *Rhodococcus opacus*, which break down complex pollutants in wastewater. In combination with polyurethane, it enhances the removal of COD by up to 95%, thanks to the synergy between microbial degradation and N–BC adsorption, which also provides essential nutrients such as iron to support microbial activity (Chaubey et al. 2024).

12.12 AS A FUEL ADDITIVE

N–BC has established itself as a sustainable and efficient alternative to traditional metal-based additives, which are harmful to health and the environment (Chaubey et al. 2024). Derived from residual biomass, such as sugarcane, N–BC has been shown to improve fuel performance in diesel engines and significantly reduce pollutant emissions. For example, studies report a reduction of up to 20.51% in nitrogen oxide (NO_x) emissions and 25–28% in unburned hydrocarbons when N–BC concentration is increased to 125 ppm. Furthermore, CO_2 emissions were reduced from 0.03 to 0.015%. The use of N–BC and fuel oil mixtures has also shown additional benefits, such as reduced CO and particulate emissions due to its high catalytic efficiency, high surface-to-volume ratio, and heat transfer capacity, optimizing combustion by lowering temperature and incomplete fuel oxidation (Anupama and Khare 2021; Chaubey et al. 2024).

12.13 AS A CATALYST

N–BC acts as a catalyst in reactions such as esterification and photoreactions, helping to improve the yield of the resulting products. Specifically, sulfonated nano-biochar showed remarkable capacity in catalyzing the esterification of oleic acid with methanol, achieving a 64% yield of methyl oleate compared to 4% without a catalyst. This efficiency is attributed to the sulfonic acid sites on its surface, as well as its porous structure and carbonaceous hydrophobic regions, which facilitate both reactant accessibility and the removal of water required to balance the reaction (Feliz Florian et al. 2024; Yang et al. 2025).

Biochar and nano-biochar-derived catalysts are sustainable options due to their biodegradability, origin from residual biomass, and their thermal and mechanical stability, as well as their reusability and ease of separation. A notable example is nano-biochar with incorporated ZnO, which demonstrated photocatalytic activity by eliminating 99% of phenol in 90 minutes under light irradiation. This compound retained 95% of its photocatalytic capacity after five cycles of use, significantly outperforming the 69% retention of pure ZnO, demonstrating the protection nano-biochar provides against photocorrosion (Chaubey et al. 2024).

12.14 OTHER APPLICATIONS

N–BC is gaining popularity in various applications due to its low cost, high surface capacity, and good dispersion. One of its newest applications is the extraction of antibiotic resistance genes (ARGs) in water bodies, which is crucial for decontamination. Traditional water treatment methods fail to remove ARGs before they are released or reused. These genes spread through the use of sewage sludge and manure on soils, reaching water bodies via infiltration and runoff. N–BC has greater mobility in the soil, allowing it to interact more effectively with ARGs in natural environments, showing a more efficient adsorption and fragmentation capacity (Lian et al. 2020).

In the rubber industry, lignin-derived N–BC has proven to be an excellent reinforcement for styrene-butadiene rubber. Its use significantly improves tensile strength and elasticity, outperforming conventional fillers such as commercial carbon black and low-rank coal. This positions N–BC as a renewable and efficient alternative for composite material manufacturing and agricultural applications (Greenough et al. 2021).

Nano-biochar, thanks to its unique and sustainable properties, offers a wide range of applications that continue to grow. Its functional characteristics position it as a key material for developing innovative solutions in multiple fields, with untapped potential still to be discovered.

12.15 BIOCHAR APPLICATIONS AS A FILLER IN RUBBER COMPOSITES

Initially, biochar garnered attention for its potential to improve soil fertility, enhance agricultural yields, and contribute to carbon sequestration, making it an attractive tool in combating climate change. Furthermore, its production from agricultural waste not only mitigates the environmental impact of conventional disposal methods, such as burning, but also yields high-value products (Anupama and Khare 2021; Jiang et al. 2023; Chaubey et al. 2024). For these reasons, biochar is positioned as a key component in the development of compounds that are more environmentally responsible and less dependent on fossil fuels.

In recent years, interest in biochar has extended beyond agriculture, shifting toward industrial applications. One prominent application is the development of rubber compounds. According to Greenough et al. (2021), the rubber formulation represents the most essential component, comprising an elastomer, a reinforcing filler, vulcanizing agents, protective additives, and processing aids. The elastomer is a viscoelastic material characterized by weak intermolecular forces, resulting in low tensile strength, high deformability, and a low Young's modulus. To enhance its properties, reinforcing fillers are incorporated into the polymer matrix, improving durability, elasticity, resilience, tensile strength, viscosity, hardness, and resistance to adverse weather conditions (McKeen 2023).

Carbon black is a traditional reinforcing filler commonly employed in the manufacturing of components for the automotive, rail, and aviation sectors (Greenough et al. 2021). However, carbon black has significant drawbacks: its production from non-renewable sources generates high levels of carbon dioxide emissions (2.4 tons of CO_2 per ton of carbon black) and poses risks to human health and the environment (Fan et al. 2020).

In response to these limitations, the search for sustainable alternatives to carbon black has intensified, with a focus on renewable materials such as biochar. Derived from resources like corn (Peterson 2012), hardwood (Jiang et al. 2020), and coffee waste (Raju et al. 2021), biochar has emerged as a promising solution. In this regard, Peterson (2020) demonstrated that coppiced biochar can partially replace carbon black in polybutadiene/natural rubber composites, highlighting its potential as a renewable and functional filler. Incorporating biochar as a filler in rubber compounds not only reduces dependence on carbon black but also enhances the sustainability and performance of the materials (Jong et al. 2014). However, to match the performance of carbon black in demanding applications, it is crucial to control parameters such as particle size, pore distribution, and ash content in the biochar. For example, a more uniform particle size can prevent stress concentrations in rubber compounds and enhance overall durability.

Additionally, factors such as particle size, structure, and surface chemistry of biochar significantly influence the mechanical properties, durability, and performance of rubber compounds (Peterson et al. 2016).

Since virtually all biomass contains carbon, the potential raw materials for sustainable biochar production are nearly limitless. Nevertheless, a significant challenge lies in its lack of purity. Biomass is considerably more heterogeneous than petroleum in terms of carbon content and introduces ash due to its inherent inorganic components. While it is feasible to produce biochar with high carbon content and minimal ash, achieving this requires careful and tightly controlled processing during the pyrolysis of the raw material (Peterson 2019).

Research has demonstrated that biochar possesses several advantageous properties, including high thermal and electrical stability, low production costs, and a composition with reduced levels of nitrogen, sulfur, and ash. These qualities make it a promising and environmentally friendly alternative (Ismail et al. 2002; Jiang et al. 2020; Sunali et al. 2022).

12.16 PROPERTIES OF BIOCHAR AS A FILLER FOR RUBBER COMPOSITES

The compatibility of carbon black with the rubber matrix is a critical factor in determining its effectiveness as a reinforcing filler in rubber compounds. Researchers attribute its reinforcing capability primarily to three essential properties: a high elemental carbon content, minimal ash content, and a small particle size. Furthermore, additional parameters such as structural morphology, surface chemistry, and surface area significantly influence the filler-matrix interactions, enhancing the overall performance of the composite material (Ulfah et al. 2015)

Biochar is a carbonaceous material produced through the pyrolysis of biomass (Torres-Sciancalepore et al. 2023) and has attracted significant attention as a potential filler for rubber compounds due to its unique properties. According to Peterson (2022), there are several reasons to develop renewable alternatives to carbon black: (1) the fossil fuels used in its production are finite; (2) the cost of fossil fuels is volatile and can be dramatically affected by unpredictable geopolitical conditions; (3) reducing the use and production of carbon black helps improve air quality. For these reasons, biochar is positioned as a key component in the development of more environmentally responsible compounds that are less dependent on fossil fuels.

Some of the key properties of biochar that make it suitable as a replacement for carbon black, as a filler, in rubber compounds include:

12.16.1 High Carbon and Low Ash Content

Carbon-rich biochar can be achieved by carefully selecting appropriate feedstocks and optimizing pyrolysis processing parameters, which involve heat treatment in the absence of oxygen. During this process, solid biochar is produced alongside gaseous (syngas), liquid (bio-oil), and ash co-products. Research indicates that hardwood is the most commonly used feedstock for biochar production due to its higher carbon content and lower ash content compared to other lower-value biomass feedstocks. This makes hardwood-derived biochar particularly suitable for applications requiring superior material properties (Peterson 2022).

12.16.2 Cost-Effectiveness

Biochar is often more economical than traditional fillers, particularly when sourced from agricultural or industrial waste, making it an appealing option for large-scale rubber production. However, substituting carbon black with biochar can negatively impact the mechanical properties of rubber compounds due to biochar's significantly larger particle size. To address this issue, pre-treatment methods such as grinding are typically required.

To enhance the reinforcing capabilities of biochar, advanced technologies and modifications are employed, including ball milling, the use of coupling agents, surface chemical modifications, high-pressure homogenization, and acid hydrolysis. While these approaches improve performance, they also increase production costs, processing time, and energy consumption, which may offset biochar's initial cost advantages (Jiang et al. 2020).

12.16.3 Reduced Environmental Impact

Biochar, derived from renewable biomass sources, offers an environmentally friendly alternative to fossil-based fillers. Its use in rubber compounds reduces the dependence on non-renewable resources and contributes to the circular economy by using waste biomass. Replacing carbon black with biochar, as a rubber filler, can reduce greenhouse gas production while reducing the dependence on fossil fuels with a renewable alternative (Fan et al. 2020).

Furthermore, biochar, as a natural and biodegradable material, presents a compelling option for developing more sustainable rubber compounds with a reduced environmental footprint compared to those utilizing synthetic fillers. Its biodegradability aligns with global efforts to reduce waste and promote eco-friendly alternatives.

In summary, biochar's unique combination of high carbon content, low ash levels, excellent thermal stability, and cost-effectiveness, combined with its positive environmental impact, positions it as a promising alternative to traditional fillers such as carbon black. Biochar offers numerous advantages, including improved performance characteristics and a reduced environmental impact, making it a sustainable and versatile option for rubber compound formulations.

12.17 BIOCHAR AS FILLER OF CARBON FIBER REINFORCED BIO-EPOXY COMPOSITES

The growing demand for sustainable materials in the industry has driven the development of eco-friendly alternatives, such as bio-epoxies, derived from renewable sources with a lower environmental impact compared to traditional epoxies (Bartoli et al. 2022). Biochar has emerged in this context as a sustainable, economical, and viable alternative to traditional carbon-based fillers for the development of polymer-based composites (Matkiewicz 2020a; Bartoli et al. 2022). As an innovative additive, biochar enhances the properties of bio-epoxy composites by increasing their rigidity and thermal stability without negatively affecting the glass transition temperature (Bartoli et al. 2022).

Biochar exhibits high thermal stability, a large surface area, and electrical conductivity. Moreover, its primary properties can be tailored by controlling the production conditions. This flexibility positions biochar as a competitor to high-performance fillers in the formulation of multifunctional polymer-based composites, contributing to high mechanical and electrical properties. Additionally, biochar can be derived from diverse biomass sources, including post-consumption agricultural waste, offering a compelling pathway toward a "zero-waste" circular bioeconomy (Bartoli et al. 2022).

Matykiewicz (2020a) investigated the effect of incorporating biochar (2.5%, 5%, and 10% by weight) into carbon fiber-reinforced bio-epoxy composites. The study revealed several significant improvements in the properties of these materials:

— Biochar-modified composites showed a notable increase in impact and flexural strength, with the most pronounced effect observed in samples containing 10% biochar by weight.

— The addition of biochar did not cause agglomeration, indicating good interaction between components and an increase in the maximum load (F_{max}) during impact tests.

— The thermogravimetric analysis demonstrated that biochar enhanced the thermal stability of the composites, delaying degradation and increasing the residual mass at 900°C.

These findings confirm biochar's potential as a functional additive to improve the performance of carbon fiber-reinforced bio-epoxy composites.

12.18 CHARACTERISTICS OF CARBON FIBER-REINFORCED BIOEPOXY COMPOSITES

Carbon fiber-reinforced bio-epoxy composites are advanced materials combining bio-based epoxy resins with carbon fibers to create lightweight, high-performance structures. Designing these composites requires careful consideration of factors such as processing techniques, component heat susceptibility, mechanical treatments, durability, and material integration.

Bioepoxy Resins are thermosetting polymers that cure through heat, altering their chemical composition to form rigid and permanent structures. Widely used across industries, bio-epoxy resins offer high mechanical strength, favorable thermal properties, stability against chemical and atmospheric conditions, and low specific weight. Available in various forms (liquid, paste, or powder), they can be customized with reactive or non-reactive additives for specific applications (Matykiewicz 2020b). The ability to utilize various hardeners allows curing at room or elevated temperatures, achieving outstanding mechanical and thermal properties.

Bioepoxies have low impact resistance and inherent brittleness, necessitating reinforcement with synthetic or natural fibers, such as glass or carbon to enhance their structural performance (Ghabezi and Harrison 2020).

Biochar is widely recognized for its ability to enhance the mechanical and thermal properties of polymers. Studies highlight its benefits in increasing stiffness and thermal stability. However, its performance depends significantly on characteristics such as particle size, porous structure, and fixed carbon content, which vary according to production processes (Dinesh et al. 2020; Matykiewicz 2020a).

12.19 SUSTAINABILITY AND CIRCULAR ECONOMY

Biochar, derived from biomass residues, aligns with the principles of the circular economy by promoting resource reuse and contributing to CO_2 emissions reduction. This porous, chemically stable material, produced through the thermal decomposition of biomass, is gaining attention as a sustainable additive in carbon fiber-reinforced bio-epoxy composites. Integrating biochar into these composites combines its inherent sustainability with the advanced mechanical and functional properties provided by carbon fibers (Mahdian et al. 2020).

This approach not only enhances the mechanical and thermal properties of composite materials but also reduces their environmental footprint. To ensure optimal performance, the design of composite material must consider critical factors such as processing techniques, the thermal sensitivity of components, mechanical treatment, durability, and efficient material integration (Andrzejewski et al. 2018).

12.20 BIOCHAR APPLICATIONS AS AN ADDITIVE TO ALUMINUM METAL MATRIX COMPOSITES

Metal matrix composites (MMCs) refer to a group of new-age materials constituted of base metal alloys mainly reinforced with ceramic particles to tailor their mechanical and thermal properties to achieve specific requirements (Sckhar et al. 2021). A comprehensive bibliometric analysis of MMC research has been presented by Sekhar et al. (2021), providing an overview of the current trends and advances in this field. Among them, aluminum metal matrix composites (AMMCs) emerged as particularly attractive and frequently preferred base metals for the MMCs. This is consistent with AMMCs' combination of high specific stiffness, good fatigue properties, and the potential for relatively low-cost conventional processing (Hooker and Doorbar 2000), plus lightweight machinability, sustanibility for aerospace and automobile applications. As a general rule, when customizing MMCs, several variables need to be taken into consideration; they include type and level of reinforcement, choice of the matrix alloy (generally standard engineering alloys), and the composite processing route. All these factors shall not be conceived in isolation when developing a novel material. In this connection, the introduction of agricultural waste, including biochar, as reinforcement in the design of AMMC has given rise to green composites.

Research on biochar use as an additive for the design of AMMC is an ongoing topic.

In Table 12.2, the design of three AMMCs is presented from the literature. To begin with, Raja et al. (2021) reinforced the commercial Al-based alloy Aluminum 5083-O by adding SiC and TiO_2 (synthetic additives), and Polyalthia longifolia seeds biochar via stir casting. The biochar composition was characterized by Energy-dispersive X-ray (EDS) spectroscopy, which revealed it was predominantly composed of $CaCO_3$. They found an increase in the Ultimate Tensile Strength (UTS) with the increase of biochar particles, which the authors attributed to strengthening mechanisms (e.g., hindering the movement of the dislocations). Besides, they explained the degradation of the tensile properties by reinforcing at 7.5 wt% biochar, which was due to the increased presence of $CaCO_3$, combined with the inhomogeneous distribution of reinforcement particles, leading to a weak particle-matrix interface. They also observed an increase in microhardness and wear rate with the increase in biochar particles in the composite. A critical point that was revealed through scanning electron microscopy (SEM) was the presence of pores, which they related to the use of biochar as a reinforcement. Porosity could lead to voids that tend to become crack-initiating points, affecting mechanical properties such as UTS. Although no further clarifications were provided, the authors concluded that the mechanical properties enhancement was mainly related to the presence of TiO_2 and SiC particles. At this point, studies that do not include synthetic additives, but only use biochar instead,become relevant. Jannet et al. (2022) combined Aluminum 6061 as the base alloy and neem seed biochar as the reinforcement additive in different percentages for obtaining AMMCs via stir casting. From biochar characterization by EDS, the authors interpreted iron as present in the additive. Mechanical tests were conducted for the three AMMCs as obtained. The authors found a tendency in the tensile strength and the Vickers microhardness with an increase in reinforcement percentage, and a reduction in the wear rate, which they linked to the grain size reduction and the role of the biochar particles in hindering the movement of the dislocations. They also verified through SEM images the absence of pores, which the authors relied on to explain the increase in strength of the tested AMMCs.

Table 12.2 Summary of the experimental sets from research work, which include biochar as an additive for AMMCs' design

	Base Alloy	Additives (wt%)	Biochar Nature	Processing Route
Raja et al. (2021)	Al 5083-O	5 SiC-5 TiO_2-5 Biochar 5 SiC-2.5 TiO_2-7.5 Biochar 5 SiC-3.5 TiO_2-6.5 Biochar	Polyalthia longifolia seeds	15 min stir casting (780°C) + 10 h cooling
Jannet et al. (2022)	Al 6061	2.5 Biochar 5 Biochar 7.5 Biochar	Neem seeds	~5 min stir casting (750°C) + cooling
Praveen et al. (2023)	Al 5083	4 Biochar 8 Biochar 12 Biochar	Red sandal, Ashoka, and neem seeds	6 to 7 min stir casting (700°C) + transfer to a heated mold (680°C) to harden for unspecified time + cooling

A variant in the research that incorporates biochar as an additive is the one proposed by Praveen et al. (2023), who investigated the machinability of AAMCs as archived by adding red sandal, Ashoka, and neem seeds biochar. The authors concluded that the incorporation of biochar into Al alloys could enhance their mechanical (e.g., hardness and wear resistance) and thermal properties (e.g., thermal conductivity), positioning them as possible candidates for numerous engineering applications.

Certainly, the incorporation of biochar as a sustainably sourced additive in AMMCs is a promising trend in the search for more sustainable and efficient solutions that combine improved mechanical and thermal properties with environmental benefits, such as utilizing biomass waste and lowering its disposal. In other words, the use of biochar in AMMCs would encourage environmentally friendly manufacturing. Nevertheless, as highlighted by Praveen et al. (2023), further research efforts are essential to optimize and standardize the fabrication processes of biochar-reinforced AMMCs and to investigate the long-term durability and stability of these composites under service conditions.

12.21 CONCLUSION

Biochar has excellent potential as an additive in composite materials, both in the field of polymers and types of cement. Its physicochemical properties, such as a high specific surface area, porosity, and the ability to modify mechanical characteristics, make it an interesting material for improving the sustainability and performance of composites. However, the variability in its properties and the need to optimize its integration into different matrices are challenges that need to be addressed in future studies and industrial applications.

REFERENCES

Aboughaly, M., A. Babaei-Ghazvini, P. Dhar, R. Patel, and B. Acharya. 2023. Enhancing the potential of polymer composites using biochar as a filler: A review. Polymers. 15: 3981.

Ahmad, A., M. Zahra, F.E. Alam, S. Ali, M. Pervaiz, Z. Saeed, et al. 2023. A sustainable approach for the multi-dimensional exploitation of mixed biochar-based nano-composites. Fuel. 336: 126930.

Andrzejewski, J., M. Misra and A.K. Mohanty. 2018. Polycarbonate biocomposites reinforced with a hybrid filler system of recycled carbon fiber and biocarbon: Preparation and thermomechanical characterization. Journal of Applied Polymer Science. 135: 46449.

Anupama and P. Khare. 2021. A comprehensive evaluation of inherent properties and applications of nano-biochar prepared from different methods and feedstocks. Journal of Cleaner Production. 320: 128759.

Barbhuiya, S., B.B. Das and F. Kanavaris. 2024. Biochar-concrete: A comprehensive review of properties, production, and sustainability. Case Studies in Construction Materials. 20: e02859.

Bartoli, M., R. Arrigo, G. Malucelli, A. Tagliaferro and D. Duraccio. 2022. Recent advances in biochar polymer composites. Polymers (Basel). 14: 2506.

Chaubey, A.K., T. Pratap, B. Preetiva, M. Patel, J.S. Singsit, C.U. Pittman, et al. 2024. Definitive review of nanobiochar. ACS Omega. 9: 12331–12379.

Chausali, N., J. Saxena and R. Prasad. 2021. Nanobiochar and biochar based nanocomposites: Advances and applications. Journal of Agriculture and Food Research. 5: 100191.

Danish, A., M. Ali Mosaberpanah, M. Usama Salim, N. Ahmad, F. Ahmad and A. Ahmad. 2021. Reusing biochar as a filler or cement replacement material in cementitious composites: A review. Construction and Building Materials. 300: 124295.

Dinesh, S., P. Kumaran, S. Mohanamurugan, R. Vijay, D.L. Singaravelu, A. Vinod, et al. 2020. Influence of wood dust fillers on the mechanical, thermal, water absorption and biodegradation characteristics of jute fiber epoxy composites. Journal of Polymer Research. 27: 9.

Dixit, A., A.K. Verma and S.D. Pang. 2021. Dual waste utilization in ultra-high performance concrete using biochar and marine clay. Cement and Concrete Composites. 120: 104049.

Fan, Y., G.D. Fowler and M. Zhao. 2020. The past, present and future of carbon black as a rubber reinforcing filler – A review. Journal of Cleaner Production. 247: 119115.

Feliz Florian, G., M. Ragoubi, N. Leblanc, B. Taouk and L. Abdelouahed. 2024. Biochar production and its potential application for biocomposite materials: A comprehensive review. Journal of Composites Science. 8: 220.

Ghabezi, P. and N. Harrison. 2020. Mechanical behavior and long-term life prediction of carbon/epoxy and glass/epoxy composite laminates under artificial seawater environment. Materials Letters. 261: 127091.

Ghaedi, S., H. Rajabi, M.H. Mosleh, and M. Sedighi. 2024. MOF biochar composites for environmental protection and pollution control. Bioresource Technology. 418: 131982.

Greenough, S., M.J. Dumont and S. Prasher. 2021. The physicochemical properties of biochar and its applicability as a filler in rubber composites: A review. Material Today Community. 29: 102912.

Gupta, S., H.W. Kua, and C.Y. Low. 2018. Use of biochar as carbon sequestering additive in cement mortar. Cement and Concrete Composites. 87: 110–129.

Hooker, J.A. and P.J. Doorbar. 2000. Metal matrix composites for aeroengines. Materials Science and Technology. 16: 725–731.

Ismail, H., M.R. Edyham and B. Wirjosentono. 2002. Bamboo fibre filled natural rubber composites: The effects of filler loading and bonding agent. Polymer Testing. 21: 139–144.

Jannet, S., R. Raja, V. Arumugaprabu, G.B. Veeresh Kumar, S. Vigneshwaran, P.S. Rama Sreekanth, et al. 2022. Effect of neem seed biochar on the mechanical and wear properties of aluminum metal matrix composites fabricated using stir casting. Materials Today: Proceedings. 56: 1507–1512.

Jiang, C., J. Bo, X. Xiao, S. Zhang, Z. Wang, G. Yan, et al. 2020. Converting waste lignin into nano-biochar as a renewable substitute of carbon black for reinforcing styrene-butadiene rubber. Waste Management. 102: 732–742.

Jiang, M., L. He, N.K. Niazi, H. Wang, W. Gustave, M. Vithanage, et al. 2023. Nanobiochar for the remediation of contaminated soil and water: Challenges and opportunities. Biochar. 5: 2.

Jong, L., S.C. Peterson and M.A. Jackson. 2014. Utilization of porous carbons derived from coconut shell and wood in natural rubber. Journal of Polymers and the Environment. 22: 289–297.

Lian, F., W. Yu, Q. Zhou, S. Gu, Z. Wang, and B. Xing. 2020. Size matters: Nano-biochar triggers decomposition and transformation inhibition of antibiotic resistance genes in aqueous environments. Environmental Science & Technology. 54: 8821–8829.

Liang, Q., D. Pan and X. Zhang. 2023. Construction and application of biochar-based composite phase change materials. Chemical Engineering Journal. 453: 139441.

Liu, J., G. Liu, W. Zhang, Z. Li, F. Xing, and L. Tang. 2022. Application potential analysis of biochar as a carbon capture material in cementitious composites: A review. Construction and Building Materials. 350: 128715.

Mahdian, M., L.Y. Huang, D.W. Kirk and C.Q. Jia. 2020. Water permeability of monolithic wood biocarbon. Microporous and Mesoporous Materials. 303: 110258.

Matykiewicz, D. 2020a. Biochar as an effective filler of carbon fiber reinforced bio-epoxy composites. Processes. 8: 724.

Matykiewicz, D. 2020b. Hybrid epoxy composites with both powder and fiber filler: A review of mechanical and thermomechanical properties. Materials. 13: 1802.

McKeen, L.W. 2023. The Effects of Temperature, Humidity, and Other Factors on the Properties of Thermoplastic Elastomers. pp. 559–603. *In*: The Effect of Temperature and Other Factors on Plastics and Elastomers, Elsevier.

Mishra, G., P. Danoglidis, S.P. Shah and M. Konsta-Gdoutos. 2023. Optimization of biochar and fly ash to improve mechanical properties and CO_2 sequestration in cement mortar. Construction and Building Materials. 392: 132021.

Peterson, S.C. 2012. Evaluating corn starch and corn stover biochar as renewable filler in carboxylated styrene–butadiene rubber composites. Journal of Elastomers and Plastics. 44: 43–54.

Peterson, S.C., S.R. Chandrasekaran and B.K. Sharma. 2016. Birchwood biochar as partial carbon black replacement in styrene–butadiene rubber composites. Journal of Elastomers and Plastics. 48: 305–316.

Peterson, S.C. 2019. Silica-milled paulownia biochar as partial replacement of carbon black filler in natural rubber. Journal of Composites Science. 3: 107.

Peterson, S.C. 2020. Coppiced biochars as partial replacement of carbon black filler in polybutadiene/natural rubber composites. Journal of Composites Science. 4: 147.

Peterson, S.C. 2022. Carbon black replacement in natural rubber composites using dry-milled calcium carbonate, soy protein, and biochar. Processes. 10: 123.

Praveen, K., R. Raja, S. Jannet and L. George. 2023. Investigation on the machinability of biochar reinforced AA5083 metal matrix composites fabricated by stir casting technique. Engineering Research Express. 6: 015016.

Pudełko, A., P. Postawa, T. Stachowiak, K. Malińska and D. Dróżdż. 2021. Waste-derived biochar as an alternative filler in biocomposites—Mechanical, thermal and morphological properties of biochar added biocomposites. Journal of Cleaner Production. 278: 123850.

Raja, R., L. George, S. Jannet, I. Prudthvi and S.R. Kandavalli. 2021. Development and analysis of SiC, TiO_2, and biochar hybrid reinforced aluminum-based metal matrix composite. Advances in Materials and Processing Technologies. 8: 1–9.

Raju, G., M. Khalid, M.M. Shaban and B. Azahari. 2021. Preparation and characterization of eco-friendly spent coffee/ENR50 biocomposite in comparison to carbon black. Polymers. 13: 2796.

Rashid, S., A. Raghav, A. Goyal, A.B. Danie Roy and M. Singh. 2024. Biochar as a sustainable additive in cementitious composites: A comprehensive analysis of properties and environmental impact. Industrial Crops and Products. 209: 118044.

Sekhar, R., D. Sharma and P. Shah. 2021. State of the art in metal matrix composites research: A bibliometric analysis. Applied System Innovation. 4: 86.

Sunali, J. Mago, A. Negi, K.K. Pant and S. Fatima. 2022. Development of natural rubber-bamboo biochar composites for vibration and noise control applications. Journal of Cleaner Production. 373: 133760.

Torres-Sciancalepore, R., M. Riveros-Gomez, D. Zalazar-García, D. Asensio, M.P. Fabani, R. Rodriguez, et al. 2023. Two-step valorization of invasive species *Rosa rubiginosa* L. husk waste through eco-friendly optimized pectin extraction and subsequent pyrolysis. Journal of Environmental Chemical Engineering. 11: 110802.

Ulfah, I.M., R. Fidyaningsih, S. Rahayu, D.A. Fitriani, D.A. Saputra, D.A. Winarto, et al. 2015. Influence of carbon black and silica filler on the rheological and mechanical properties of natural rubber compound. Procedia Chemistry. 16: 258–264.

Verma, L. and J. Singh. 2023. Production of biochar-based nanocomposites from chemical and biological methods. pp. 47–56. *In*: D. Mishra, R. Singh, and P. Khare (eds). Biochar-Based Nanocomposites for Contaminant Management. Advances in Science, Technology and Innovation. Springer, Cham.

Wen, J., B. Wang, Z. Dai, X. Shi, Z. Jin, H. Wang, et al. 2023. New insights into the green cement composites with low carbon footprint: The role of biochar as cement additive/alternative. Resources, Conservation and Recycling. 197: 107081.

Yang, Y., G. Li, X. Yue, K. Zhang, Z. Zhang, H. Zheng, et al. 2025. Advances in biochar composites for environmental sustainability. Advanced Composites and Hybrid Materials. 8: 74–90.

Zhang, P., W. Duan, H. Peng, B. Pan and B. Xing. 2022. Functional Biochar and Its Balanced Design. Environmental Au. 2: 115–127.

Zouari, M., D.B. Devallance and L. Marrot. 2022. Effect of biochar addition on mechanical properties, thermal stability, and water resistance of Hemp-Polylactic Acid (PLA) composites. Materials. 15: 2271.

Index

For Product Safety Concerns and Information please contact our EU representative GPSR@taylorandfrancis.com Taylor & Francis Verlag GmbH, Kaufingerstraße 24, 80331 München, Germany